A RACE FOR THE FUTURE

A RACE FOR THE FUTURE

Scientific Visions of Modern Russian Jewishness

MARINA MOGILNER

HARVARD UNIVERSITY PRESS

Cambridge, Massachusetts
London, England

2022

Cataloging-in-Publication Data is available from the Library of Congress
ISBN: 978-0-674-27072-5 (alk. paper)

To my father, Boris Mogilner (1941–2020)

CONTENTS

CONTENTS

A RACE FOR THE FUTURE

Introduction

Jews, Race, and the Challenge of (Post)Imperial Modernity

> That race is invoked by, and coexists with, a range of political
> agendas is not a contradiction but a fundamental historical
> feature of its multiplex political genealogies.
>
> —Ann Laura Stoler, "Racial Regimes of Truth"

A T THE TURN OF THE twentieth century, American Jews whole-
heartedly resisted their categorization as a distinctive race, which in
the context of the Progressive Era's rising antiemigration attitudes and
policies threatened to problematize their much-coveted status as white.
American Jewish race scientists in particular challenged the racial othering of recent Jewish immigrants from Eastern Europe by applying
methods of physical anthropology—the science of race—to study Jewish
assimilation. An immigrant from the Russian Empire, the New York–
based physician and scholar Maurice Fishberg, in his monumental monograph *The Jews: A Study of Race and Environment* (1911), denied that
Jews constituted a race at all. At the same time, the doyen of modern anthropology in the United States, Columbia University professor of anthropology Franz Boas, a Jewish emigrant from Germany, conducted at the
request of the US Immigration Commission an exemplary anthropometric

survey of Jewish and Italian immigrant communities in New York. First published in its entirety as a 573-page volume by the commission and then as an article in a 1912 issue of *American Anthropologist*, the study presented anthropometric statistics and calculations that proved the plasticity of racial traits, including the same cephalic, or head, index previously seen as the main and most stable anthropological indicator. Boas's study revealed that American-born children of Eastern European Jews in New York differed from their siblings born in Eastern Europe, as well as from their parents. They resembled the surrounding population physically and thus, it was assumed, in terms of their mental makeup. Boas did not explain the exact causes and mechanisms of this racial malleability, pointing vaguely to the role of "environment." However, his scholarly methodology decidedly undermined the perception of race as something enduring. Politically, his study denied the widespread eugenic and sociological speculations about the degenerate and unassimilable nature of Eastern European Jews as a particular race.

This academic politics of Jewish race stood in stark contrast to a trend that was simultaneously gaining momentum in the Russian Empire—home to the absolute majority of the New York Jews studied by Fishberg and Boas, and of world Jewry in general. Russian Jewish race scientists interpreted Boas's study as a call to double down on Jewish self-perception as a distinctive race, and to greatly expand the collection of anthropometric statistics on Russian Jews. This misinterpretation was backed by the authority of one of the leading cultural anthropologists in the Russian Empire, a personal friend of Boas and his collaborator in North Pacific ethnographic studies, Lev Shternberg. For a variety of reasons, Shternberg succeeded in turning Boas's conclusions upside down and presenting his study of New York immigrant families as proof that Jews constituted a pure race, easily adaptive to a better environment but retaining its purity under external influences. Departing from this assumption, Shternberg articulated a call to educated Jews of the Russian Empire to collect anthropometric data from hundreds of thousands of Jews in the Pale of Settlement and beyond, across the empire, in traditional and modern settings.

The goal of this grand anthropological self-exploration was to develop the Boasian project by explaining how exactly the Jewish race was adapting to a more modern, industrial, and democratic ("American") environment while at the same time avoiding assimilation. This project was put into practice by self-mobilized Jewish intelligentsia and professionals who were deeply concerned with the postimperial future of the nonterritorial Jewish nation. Like Shternberg, they perceived individual Jewish bodies and the collective body social as the true and lasting "living space" of the otherwise territorially scattered and internally culturally diverse Jewish nation. Together, physicians, teachers, lawyers, journalists, and other educated and semieducated readers of the Russian Jewish press, students in imperial universities and private colleges, patients who allowed physicians to take their anthropometric measurements, communal and relief organization activists, and relief recipients formed a movement that shared a social vision of Jewishness as a "race." The consensus around the Jewish body as a basis of nationality did not cancel political disagreements within the community of Russian Jews. At the same time, it generated a space for race science and biopolitics focused on a collective physical Self. This reformist politics and social movement survived, in changed form, throughout the early Soviet period and influenced not only the lives of Russia's Jews but also Soviet general population policies.

A Race for the Future tells the story of this massive and far-reaching but surprisingly forgotten phenomenon as a story of Jewish imperial and postimperial modernity. The urge to racialize Russia's Jewry was prompted by the same desire to normalize Jews and present them as moderns that compelled their American counterparts to advance the opposite argument, and, as John M. Efron, Mitchell B. Hart, and others have shown, the same desire that inspired Jewish race science and racialized politics in Germany, England, and France.[1] What differed were the contexts in which the enthusiasts of Jewish self-racializing acted and the future modernities they could imagine for themselves. Russian Jews turned to "race" in the context of the nationalizing empire, and hence empire is a major protagonist and a key context-setting category in this book.[2]

Empire matters here as a historical Eurasian state with the largest Jewish population at the time—the Russian Empire, an anthropological "field" embraced by Russian Jewish race scientists, a name for a supranatural polity as imagined by the nationalist Jewish activists of self-racializing, and an analytical model informing my own analysis.[3] In the latter sense, I prefer to think about the phenomenon of empire not in a conventional sense, as a large, authoritarian, expanding state ruling over a variety of populations, but in poststructuralist terms, as an "imperial situation" characterized by the multilayered and unsystematic overlapping of incongruent taxonomies, principles of group affinity, hierarchies of authority, and regimes of hegemony. The essential point here is that these regimes of difference and hierarchies of inequality are synchronic, they do not supplant each other, and they only partially overlap.[4] In other words, one's place in a given hierarchy of diversity and difference is not always regular and fixed, and it never neatly coincides with externally well-bounded and internally homogeneous ethnicities or races (or nations and classes)—the overlap is always contextual and never complete. The imperial situation optics make the meaning and boundaries of any universal category—be it a Jew, a Russian, or race—a product of imperial "strategic relativism" that problematizes monological explanatory narratives and simplistic colonial binaries.[5]

By embracing this approach, *A Race for the Future* aspires to a place among studies in new imperial history that explore imperial heterogeneity and adaptability and are focused on explicating the logic of every imperial situation. New imperial history prioritizes exposing the constant negotiation and rearrangement of difference, degrees of sovereignty and subjecthood, and terms of exclusion and inclusion.[6] As Pieter Judson, the premier historian of the Habsburg Empire, has outlined this approach:

> We are not looking for "the thing," we are looking for the situation that produces the phenomenon, the thing. So that when we are looking for categories it seems to me we are looking for phenomena, and we as historians who examine change over time should be looking for what much more often is the situation that produces this phenomenon.[7]

This book also "looks for the situation"—the imperial situations that triggered Russian Jewish self-racializing and informed the meanings and applications of the "Jewish race" trope in Jewish imperial and anti-imperial science and biopolitics. Thus, *A Race for the Future* may be best described as the first attempt to write a comprehensive story of "race" as an empowering discourse embraced by Jews in the Russian Empire in response to the specific challenges of their imperial situation.

Russian Jews living in the Pale of Settlement, in central Russia, in Siberia, or in the Caucasus and Central Asia often had little in common. Many of them embraced hyphenated identities (Russian-Jewish) and Russian language and culture, which by the turn of the century were ceasing to function as an imperial and civilizational medium and were being progressively claimed by ethnic Russians and the imperial dynastic regime for a modern Russian national project. By the standards of social and political sciences of the time, Jews did not constitute a nation at all, because they did not represent a community of language, traditions, and territory. All they could realistically have hoped for in the postimperial future was a minority status at best and accelerated assimilation at worst, which in any case meant confinement to what Dipesh Chakrabarty famously called "the waiting room of history."[8] By grounding Jewish nationality not in culture and territory but in physicality and biological heredity, "race" offered a scientifically sound and politically effective answer to this predicament of inescapable ahistoricity, a remedy for being defined "in terms of a lack, an absence, or an incompleteness that translates into 'inadequacy.'"[9]

Some Russian Jewish race scientists viewed their studies as explicitly political, practiced by Jews for Jews through self-organization outside of institutionalized academia, which excluded Jews anyway. The science of Jewish race was thus about producing independent "authentic" knowledge about the Jewish physical body. This kind of knowledge was expected to afford Russian Jews discursive and biopolitical control over their presently colonized nation and over its postimperial future. Other self-proclaimed students of the Jewish race preferred to carefully differentiate between Jewish politics and Jewish science. Still others refrained from politics altogether and expected the language of modern science to bring about the desired social and political change. Russian Jewish race scientists differed on whether Jews were Semites or not; on whether Jews were a mixed or a pure race; on the place of "stagnating" Jews of the Russian Caucasus and Central Asia in the scientific constructs of the Jewish race/nation aspiring to a place in modernity; and on many other issues. They also held different political views. These differences present a powerful lens onto the turn-of-the-century entangled crises of the sense of Jewishness as a modern nation and of the old particularistic imperial order. The language of "race" seemed to offer a solution by stabilizing groupness in *longue durée* structures deemed objective (peoples, nations, and creeds) and helping to differentiate population groups emerging from the imperial mix. Nominally differentiated as cultural or biological, in practice these hierarchical or horizontal visions of racialized groupness were always hybrid.[10] The language of race was globalizing; it enabled broad comparisons and claims. In the Russian Empire, race as an argument helped to frame the imperial regime as "archaic" because of the regime's reluctance to deploy the modern concept of race consistently in its own scientific population politics (instead relying on traditional categories such as mother tongue, legal estate, or regional belonging).

Embracing such a "cognitive" perspective on race helps us understand how the turn to self-racializing could be empowering and liberating for

Russian Jews. In this, *A Race for the Future* resonates with other studies of late imperial and postimperial transitions, and a question formulated by the historian Peter C. Perdue on Chinese early twentieth-century anti-imperial nationalism looms large in this book, too: "Why did the passage from empire to nation produce such a violently racist ideology?"[11] As we will see, Russian Jewish self-racialization was not racist, although it did possess powerful exclusionary potential. Rather, this book argues in favor of the intellectual productivity of thinking about Jewish self-racializing as an original attempt to elaborate a nonterritorial and culturally hybrid notion of legitimate nationhood in the scientific and biopolitical language that held authority at the turn of the twentieth century. To put it better yet, the Russian Jewish engagement with "race" is interpreted as an anticolonial strategy for establishing national "authenticity" and safeguarding it from colonial domination attributed to the Russian Empire, and after the Bolshevik Revolution of 1917—to the Soviet "affirmative action empire" (to use Terry Martin's metaphor).[12]

The book's chronology transcends the historical divide of 1917, separating the old empire from the Bolshevik multinational state. The story begins in the context of the imperial regime's liberal reforms of the 1860s and 1870s, when Russian Jewish intellectuals began showing steady interest in exploring the concept of race as grounded in Darwinian/Lamarckian epistemologies, expressed through social, medical, and anthropometric statistics, embraced by theories of nationalism and easily identifiable in the global archive of colonial knowledge.[13] The story ends in the USSR of the early 1930s, when the formulation and official adoption of the doctrine of historical materialism in 1934 put an end to the proliferation of the old-type sociobiological discourses of groupness, and Soviet authorities made a decisive choice in favor of territorial Soviet national modernity for the Jews, with the Jewish Autonomous District (since 1934 the Jewish Autonomous Region). The book is organized in two parts according to the mode of historical narration (intellectual and social histories): "The Science of Race" and "The Biopolitics of Race." Each part features a broad range of protagonists: scientists, politicians, writers, and professionals (physicians, lawyers, teachers); men and women; and Jews from imperial centers and the borderlands. Some chapters, especially in Part I, are centered around individual "imperial" biographies. Other chapters reconstruct the social contexts of the most abstract ideas and the intellectual roots of the most practice-oriented social movements and initiatives (for example, the project of Jewish statistical self-exploration that well outpaced in scale and scope the official imperial statistics).

The result is a complex and multifaceted picture of one of the most comprehensive projects of self-racializing in pre–World War II European history. Unlike many similar contemporary projects, the Russian Jewish project never relied on the power of an interventionist national state. It was a project of grassroots self-organization and decolonization, of overcoming one's perceived national incompleteness, of solidifying collective identity through purging imperial hybridity, and of claiming biopolitical and political citizenship as a colonized nation. Decades later Frantz Fanon summarized the goal of such projects in one aphoristic sentence: "Decolonization unifies this world by a radical decision to remove its heterogeneity, by unifying it on the grounds of nation and sometimes race."[14]

Part I opens with a broad overview of Jews mastering the idiom of race in the context of the Russian Empire, tackling the challenge of nationalization from above and below, and entering the epoch of mass politics. This was a truly novel moment not just in the history of the empire but also in Jewish modernism, signifying a break with the previous tradition of Haskalah (Jewish enlightenment). Derek J. Penslar's comparison of Haskalah to anticolonial thought among Asian intellectuals underscores the centrality of "authentic" textual and spiritual canons for both projects of modern subjectivity: both "used Western methods to study their civilizations' classic texts" and located the true national selves "in the realm of spiritual and literary creativity."[15] Svetlana Natkovich registers the break with this self-referential tradition of cultural self-exploration in the Russian imperial context by exposing the growing centrality of the physical body in literary works in Hebrew and Yiddish produced by the mid-nineteenth-century Russian maskilim (proponents of Jewish enlightenment). While emerging as a space of articulation of new, modern Jewish subjectivity, this Jewish body, Natkovich writes, still remained unattached to any specific scientific worldview: "Lacking the opportunity to be accommodated into their designated social milieux, these reformed bodies found themselves alienated from both their Jewish and Russian surroundings, shuttling between them as ungraspable, 'floating' signifiers." The result was a view of physical bodies as stripped of a coherent framework of signification, or an emphasis on consciousness as being uncoordinated with its physical receptacle.[16] The turn to scientific race accommodated this anticolonial act of using the physical body as a subaltern language by grounding it in a distinct framework of signification and connecting individual and collective planes of modern Jewishness in a scientific national epistemology. Moreover, the act of embracing scientific race put Jews at the forefront of the empire's self-modernization as an empire of knowledge.

The remaining chapters in Part I consider the biographical, professional, and political trajectories of three leading Russian Jewish race scientists: Samuel Weissenberg (1867–1928), Arkadii El'kind (1868–1921), and Lev Shternberg (1861–1927). Their imperial biographies and intellectual projects represent three distinct paradigms of Jewish race science and three political visions of the Jewish modern postimperial future, each of them contingent on preserving empire in the region as a form of democratized supranational polity. They also illustrate three possible positions of a Russian Jewish race scientist in the empire—the colonial periphery; the all-imperial network of "citizen science" calling for scientification and modernization of the Russian imperial regime; and official academia. Shternberg's unique position within the official, otherwise Jew-free, imperial academia necessitated a complete separation of his normal science of ethnography of Siberian indigenous peoples from his racialized and politically militant Jewish science—a move interpreted in the book in light of the larger problem of "subaltern science." As a founder of Soviet ethnography and Soviet Jewish science, Shternberg appears in many ways central to both parts of the "race for the future" story.

Amazingly, Shternberg's Jewishness and his Jewish science have remained until now the least studied aspects of his otherwise well-known academic and political persona.[17] This fact alone explains why Part II opens with an introduction, "Jewish Biopolitics as a Victim of Aphasia," illuminating why this particular aspect of Russian Jewish imperial and postimperial modernity was never incorporated into big historical narratives, whether Jewish, imperial, or Soviet. Colonial aphasia, as Ann Laura Stoler reminds us, is not an innocent forgetting, "a matter of ignorance or absence." It requires "the loss of access and active dissociation": "Aphasia is . . . a difficulty in generating a vocabulary that associates appropriate words and concepts to appropriate things," which is precisely the case with the "loss" of historical or memory narratives of Russian and early Soviet Jewish racial biopolitics.[18] This book offers a framework of the subaltern politics of self-racializing as a means of overcoming the structural limitations of aphasia, being able to ask proper research questions and to "associate appropriate words and concepts to appropriate things." My discussion in Part II is structured as a social history of Jewish biopolitical undertakings, which I treat as forms of apolitical grassroots politics that claim control over the biological, social (and, by extension, political) body of the nation.

Chapter 5 offers a comprehensive investigation of professional associations of Jewish physicians (medical societies) in the Russian Empire, their secession from the all-imperial progressive medical movement, and consolidation in 1912 into the Society for Protection of the Health of the Jewish Population (Obshchestvo okhraneniia zdorov'ia evreiskogo naseleniia, OZE), which evolved into a broad public movement of adepts of Jewish biopolitics. The OZE embraced biopolitics as a form of nonparty and nonideological politics that seemed a realistic alternative to a political revolution. In some sense, these societies and associations fit Jeffrey Veidlinger's definition of modern Jewish secular public culture as a space of remaking an imperial Russian Jewish society by means of "construction of an entirely new system of myths, the articulation of new legends, and the enactment of new ritual behavior."[19] The content of these new myths and the practices of reenactment and performance of "race" are explored next. Chapter 6 discusses the development of Jewish statistical and eugenic discourses and initiatives, and the medicalization of the problem of Jewish childhood in the years between 1912 and the 1920s. Chapter 7 traces this story into the Civil War period, problematizing in particular the persistence of empire in the OZE biopolitical vision and exploring how this vision was altered by the anthropology of violence and genocide carried out by OZE activists in pogrom-stricken Ukraine. This chapter reconstructs the afterlife of the OZE in the wake of its formal ban by the Soviet authorities in 1921, in a gray zone between legality and illegality, and under the umbrellas of the American Relief Administration and the American Jewish Joint Distribution Committee or behind the facade of Soviet statistical and medical institutions. The story culminates in Chapter 8 with a detective-like investigation of a wide semilegal network of former OZE activists in central Russia, who in 1921–1922, under the conditions of famine and the regime of Red Terror, continued collecting and analyzing Jewish racial statistics. They perceived the situation through the model of Jewish accommodation to modernity, which Shternberg had adopted from Boas in 1912, and worked hard to demonstrate Jewish adaptability to the changing postimperial environment and Jews' biopolitical right to negotiate the terms of Jewish national modernity with the new regime.

The negotiations, indeed, continued throughout the 1920s, having received, on the one hand, a new impetus from Soviet eugenics, affirmative nation-building, and scientific population politics and, on the other, ascending anti-Semitism in the postimperial nation-states of East-Central Europe (and across the West). This last part of the "race for the future" story up to the early 1930s is considered in the Conclusion. The Soviet imperial situation created conditions for the simultaneous empowering of the Jews and their new primitivization as a "nonproductive" and nonterritorial nation, and for their inclusion as a community of the emerging Yiddish proletarian culture and exclusion as an old and irredeemable community of origin (race). Andrew Sloin, whose innovative book explores the latter aspect of Soviet political transformation, added to the list the persisting semantic overlap between Jews as a nationality and Judaism as a religious congregation: "While party and ethnographic institutions rejected the conflation of religious and national categories for classifying Muslims, Christians, and other religious groups, the terms 'Jew,' and 'Jewish,' . . . were recognized in all of their ambiguity as acceptable national designations."[20] This ambiguous and somewhat liminal nature of Soviet national Jewishness before the Cultural Revolution and the First Five-Year Plan left room for self-racializing as a basis for scientific and materialistic Jewishness. And although the Jewish "race for the future" did not succeed in the end, its story suggests a new perspective on "race" as a language of self-description and resistance to hegemony in various imperial formations, tsarist and Soviet.

PART I

THE SCIENCE OF RACE

CHAPTER ONE

The *Dawn* of the Jewish Race in the Late Nineteenth-Century Russian Empire

The Jewish race has rendered the greatest services to the world. Assimilated into the different nations, integrated into the various national units, it will continue to do as it has done in the past, and, by collaboration with the liberal forces of Europe, will contribute greatly to the social progress of mankind.

—Ernest Renan, 1883

T HE MODERN SCIENTIFIC repertoire of race concept had formed as a reaction to the eighteenth-century defamiliarization and secularization of the world that accompanied the demise of the ancien régime, with its multiple social and political dislocations and ruptures. The notion of race promised to restore orderliness and continuity to the perception of historical development and social order in the emerging, increasingly volatile modern social imaginary. As a synthetic category, race was expected to explain the processes and forces, both biological and historical, that shaped universal humanity in the past as well as observable human differences in the present, and hence it elucidated progress too—another new concept that had emerged by the end of the eighteenth century.[1]

The concept of race drew its legitimacy not from some particularistic intellectual tradition but from a new scientific paradigm that embraced criteria of objectivity that were typical of the natural sciences along with a truly global worldview and classificatory approach. Thus understood, race was firmly inscribed in the epistemological and political repertoire of modernity claimed by the West.[2] The specific genealogies of this race can be traced to the early nineteenth-century philological theories of language families that conflated ideas of linguistic and social groups (later conceptualized as ethnic/racial communities); or to Johann Friedrich Blumenbach's classification of five human races—Caucasian, Mongolian, Ethiopian, Malay, and American—based on cranial typology; or to Charles Darwin's 1871 *The Descent of Man, and Selection in Relation to Sex,* in which sexual selection emerged as a secular mechanism of race production.[3] But regardless of a chosen genealogy, all of them advanced a narrative of progressive, if uneven, development and all assumed that the human species was a product of natural and social processes. Humans could thus be examined and understood with the help of modern science, and controlled and regulated without recourse to divine authority. "Race" was, therefore, not just another term; it functioned as a scientific, positivistic, and anthropocentric discourse that deployed a natural-science approach to the social world. Being intrinsic to modernity, it inspired political and social theories and practices of modern groupness, from nationalism to colonialism and from eugenics to self-help, and served as an underlying code of mass culture (tightly intertwining with gender and class).[4] The semantics of race and the term itself were meant to be universally applicable.

The Jews of the Russian Empire encountered this concept of race in the mid-nineteenth century as members of the pan-imperial public sphere, and they continued to discuss it in the Russian language over the course of several decades. Obviously, Yiddish or Hebrew concepts were available that conveyed the semantics of genealogical lineage and blood kinship, but the term "race" that interests us here was embedded in a very specific scientific paradigm and cultural moment. The German, French, Italian, and English contributions to this paradigm were so extensively and profoundly engaged by Russian imperial scholarship that Russian science became a natural framework and necessary context for the conceptualization of race by Jewish intellectuals.

In the Russian Empire of the late nineteenth century, the Russian language served as the main medium for communicating such a paradigm.

Since the 1850s, scholars residing in Russia who used to publish their works in Latin, French, or German began to see in Russian a legitimate language of science. In the 1860s, Russian was already the main language of academic science and of major political and philosophical debates, the language that made global, especially Western, knowledge available (in translation) for imperial subjects.[5] Throughout most of the nineteenth century the parameters of Russianness remained unclear and contested, and Russian language did not automatically signify Russian ethnicity. It retained the status of a cosmopolitan imperial medium—the language of high literature, science, and philosophy—and a channel for transmitting "Western" modernity. In some contexts, this role of Russian language produced epistemological domination and colonial disbalances, whereas in others its national indifference facilitated hybridity and strengthened all-imperial political and intellectual networks. Finally, Russian as an imperial medium enabled critical discourse by imperial subjects toward their own communities and institutionally reinforced, from within and from outside, "traditions" and identities.

The imperial situation of underrationalized diversity and contextually determined meanings of social categories and relationships in which Russian Jews existed was responsible for the presence in their lives of all these types of engagement with the Russian language. Sustained in different languages on a vast territory from Central Asia to Poland, and from the Caucasus to St. Petersburg, Russian Jewish culture was a hybrid phenomenon. It operated through parallel and often coinciding public spheres that partially overlapped with the Russian imperial public sphere and various national, ethnically and linguistically non-Jewish ones. Their co-existence and overlap ensured the multilingualism of Russian Jewish culture. Jeffrey Veidlinger has noted the intrusion of Russian-language words in the Yiddish vocabulary for scholarly associations and educational groups: *vechernik* (evening event) and *zasedaniia* (meeting); *liubiteli* (amateurs, aficionados or lovers, as in the Moscow-based Imperial Society of Lovers of Natural Sciences, Anthropology and Ethnography), used alongside Yiddish *libhober,* and even Yiddish words for a circle (group of intellectuals)—*redle, ringl, tsirkl*—were constructed and used as the Russian *kruzhok*.[6] It was not accidental that the leading Russian Jewish historian and ideologue of diaspora nationalism, Simon Dubnov, first formulated the parameters of his modern narrative of Jewish history in Russian, and that S. An-sky (Shloyme Zanvil Rappoport) made the same choice in the case of modern Jewish ethnography. Their politics of knowledge was inspired by Jewish nationalism and at the same time predicated on the universality of the embraced scholarly episteme, which activated a very specific archive of distinctly modern philosophical, political, and scientific knowledge.

Even though Yiddish was spoken by the majority of Russian Jews at the turn of the century (by 97 percent of those who identified with Judaic faith), there were few incentives to translate the entire body of modern scholarship into the "Jewish language" (as Yiddish was identified in the 1897 imperial census).[7] The first translations of works of modern science for Eastern European Jews started to appear in the mid-eighteenth century. However, their publication was limited and they circulated in a restricted social and intellectual sphere.[8] In the nineteenth century, Haskalah gave rise to maskilic scientific literature that "constituted a foundation for public discourse, promoted change, and shaped the outlook of broad social circles."[9] In this early period, scientific ideas, especially in fields such as medicine and exact sciences, circulated in Hebrew translations, but they aimed primarily at initiating young maskilim to the new Jewish cultural world, still largely defined by its particularism and unique textual tradition.[10] Darwinism, for example, generated little enthusiasm in this world. Joakim Philipson concludes that it "generally did not cross the border into the popular scientific Hebrew literature."[11] Moreover, in the world influenced by early Haskalah, modern natural sciences did not define the language of social analysis, whereas in the Eastern European Jewish world at large secular education in natural and social sciences, which could potentially undermine the foundations of the Jewish tradition, was often represented as a conduit to heresy and associated with conversion.[12]

That being said, in the second part of the nineteenth century we see the proliferation of discourses grounded in "Western" natural sciences and biosocial theories (a development that also evokes multiple parallels with Asian colonial intellectual and political trends).[13] They reached Jews primarily in Russian and facilitated the emergence of a modern national idiom of Jewishness. The process also coincided with the Great Reforms of the 1860s–1870s that liberalized Russia's legal, economic, and public life and produced more opportunities for Jewish selective integration. This enhanced the appeal of the state's language as an important channel of integration, including direct and immediate integration into the world of modern science and scientific politics. In 1912, multilingual Jewish university students in Moscow described their linguistic preferences as being dependent on a context and a situation. Their answers exposed the exclusive connection between modern science and the Russian language that had continued since the nineteenth century:

> When I am spending a long time at home [that is, in the Pale of Jewish settlement], I tend to think in Jewish [that is, Yiddish]; when I am not at home, I think in Russian; I always [think] only in Russian about scholarly questions;
>
> The language of thinking depends on a subject: it would be strange to think in Jewish about the osmotic pressure and similar things. I think in Jewish about matters of everyday life;
>
> When I am thinking about people and phenomena of everyday life I often do this in Jewish language; in all other cases I use Russian, because I've learned about science and problems of public life from the Russian-language books.[14]

Unlike other collective subjects of the Russian Empire with sizable educated elites, Jews could not pursue official academic careers without being baptized. This fact influenced the politics of Jewish knowledge by shifting its centers from universities and affiliated intellectual networks to the Jewish press in Russian and, to a lesser degree, Jewish languages. After 1905, a few newly established Jewish scholarly societies joined the cause, but this could not compensate for the absence of Jews in academic science, especially since despite the sporadic ideological pushes from the government, Russia's universities generally failed to perform as agents of ethnocultural Russification. As one study asserts, "Even in the early twentieth century within the walls of Russian universities, the old enlightenment politics of knowledge continued—although in a new, liberal-positivistic, décor. This politics in fact explicitly promoted the study of local differences."[15] Regardless of the changing priorities of the regime, imperial universities hired people who identified as "Little Russians" (Ukrainians), Poles, Germans, and even Muslims, or simply did not identify with ethnic Russianness, and they were teaching students whose ethnic and confessional composition was relatively diverse.

Under these conditions, the continuation of the "old enlightenment politics of knowledge" that tolerated research into local ethnography, languages, folklore, or literary traditions (as part of the Oriental or Slavic studies curriculum) produced unintended effects. It could, for example, foster the emergence of networks of intellectuals invested in applying academic knowledge to nurture modern national identities (Ukrainophilism in the nineteenth century is the best case in point). Or it could promote the version of Oriental studies that criticized European Orientalism and colonialism.[16] Russian Jews were special not only because their access to academic jobs depended on baptism but also because they did not in principle qualify as legitimate objects of academic research. They were not recognized as one of the empire's indigenous territorial ethnographic cultures, its own Orientals, or a Eurasian linguistic group whose presence in the region since early times could be traced through culture as well as historical and archaeological records. Only the Russian Jewish press could partially compensate for the omission of Jews by the institutionalized sphere of modern knowledge production. Of course, the Russian language of this press can be interpreted as a sign of assimilation, but it, in fact, empowered Jews to develop their own, original biological, medical, statistical, economic, and political expert discourses of Jewishness.

In the 1860s and early 1870s, the minuscule print run of Jewish periodicals in the Russian language reflected the actual size of the target audience for these publications—"a handful of Odessa and Petersburg Jewish intelligentsia."[17] Even this readership had to be won over, for the Russian Jewish intelligentsia tended to place their hopes with the Russian liberal press, which they viewed as a direct pathway to the all-imperial public sphere of *obshchestvennost'*—a surrogate modern political nation held together by common values and a shared mode of social thinking rather than ethnic origin. Measured against the appeal of the Russian liberal press, the popularity of Russian Jewish periodicals such as *Rassvet* (Sunrise) or *Den', organ russkikh evreev* (Day, the organ of Russian Jews) was low.[18] In 1884–1899, *Voskhod* (Dawn; published in St. Petersburg 1881–1906) and its supplement, *Nedel'naia khronika Voskhoda* (Weekly chronicle), remained the only Russian-language Jewish periodicals in the empire, but the increased number of *Voskhod* subscribers from 950 in 1881 to 4,397 in 1885 reflected the growing influence of the model of Jewish modernity it advanced. Dmitry El'iashevich calculated that in the 1890s, the number of potential readers of the Russian Jewish press had already reached one million.[19] A constellation of seemingly opposing trends produced this change: growing Jewish integration; the diminishing philosemitism of the Russian liberal press in response to the Jews' resistance to full assimilation and the upsurge of Russian nationalism in the wake of the emancipation reform of 1861; and, of course, the pogroms of 1881, which stimulated critical reflection among Jews on earlier scenarios of integration and self-modernization.[20] In the words of Dubnov, by the mid-1880s, *Voskhod* had become a "mouthpiece of our internal indignation."[21] It offered a platform and a new scientific language for conversations about modern Jewishness.[22]

Race belonged to this new conceptual repertoire grounded in natural sciences, demography, medicine, and statistics. Obviously, racial anti-Semitism was a factor stimulating Jewish race thinking, but it does not explain or exhaust Jewish self-racialization as we observe it in *Voskhod* and other similar venues. The national inferiority complex that underlined Jewish race thinking did not simply mirror anti-Semitic invectives. Rather, it followed from the application of the most advanced political and sociobiological theories to the Jewish people. Simply put, Jews did not conform to the widely accepted scientific understanding of nation as defined by the commonality of language, traditions, historical experience, and national territory.

"In its present state, Jewry is an incomplete and abnormal nation," declared the prominent Jewish lawyer and intellectual Oskar Osipovich Gruzenberg, in *Voskhod* in 1894.[23] At this point he was merely summarizing the general sentiment that had already been in the air for a few decades but was expressed primarily in the Russian press. At least as far back as 1861, R. Goldenviezer complained in the liberal daily *Moskovskie vedomosti* (Moscow News) that having lost their political independence, national territory, native "soil," and "historical-political individuality," Jews ceased to be a nation. As a typical midcentury positivist, Goldenviezer chose the organicist metaphor of a plant transferred to an alien terrain. In order to survive and blossom, it had to be firmly planted in its new soil and it had to adapt to new conditions. For Goldenviezer, this example translated into an assimilationist political program.[24] Indeed, if Jews were deficient as a modern nation because they resembled a plant growing in alien soil, then according to scientific laws, they had no other choice but to adapt, that is, to join a nation on whose "soil" they proliferated. A response to Goldenviezer in the Russian Jewish weekly *Sion,* formulated from the antiassimilationist position by Julius Iosifovich Goldendakh, used biological metaphors, too: "I would ask Mr. Goldenviezer: does a plant that is implanted on a new terrain really change, that is, cease to be itself? It will take root [*primetsia*] and blossom—agreed, but it will still preserve (if only with some changes) its [original] nature."[25]

Goldendakh was known as the translator into Russian of Thomas H. Huxley's *Evidence as to Man's Place in Nature* (1863) and works by German natural scientists, so one can easily pinpoint the sources of his own scientific language.[26] He called on Jews to continue their "struggle for existence," but primarily through familiarizing the Jewish masses with "contemporary thought" and "awakening them from their apathetic drowsiness." The resemblance of this rhetoric of awakening to the Haskalah sermon should not mislead us. Goldendakh's language was explicitly Darwinian, and his understanding of the historical moment had been formed under the influence of nationalism. Jews had no choice but to awaken to their true nature because "blood [was being] spilled in the name of nationality" and "nationality constitute[d] the alpha and omega of all political events."[27]

Twenty-three years later the rhetoric remained the same: "From all sides we hear . . . that modern Jewry is not a nation, but only a religious congregation of the 'Mosaic faith.' . . . That to be a nation, presumably a common political order, common territory, and common spoken language are required, but Jews do not have them."[28] In the 1880s, when such complaints became commonplace, imperial particularism in politics was already under attack, including from the representatives of growing Russian nationalism within the ruling regime and state bureaucracy. It was gradually giving way to a more systemic and rational politics of groupness that negated all kinds of exceptional cases (of both "positive" and "negative" treatment), hybrid identities, and multiple loyalties—those foundational pillars of Russian Jewish identification. Under these circumstances, pressure was growing on the Russian Jewish intelligentsia to formulate an objective and scientifically verifiable justification of Jewish groupness as the basis for modern political subjectivity in the nationalizing empire.

Being a key concept of prestigious "Western" science and a hegemonic political epistemology of difference, "race" helped to meet this challenge. "Race" referred to the kind of primordial unity that preceded the commonality of language, traditions, and even territory, and thus presented a chance to overcome the national inferiority complex. Mastering the science of race, equally new to all the participants of the public all-imperial sphere, afforded Jewish intellectuals and professionals a Foucauldian type of expert power, including the power to perfect, control, and protect their own "race." The anthropocentric logic of "race" endowed them, the secularized Jewish intellectuals, with this power, while the nationalizing moment in the empire gave them the chance to define Jewish groupness in new terms, compatible with different visions of imperial and even a projected postimperial modernity.

A Philosemitic Science: Ernest Renan

The range of the main responses to the conundrum of Jewish national deficiency included Zionism, autonomism, socialist universalism (albeit gradually turning national as well), and assimilationism, or various combinations of the above.[29] Of these, Palestinophilism/Zionism and assimilationism were equally alien to the philosophy of *Voskhod:* the first as the most fantastic and remote idea, the second as a false reaction to the trauma of collective modern Jewish subjectivity. However, even these two extremes found their way to the intellectual marketplace of modern Russian Jewishness as exemplified in *Voskhod.* Assimilation in particular resided in the works of Ernest Renan, who enjoyed great popularity among Russian Jewish readers. Renan denied the existence of Jewish nationality in both the political and racial senses, but recognized Jews' lasting contribution to human civilization. His speeches on Jews and Judaism were promptly translated and published in Russian for the Jewish audience, carrying with them the authority of academic science.[30]

One of the earliest translations appeared in *Den'* in 1870. This was a swiftly reproduced article originally published in the November 1869 issue of the magazine *Revue des deux Mondes* under the title "Renan on the Tasks of the Jewish Nation." The "tasks" consisted in contributing to the progress of humankind. While denying national Jewishness as a modern identity, Renan flatteringly included ancient Jews in the selected club of nations that had changed the world (alongside ancient Greece, Renaissance Italy, and modern France).[31]

In 1882, *Nedel'naia khronika Voskhoda* (no. 51) disseminated Renan's letter defending Jews from the blood libel accusation—"the most ridiculous slander ever to be fabricated by hatred and fanaticism."[32] In January 1883, the weekly *Russkii evrei* (Russian Jew) followed suit and published Renan's authoritative historical denunciation of the blood libel. Not only did Renan remind his readers that the blood libel accusation was leveled against the first Christians in Rome. He also exposed the sheer absurdity of the blood libel being ascribed to Jews, for whom "the consumption of blood would be a crime against the prescriptions of the Jewish religion."[33] The article ended with Renan's appeal on behalf of the Jews accused in the Tiszaeszlár affair (the blood libel case in Hungary that was resolved in 1883 with the acquittal of all those accused).[34]

Later in 1883, *Russkii evrei* allocated eight full pages to the Russian translation of Renan's speech on Jewish science for the Jewish Scholarly Society (Société des études juives) in Paris, of which Renan was a member and which was presided over by Alphonse de Rothschild. Again, the speech culminated with an optimistic assessment of the Jews' role in modern history: "Every Jew is a liberal. (General applause.) He is a liberal by his very nature. By contrast, the enemies of Judaism—just look at them attentively—are enemies of the spirit of modernity."[35] Renan lectured that the Bible was the product of Jewish thought that brought East and West together and laid the foundation for the idea of equal rights. The French Revolution had consummated this idea through the emancipation of the Jews.

Simon Dubnov was so impressed by this speech that in 1886 he referred to it in *Voskhod* to counterbalance Renan's provocative statements about Jews and Judaism in *Antichrist* (1873).[36] Dubnov warned his readers that "in the book under review [Renan] claimed that after the rise of Christianity Judaism had lost its right to exist, while in his later speech to the Jewish Scholarly Society (1883) he said that Judaism's service to humanity in the past has been so great that it should continue into the future." This incongruity was hard to square, but Dubnov claimed that the speech better represented Renan's real views.[37] Renan's next speech, delivered on January 27, 1883, in Paris at the Saint-Simon Circle (Cercle Saint-Simon) and promptly published in translation in *Voskhod* (April 1883), "Judaism as Race and as Religion," seemingly confirmed Dubnov's optimistic assertion.[38] Although Renan's rhetoric was replete with such terms as "race" and "blood," "his historical erudition resisted the dominant verbal conventions" that constructed Jews as a pure race distinct from all others races.[39]

> It is my conviction that in the Jewish population as it exists today there is a considerable element of non-Jewish blood—even so much as to make this race, which is considered as the very ideal of pure *ethnos* and which has survived the centuries by means of prohibition of mixed marriages, appear as penetrated by foreign blood to only slightly less an extent than all the other races.[40]

Only in antiquity did Judaism function as the religion of a pure race, Renan explained. Since then, the original race was transformed by racial mixing, and Judaism could not continue as a national religion. Dubnov preferred to interpret this conclusion as a call for Judaism to reclaim its status as a "universal religion of the civilized world."[41] However, one could as easily see in Renan's speech a call to Jews to stop identifying their national belonging with Judaism. The latter reading accorded with Renan's stated preference for Jewish assimilation: "The Jewish race has rendered the greatest services to the world. Assimilated into the different nations, integrated into the various national units, it will continue to do as it has done in the past, and, by collaboration with the liberal forces of Europe, will contribute greatly to the social progress of mankind."[42]

Dubnov's "misreading" of this and similar statements reflected the profound ambiguity of his position vis-à-vis modern academic philosemitic science, personified by luminaries such as Renan.[43] Like other contributors to Russian Jewish publications, Dubnov always stressed Renan's high standing in French academia (Grand Officer of the Legion of Honor, member of the Académie française, and administrator of the Collège de France) as normalizing the philosemitic approach he represented. At the same time, Dubnov rejected the scenario of Jewish national disappearance due to the fulfillment by Judaism of its unique historical mission. As an autodidact in social sciences, Dubnov placed Renan among his main teachers whose ranks also included John Stuart Mill, Herbert Spencer, and Hippolyte Taine. But in addition to admitting his intellectual debt to Renan, Dubnov saw in him a kindred soul: "He, a graduate of Catholic seminaries, had accomplished the same initial journey [as Dubnov himself had] from dogmatic theology to scientific philosophy . . . and then to the study of religious movements."[44] Similar sentiments must have been shared by many educated Russian Jews who appreciated Renan's personal evolution and his willingness to contribute to Jewish science and protect Jews from dehumanizing accusations. One cannot underestimate the fact that in the wake of the 1881 pogroms, Renan, together with eight other gentile French scholars, signed a petition drafted by the committee led by Victor Hugo on behalf of the Russian Jews.[45]

When viewed through the lens of *Voskhod* and other Russian Jewish periodicals, Renan did not emerge as a source of definitive answers to the problem of Jewish national deficiency. Rather, his works in Russian translations helped to legitimize the serious scrutiny of modern Jewish national groupness in scientific terms that had nothing to do with any specific Jewish tradition or worldview. The science he represented was interpreted as global, unequivocally philosemitic and possessing the expert license and moral authority to voice a political opinion.

When it came to the actual "answers," even Renan's most loyal devotees could disagree. Dubnov's criticism of Renan's five-volume *History of the People of Israel* (*Histoire du peuple d'Israël,* 1887–1893) for its "extremely tendentious" presentation in the last volumes is a good case in point.[46] But this did not diminish the significance of Renan's humanist and universalist scholarship for the Russian Jewish intelligentsia—so much so that the most problematic aspects of Renan's scholarly views of Jewishness were selectively censored in the Russian Jewish press. His lectures directly addressing Jews and Judaism circulated broadly in Russian, but his earlier works in which he created a negative image of Semites, especially his fundamental academic monograph on Semitic languages, *Histoire générale et système comparé des langues sémitiques* (1855), were hardly mentioned. Furthermore, Renan's Russian Jewish popularizers tended to ignore his endorsements of European colonialism that markedly contrasted with his reputation as a humanist and universalist. Thus, in the aftermath of the revolutions of 1848 Renan argued that, without long preparation, liberty could not be extended to "savage races" like Black Africans. *La réforme intellectuelle et morale,* his antidemocratic manifesto in the wake of the Franco-Prussian War, argued for European colonialism as a "political necessity of the absolutely highest order." As Robert D. Priest observes, according to Renan, Europeans—a "race of soldiers" (that did not explicitly exclude Semites)—had a historical right to conquer "inferior races" such as the Chinese, a race "lacking almost any sense of honor," and "negroes," "a race of tillers of the soil" and nothing more.[47] Being highly sensitive to "inferior race" language in scientific constructions of Jews, Russian Jewish late nineteenth-century intellectuals preferred to ignore instances of Renan's applying the same language to Africans or Chinese. Moreover, as we will see, his racism and colonialism resonated with the way some Russian Jewish intellectuals imagined their own role in their own empire.

Enthusiasm about Renan tended to feed two schools of thought, both of which manifested themselves in *Voskhod* and other Russian Jewish venues. One school is particularly well known today due to its articulation by Dubnov in the model of Jews as a spiritual and historical, but not political, nation. The other school of thought was never extensively studied in the Russian Jewish context. It reflected a medicalized view of national groupness and presented Jews as an objectively existing biosocial entity (race-nation). Both approaches engendered numerous interpretations and iterations, and combined specifics of the Russian imperial situation with the "Renanian" discourse of the civilized European modernity that incorporated ancient and modern Jews.

Simon Dubnov's Social Darwinism

Dubnov's programmatic articles ("letters") in *Voskhod* on "old and new Jewry" are well known, as are his later works on Jewish history and politics.[48] He famously located the current "hegemonic center" of Jewish history in Russian Jewry—the only modern-day Jewish group that preserved authentic Jewish social organization and spirituality, and whose very existence had to defy Renan's assimilationism. The Jewish spiritual and historical nation fulfilled itself through communal existence and institutions, as well as Judaism understood as a common and lasting cultural text of the nation. In addition, it could claim a shared historical experience in a particular time and place. In short, Dubnov's nation did not require national territory and full political self-determination to complete itself. His conceptual thinking about national groupness experienced multiple influences. A list of the most obvious would include Renan, Mill, Austro-Marxism, Polish nationalism, Russian *obshchestvennost'* progressive mobilization for local causes, and the example of the zemstvo as a form of representative self-government. The same set of influences inspired Dubnov's political project of Jewish cultural-administrative autonomy within a multinational polity.[49]

Dubnov reinterpreted Jewish national deficiency as an advantage, a sign of the high stage of national evolution that had been achieved, when "political victories, territorial annexations or subjugation of other peoples" stopped being relevant:

> When a people loses not only its political independence but also its land, when the storm of history uproots it and removes it far from its natural homeland and it becomes dispersed and scattered in alien lands, and in addition loses its unifying language; if, despite the fact that the external national bonds have been destroyed, such a nation maintains itself for many years, creates an independent existence, reveals a stubborn determination to carry on its autonomous development—such a people has reached the highest stage of cultural-historical individuality and may be said to be indestructible, if only it cling forcefully to its national will.[50]

This passage from Dubnov's first letter, published in *Voskhod* in 1897, is strikingly revealing. He starts with the normative definition of a nation and continues with the affirmation that Jews, who had lost their territorial home, common language, and ethnographic culture, did not satisfy its criteria. But then a highly original claim follows, presenting Jews as the winners in the struggle for existence, in which they engaged with "stubborn determination." As it turns out, protection of the nation's autonomous inner self was a successful compensatory strategy that elevated Jews to "the highest stage of cultural-historical individuality."

In addition to the classical postcolonial concern over a lost versus preserved national authenticity, the overtones of social Darwinism in this passage are undeniable. Their presence supports Jonathan Frankel's observation that the doyen of Russian Jewish history believed in the law of survival of the fittest and "applied [it] to nations no less than to species."[51] Social Darwinism was an essential element of Dubnov's science of the colonized, articulated from outside of official academia and in terms that were not politically neutral but implied struggle and resistance. Dubnov blended universalist scholarly idioms with "authentic" Jewish self-knowledge. Thus, the first letter cited above combined a typical Jewish survival trope with the Darwinist-Malthusian concept of the "struggle for existence." The ensuing semantic transformation replicated a more general pattern of the adaptation of Darwinism by Russian intellectuals, who tended to de-emphasize Malthusian themes in Darwinism and diminish the role of the "struggle for existence" mechanism in favor of the idea of cooperation.[52] Although in Dubnov's works the "struggle for existence" was never explicitly downplayed, he also emphasized cooperation as a winning strategy that helped Jews in their collective struggle for existence and elevated them to the highest stage of social evolution.

Like many of his contemporaries, Dubnov embraced the evolutionary model of natural history in his historical narrative, but he imagined the object of evolutionary progression as a spiritual rather than political or territorial community. Race solved for him the problem of designating the limits of this community's social body, scattered all over the globe. Race also helped to locate this imagined community "vertically, with respect to the four-thousand-year-trajectory of Jewish history."[53] The core of Dubnov's Jewish nation remained intact regardless of how the nation expressed itself in a particular period and locality—through statehood (as during the biblical period), religion (premodern times), or historical consciousness (in the modern era). In the universal struggle for existence, Jews could rely on their superior spirituality and advanced social practices, but—even more so—on the "instinct of national self-preservation" that was "unusually developed among all members of the nation."[54] This primitive racial instinct preceded religious unity and historical solidarity. The first letter on "old and new Jewry" in *Voskhod* introduced these forms of bonding as successive evolutionary developmental stages.

The standard English edition of the letters I use in this book translates "race" as "tribe," which is misleading.[55] In the original Russian text, Dubnov uses "race" consistently and in the very direct sense of biological species. He distinguishes three "types of the nation," each characteristic of a specific historical epoch: (1) "racial type" (of the nation [translated in the English edition as "tribal type"]); (2) "territorial-political or socially autonomous" type; and (3) "cultural-historical or spiritual" type. As history progresses, each type gets subsumed into the succeeding type, producing a "mechanical or chemical combination." To convey the complexity of this evolution, Dubnov invented the term "*natio*-species" or even "*natio*-breed" (*natio-poroda*) and referred to the law of heredity that enabled "the fundamental stratum, the tribal substrate of nationality" to consolidate multiple generations.[56] As history progressed, the feeling of national belonging was becoming more and more complex. Fortunately, the fundamental primordial racial kinship and the inherited instinct of survival neutralized the negative effects produced by transitions from one hegemonic center of Jewish history to another. Dubnov's own summary of the first letter is also the best summary of this dialectic of the biological and historical.[57]

FIRST LETTER: "THE DOCTRINE OF JEWISH NATIONALISM"

   I. The course of the development of the national type: racial aspect [*rasovyi moment*—translated in the English edition as "tribal foundations"], territorial, political, and cultural-historical or spiritual [aspects].—The transition from material to spiritual culture in the growth of a nation.—The test of the internal strength of a nation: its loss of statehood and territory.

   II. Corresponding evolution of Jewry [translated in the English edition as "this development in Judaism"—an incorrect translation of Dubnov's secular understanding of the phenomenon of the Jewish nation]: embryo [*embrion*—this biological concept is translated as "the early beginnings of the growth"] of the spiritual nation in the period of the Prophets, its ["continued" in the English translation] growth during the era of the Second Commonwealth and its final realization [Dubnov used the Russian *zakal*—"hardening"] in the Diaspora (autonomism, the system of hegemonies [translated in the English edition as "Diaspora centers"]).—The nation as both the creator and the product of its history; the accumulation of national typicality [*narastanie natsional'noi tipichnosti*—translated in the English edition as "the crystallization of national types"].

III. The transition from objective to subjective criteria in scientific definitions of the term "nation."—Spiritual affinity is stronger [*sil'nee*—"more important" in the English translation] than blood relationships; the essence of the nation [*kvintessentsiia natsii*—translated in the English edition as "the essential basis of the nation"].—The negation of Jewish nationality [*otritsanie evreiskoi natsional'nosti*—translated in the English edition as "The negation of Jewish nationalism"].—Confusion of national and religious foundations and the secularization of the Jewish national idea; the assimilationist current; religious *nation* [*religioznaia natsiia*—translated in the English edition as "religious nationality"] and religious *group*.—Minimum of Judaism and the limits of national desertion [*granitsa natsional'nogo renegatstva*—translated in the English edition as "the limits of nationalist rejection"].

IV. National individualism and national egoism [*natsional'nyi individualism i natsional'nyi egoizm*—translated as "national individuality and national corporate personality"].—The Jewish national idea is an expression of pure individualism [*chistogo individualizma*—translated as "enlightened individualism"].—Between individual and social stands national [*mezhdu individual'nym i sotsial'nym stoit natsional'noe*—translated in the English edition as "The national foundation as midway between the individual and the social foundations"].—Theses.

This very detailed summary traces the evolution of the race-nation through multiple stages into a spiritual nation. But even at the highest stage of national evolution we see race-nation persisting in a subsumed form. Dubnov's analytical language in the summary undergoes a similar transformation: we see how terms such as "race," "embryo," "evolution," "growth," and the accumulation of "physical typicality" become gradually subsumed within the psychological and sociological rhetoric. Like Hegelian thesis and antithesis, Jews as a historical-spiritual nation and Jews as a race produce a synthesis in which they reach the ultimate evolutionary success.[58]

What matters is the role assigned to race in this synthesis at different stages of evolution. Whereas Dubnov placed "race" at the bottom of the hierarchy of national forms, so that it performed almost no active role in the modern life of contemporary Jews, many of his contemporaries understood the synthesis of biology and history differently, advancing "race" as the main argument in claiming the status of a normal nation. But one way or another, both schools of thought needed "race" for their secular, scientific constructions of Jewish nationality as residing in the common national body, sharing a collective spirit and history, struggling for survival against imperial and nationalizing regimes and other nations, and sustaining the threat of assimilation.

Statistical Jewishness

From this perspective, there was no contradiction between Dubnov's concept of historical-spiritual nation and the medical-statistical discourse of Jewishness, also popularized in *Voskhod*. The latter, however, lacked the conceptual bravery and originality so characteristic of Dubnov's approach. Most medical-statistical publications of the 1870s and 1880s emulated foreign examples and used statistics collected for Western Jewry, as comparable statistics on Russian Jews were still insufficient.

The article "From the World of Anthropological Statistics" by V.R. (*Voskhod*, August 1881) was a typical example of the genre. It offered a statistical snapshot of the Jewish biosocial organism and advanced the concept of Jewish "vital energy" (*zhiznennaia energiia*), understood as racial endurance. V.R. cataloged selected Jewish anthropometric measurements together with what could be better described as sanitation practices and life statistics. He considered Jewish marriages to be moderately fertile and registered the high rate of infants' survival; low numbers of illegitimate and stillborn children; a higher percentage of male births and higher population growth than that among Christians; higher life expectancy; immunity to some epidemic diseases; and so on.[59] The article's conclusion affirmed the unique power of the Jewish race, that is, the Jews' unique racial vitality, "which in all its biological manifestations gives Jews obvious advantages over other peoples."[60] This quotation, summarizing almost verbatim V.R.'s main conclusion, in fact, comes from a similar medical-statistical article, "Biostatistical Advantages of the Jewish Race in Europe," published two years earlier in *Russkii evrei*.

One can indeed use interchangeably quotations from these and other early articles experimenting with Jewish racial statistics. The author of the article in *Russkii evrei,* in turn, borrowed a quotation from the French statistician and director of the French statistical bureau since 1852, Alfred Legoyt, who oversaw three French censuses in 1856, 1861, and 1866. Being introduced as "a famous scholar who also is not Jewish—a double guarantee of the impartiality of his conclusions," Legoyt presumably said that "in the countries of exile" Jews preserved "the purity of race, purity of morals, . . . and finally a special energetic vitality [*energeticheskuiu zhivuchest'*] that ensured their quick acclimatization everywhere on the globe."[61] In another article that was typical for its genre, the author, Iosif Tsederbaum, turned for inspiration and statistical data to the French physiologist and anthropologist Jean Boudin. The authority of Boudin, "one of the prime physiologists and statisticians from the best medical department in France," was needed to reinforce the claim that the Jewish race has exhibited an "innate propensity for acclimatization," and high energy and vitality.[62] The role of Western European non-Jewish scientists in this discourse replicated that of Renan in the domain of humanities and social sciences—they symbolized the highest objectivity of modern science that studied Jews.

Just as it was with the appropriation of the idealized Renan, such a view required a great deal of learned ignorance and self-censorship. For example, Russian Jewish enthusiasts of the statistical approach never questioned the tendency of European race statisticians and medical doctors to evaluate Jewish vitality exclusively in colonial contexts, in their minds clearly separated from European societies. Only there the "physiological advantages that protected the Jewish race, more than other peoples, from harmful influences of climate, soil, and poor hygienic, moral, and economic conditions" were becoming visible.[63] Early Russian Jewish commentators accepted this colonial gaze as an essential rule of the "game." V.R., for example, used scattered data from statistical reports composed in English colonies in Australia, India, and Africa, which, in his view, showed that "Jewish race plays quite a substantial role there." In addition, V.R. quoted the well-known British physician Robert Lawson Tait, who "in a paper presented in September 1864 to the London statistical society and addressing mortality among mestizos from the marriages of European colonists with Indian women" placed "Jews ahead of all other peoples that have been colonizing India."[64]

Some Russian Jewish authors found such a colonial approach resonating with realities in the Russian Empire, where Jews could also perform (and did perform) the role of potential agents of colonization and Western civilizational influence.[65] The sad irony was that in the Russian imperial context, the overlap between the perception of Jews as modern and European and their assessment as primitive and alien was too obvious. In *Voskhod*, articles on Jewish vitality based on Western colonial statistics coexisted with many more articles on Jewish discrimination and degeneration in Eastern and Central Europe—a dissonance that complicated the unproblematic transfer of Western philosemitic colonial science and generated incoherency in the statistical discourse of the racialized Jewishness itself. In a study that approaches publications in *Voskhod* (1885/1886) on different aspects of the "Jewish Question" as single text, its author, Joakim Philipson, catches this incoherency particularly well:

> The text articulates an admixture of enlightenment, religious, national and evolutionary discourses, employing such key concepts (signifiers) as "civilization," "natural law" (*zakon prirody*), "unity of the human race" (*edinstvo chelovecheskago roda*), racial origin (*plemennoe proiskhozhdenie*), Talmud, "ethnographic," "universal," "strictly scientific observations" (*strogo-nauchnykh nabliudenii*) etc. A great number of references are invoked, many of which are external non-Jewish, to testify in defense of the Jews, e.g. Macaulay, Renan, Gregoire and Virchow. . . . Further external evidence is presented through some six pages of statistical figures of crime rates, comparing Jews favorably with Christians and other non-Jews. The conclusion is clear: the statistics of crime clearly shows Jews in a positive light. . . . Rudolf Virchow is invoked, as a voice of reason, speaking out in the German Reichstag in 1880: "To reproach the Jews for their level of education [*obrazovannost'*], making of it an excuse for what Darwin calls the struggle for existence—this means setting a limit to every peaceful development; to reproach the father for sending his children to schools at the highest level—this means making peace between Jews and Christians completely impossible."[66]

If there was any method in this madness, it was the general acceptance of the authority and prestige of modern natural and human sciences and the creation of a new canon that made the conversation about Jewishness truly scientific. This canon included scholars as different as the philologist Renan, the physical anthropologist Rudolf Virchow, and the biologist

Darwin. The discourse that their names authorized operated with concepts such as civilization, race, and natural laws. But the new frames of reference and the new language of debate required a new kind of data—mass, verifiable, and objective data on the state of the Jewish social and racial body, whereas the logic of Jewish self-racialization implied a kind of scientific population politics that at the time was not yet practiced consistently in the Russian Empire, including by the Russian imperial state.

Seeing like a State?

The Russian imperial state came to introduce systematic population statistics as a potential basis for practical governance in the context of preparation and implementation of the Great Reforms of the 1860s and 1870s. In the late 1870s to the early 1880s, statistics became a regular university discipline and a mandatory subject in the General Staff Military Academy.[67] In terms of philosophical influence, the German Verein für Sozialpolitik held the most prestige among Russia's statisticians. The Verein's members approached individuals as elements of social collectives—family, class, nation, and, as Andrew Zimmerman has conclusively shown, race—a view that could have been adapted successfully to populist, socialist, or statist goals.[68] The founder of the first statistical seminar at Moscow University, the professor of political economy and statistics Alexander Chuprov (who joined Moscow University's Department of Political Economy in 1874), advocated this approach.[69] Zemstvo statisticians working for local self-governments (created in the 1860s–1870s) shared the same philosophy. They accumulated enormous amounts of data pertaining to all spheres of life of Russia's rural population, thus assisting not only a state-sponsored modernization but also progressive social engineering from below.[70]

When the esteemed member of the Verein für Sozialpolitik Max Weber discovered Russia's zemstvo statistics in 1906, he was struck by its scale: "The foreigner who sighs as he contemplates the ocean of zemstvo statistics will at times feel that the ability to distinguish between the important and the unimportant is lacking."[71] But this was indeed an assessment of a biased observer whose understanding of what is "important" necessarily included, among other things, nationality and race.[72] Russian zemstvo statisticians studied individuals as members of their immediate and extended families, peasant communes, regional economic networks, gender groups, and social classes, thus ascribing to each group its own economic or political rationality and producing the picture of a complex corporate society. At the same time and quite contrary to Weber's expectations, they almost never considered individuals as participants in national (or racial) collectives. Operating under the progressive ethos of grassroots self-modernization, zemstvo statisticians tended to ignore particularistic imperial diversity as an impediment to universal improvement. They held this position until World War I, which unleashed full-force nationalizing processes everywhere, including the Russian Empire.[73] But even when data on national diversity were registered by zemstvo statistics, this information could not satisfy the needs of Jewish statisticians. The empire did not introduce zemstvo self-government in its western provinces, where the Pale of Jewish settlement was located and the majority of Russia's Jews—the potential objects of national statistics—lived.

A few reformist state agencies competed with zemstvo statisticians for the collection of data and their interpretation, especially in the fields of military and medical statistics.[74] Military statisticians strove to reinforce a racialized vision of the different ethnicities of the Russian Empire and, especially in the early twentieth century, tended to single out Jews as unreliable and unfitting by definition.[75] Unlike military experts, civic medical authorities recognized just one form of collectivity—that of the country's population. The government's statistical *Herald of Public Hygiene, Forensic and Practical Medicine* (*Vestnik obshchestvennoi gigieny, sudebnoi i prakticheskoi meditsiny*), launched in 1865, never grouped its data separately for Jews, Russians, or any other ethnicity in the empire. It operated with the term "population"—of cities, provinces, and regions—without differentiating it by ethnicity or race.[76] Hence, Jews largely remained invisible here, too.

This brief overview summarizes the range of paradigmatic choices available to Jewish advocates of a statistical solution to the perceived problem of Jewish national deficiency. Their choice in favor of national medical and anthropological statistics went beyond these options and reflected a radically different social vision. Jewish statisticians showed the desire to nationalize and racialize population politics in the empire before the most modernist agencies within the imperial state, such as the War Ministry, started doing this themselves. In this, the late nineteenth-century amateur Jewish statisticians were ahead of both the state and zemstvo statistics. More importantly, their message and their normalizing statistical paradigm explicitly served the Jewish national project. In this sense, it was an internal Jewish conversation. And when Jewish statisticians addressed the gentile society or the state, they often did it from a position of power, as experts in the field of national statistics exposing the incompetence of their opponents.

A series of characteristic essays titled "Wrong Conclusions from Correct Numbers" (1884) in *Russkii evrei* offers an example of such a critique directed at a government official who misinterpreted the demographic statistics for Vilna province. Referring to the numbers of deaths and births for all religious groups in the province, this official claimed that the Jewish rate of population growth was lower than that among their gentile neighbors. In the case of Vilna province with its distinct population profile, the nationalization of confessional statistics was a relatively easy operation. Judaism here always served as a code for Jews, Catholicism for Poles (but also Lithuanians), and Orthodoxy for all those other groups the government was ready to designate as Russians. Having arranged his statistics accordingly, the author of the article under criticism claimed that local Jews were dying out due to their failure to adjust to modern life. To protect them from unfair competition with more healthy and better integrated peoples, the Pale of Jewish settlement had to be preserved.

In response, a polemical article in *Russkii evrei* challenged the state official's credentials as a statistician, exposing for example, the false correlation between racial degeneration and the numerical decline of a group. The reasons for population decline, the response detailed, could be numerous, including mass conversion to a different confession or the labor migration of younger men. All the explanations had to be tested statistically and correlated with other factors as well as with general population statistics. Indeed, statistical data showed a high rate of economic migration from Vilna to other regions within the empire and abroad, which correlated with the decline of births in the Jewish community. Using such examples, the Jewish author, step-by-step, destroyed the scientific credentials of his opponent. Only then did he allow himself to shift to an explicitly ideological polemical mode:

> The expression "the extinction of the Jewish population of Vilna province" is wrong. Even if, in the course of eight years, from 1863 to 1871, the absolute number of Jews in the population had dropped by 16,000, the cause of this reduction is not the extinction of the Jewish race. . . .
>
> Only our homegrown quasi statisticians are capable of such a statistical forgery—to say that Jews are dying out; [to say this about] Jews—a na-

tion that is the most vital in the world and the most adjusted to life; the oldest but never-eldering people. As if we are talking about the extinction of the Kamchadal, Aleut, Lopar, Chukchi, or Redskins.[77]

The opposition of never-eldering, civilized Jews to the "primitive" aboriginal peoples of the Russian North was a vivid example of the intraimperial comparison, where one's otherness could always be relativized against someone else's primitive condition. The choice of this rhetorical strategy is telling if we consider that at the time the imperial authorities still lumped Jews together with the "Kamchadal, Aleut, Lopar, and Chukchi" into one legal category of *inorodtsy*—"aliens," literally born to a different kin. The author of *Russkii evrei* assumed an especially difficult expert position: while rejecting both imperial particularism and colonialism, at the same time, he advanced hierarchical differentiation of the victims of colonialism. "Primitive" peoples were unfortunate victims who stood on the verge of extinction not being able to "struggle for existence" and resist the inevitable. Their primitiveness was an objective scientific "fact" and their place in modernity was uncertain. Jews, on the contrary, demonstrated unprecedented biological vitality and endurance in the "struggle for existence"—and these were also considered objective scientific facts. Hence Jewish advocates of medical-anthropological statistics were ready to work hard to find a formula of nationhood that would secure a place in modernity for the Jews, leaving the "Kamchadal, Aleut, Lopar, and Chukchi" behind.

Balancing Culture and Race

For obvious reasons, physicians were the most qualified and active participants in the emerging medicalized discourse of modern Jewishness. Samuil Osipovich Gruzenberg, a graduate of the Petersburg Medical-Surgical Academy, since the 1890s worked as an editor of *Voskhod*. In his writings, Gruzenberg tried to reconcile a medical-statistical approach with a culturalist view of Jewishness as based on a rationalized Jewish tradition (a system of moral values and rational regulations). Whereas Dubnov, another editor of *Voskhod,* attempted a similar synthesis as a historian, Gruzenberg did it as a physician.

Gruzenberg, then still a medical student, began publishing on Jewish medical statistics before becoming a permanent *Voskhod* contributor and a member of its editorial staff.[78] One of his early essays, "On the Physical State of Jews in Light of the Conditions of Their Life," came out in 1884 in three consecutive issues of the *Jewish Review* (*Evreiskoe obozrenie*).[79] As one would expect from a work in this genre, it offered a catalog of medical and demographic facts about Jewish life that made Jews distinct (the small number of stillborn; the relative ease with which Jewish women gave birth; low numbers of those infected with syphilis; the physical feebleness of adult Jews; their high nervousness). The negative physical traits and practices on the list were the results of difficult conditions of Jewish life and mental exhaustion. Positive characteristics had formed under the influence of rational regulations of Judaism and reflected Jewish moral virtues such as chastity and sobriety, as well as Jews' "general intelligence and exceptional love for children."[80] In other words, Gruzenberg depicted Jewish negative specificity as cultural and transient (negative factors would surely disappear as life conditions improved), whereas he depicted the positive Jewish traits as timeless. They implied a very different temporality, an eternal Jewishness defined by a stable set of traits and practices that revealed themselves in the "Jewish national physical type," often used by Gruzenberg as a synonym for race.[81] He discussed the "Jewish national physical type" alongside the racial types studied by Western and Russian physical anthropologists, especially Paul Broca, Paul Topinard, and Russia's own Bernhard Blechmann, a student of Ludwig Stieda at Dorpat University, who pioneered the anthropology of the Jewish population in the Ostsee provinces of the Russian Empire.[82] Blechmann was the only anthropologist in the Russian Empire known to Gruzenberg as a scholar interested in Russian Jews—the "weakest nation in European Russia," according to Blechmann. The Russian Jewish press duly reviewed Blechmann's dissertation and endorsed many of his conclusions, especially on Jewish physical weakness and Jewish racial purity and stability. Gruzenberg also agreed with his findings, conveniently summarized in one paragraph in an 1883 issue of *Russkii evrei:* "Due to the purity of their type and stability of their racial distinctions, Jews provide an anthropologist with exceptionally instructive material for scholarly research. If their centuries-long and constant contacts with all peoples of the globe without exception might have partially influenced their language, costume, morals, and traditions in different countries, their racial traits and peculiarities remained untouched."[83]

Although Blechmann's assessment of Jewish weakness contradicted Gruzenberg's favorite Jewish "vitality" argument, he confirmed that "Jews have too much soul but, at the same time, too little body."[84] In his own works Gruzenberg aspired to keep soul and body connected and to marry the discourses of Jewish racial stability, physical and moral vitality, degeneration caused by social factors, superiority in the struggle for existence, and Jewish cultural traditionalism as the source of this superiority.

One of Gruzenberg's articles in *Voskhod*, "A Chapter from Jewish Anthropological Statistics: On the Distribution of Sexes during Birth" (1887), discussed the prevalence of male births among Jews. In the medical literature of the time this fact received two conflicting explanations: one was purely racial, treating the prevalence of male births as a feature of the Jewish race; the other was purely culturalist and Russia-specific. It posited that state rabbis, who under the Russian "confessional state" regime were tasked with registering births and deaths, documented Jewish boys more diligently. Boys were eligible for military conscription and potentially more likely to deal with the state in the future, whereas girls, the argument went, could remain in a domestic gray zone where official registration of birth did not matter.[85]

Gruzenberg found both explanations unsatisfactory. As he pointed out, Western European Jewish statistics showed the same pattern of prevalence of male births, but the politics of birth registrations in Europe clearly differed from that in the Russian Empire. Gruzenberg also cited statistical data for the imperial capital, St. Petersburg, where, unlike in the Pale, "registration books were kept extremely accurately" and doormen (*dvorniki*) reported to the police every case of birth and death. Still, even in Petersburg male Jewish newborns outnumbered female infants.[86] Such evidence, seemingly, pointed to a biological explanation. Gruzenberg, however, was looking for a synthetic formula. Medical science, he claimed, had established that younger women were more likely to conceive boys, and that abstinence from sex during menstruation also facilitated male births. Observant Jewish girls tended to marry young; they also followed Jewish religious regulations prohibiting sexual activity during and nine days after menstruation. Finally, they were inherently moral and rarely engaged in premarital sex. Gruzenberg related this fact to the gender-biased medical consensus of the time regarding births out of wedlock that, presumably, produced more female births than male. All these explanations belonged to the realm of Jewish "traditional" culture rather than biology, and Gruzenberg insisted that "one does not need to invoke Jewish 'racial individuality,' which is anyway [a] quite problematic [concept]."[87] However, while avoiding the language of race, he embraced a form of race thinking that treated Jewish moral virtues as racial traits and ascribed them to all Jews, at all times and everywhere.

Such ambivalence about race permeated all Gruzenberg's publications. Take his discussion of consanguineous marriage among Jews. Having been incorporated into the narrative of Jewish degeneration and widely publicized by the anti-Semitic press, this topic became especially controversial in the light of the new science and eugenics-inspired medicine. In the Russian case, it alarmed the state, too. In 1893, the Ministry of Internal Affairs specifically inquired with the Rabbinical Committee (established in 1848 as an advisory body for the government) about the existence of the Jewish laws that sanctioned marriages between blood relatives. According to ChaeRan Y. Freeze, the inquiry "resulted partly from the outcry in the science and medical community against the high number of consanguineous marriages among the Jews, and the discussion spilled over into the Jewish press as well."[88] Gruzenberg joined the discussion that "spilled over" into the Russian Jewish press to rebuff, from the point of view of modern science, the association of consanguineous marriage with savagery, primitivism, and degeneration. However, his defense strategy was as inconsistent as his entire medicalized approach to Jewishness. Gruzenberg accepted the existing statistical data as valid and admitted the persistence of consanguineous marriage among the Jews. To counter the claims of other scientists about its detrimental influence, he resorted to the racialized rhetoric of Jewish physical vitality: "It has been recognized that Jews possess a particular 'inborn vitality' (*Vitalité inhérente*), a particular inborn power that saves them from many harmful influences, including the fatal consequences of consanguineous marriages."[89]

He went as far as questioning whether consanguineous marriages indeed produced any fatal consequences. He answered in the negative. Instead, Gruzenberg blamed "heredity"—an "external circumstance that often accompanies consanguineous marriages." He thus protected consanguineous marriage as a custom, a part of the Jewish traditional way of life, while denigrating a seemingly accidental defect that was only transmitted through consanguineous marriage. In this logic, both the unique Jewish vitality and moral virtues and a weak "national physical type" could be equally passed over within a race that practiced marriages between blood relatives.[90] Consanguineous marriage was just a vehicle; the healthier, the stronger, and the more virtuous a race / nation was, the more useful a job this vehicle could perform, and vice versa. If pursued to its logical end (which Gruzenberg avoided doing), his argument amounted to an apology for racial endogamy on biological grounds. Gruzenberg protected one controversial Jewish custom by placing it in the context of the secularized Jewish "tradition," which he saw as the source of Jewish vitality and eventual success in the struggle for existence. But he did this in the language of medical statistics and within the paradigm of exact, verifiable, and universalist science. The contradiction between the two approaches exposed his fundamental dependence on race thinking: only race as a sociobiological concept seemed to allow the articulation of the desired model of Jewish nationhood. As diverse and dispersed as Jews were, they still shared one biological history, which gave rise to the Jewish "tradition"—a common way of life and a set of moral and cultural references for the entire nation.

Gruzenberg's model was obviously very confusing. A leading figure at *Voskhod* and hence an important Jewish voice, he was nonetheless often criticized for a lack of consistency. Contemporaries blamed Gruzenberg for propagating reductionist medical positivism and even cultural assimilationism (in his later publications). Trying to defend his colleague, Dubnov suggested that Gruzenberg's mindset had changed once he started writing for *Voskhod* on foreign life. Presumably, he "had adopted the ideals of the same Western Jewry, whose fate he followed in his weekly reviews."[91] Whether this was true or not, Gruzenberg's writings reflect the inconsistency of his understanding of national Jewishness. He failed to offer a scientifically convincing balance between race and culture. However, his intuition that the two were compatible in a single discourse and that the answer to the conundrum of Jewish national deficiency would emerge as a synthesis of both proved right.

Reformulating the "Jewish Question"

If anything, the discussion of modern Jewish national selfhood in the late nineteenth-century Russian Jewish press was a creative process of trial and error. It fed enthusiasm for the special Jewish science of race and for the collection of Jewish statistics. It helped a distinctly modern language of Jewish nationhood and modernity to receive public visibility and recognition. And it enabled a new Jewish expert discourse that could challenge the authority of other experts. Thus, the dawn of the modern Jewish discourse of race was much more than just a polemic with racial anti-Semitism. It registered a truly postcolonial moment when the Jewish collective Self was problematized in view of the nationalization of politics in the empire, on the one hand, and the universal modernity project, on the other. All the authors mentioned in this chapter approached "race" as a key scientific concept capable of unlocking for them a range of important epistemological and political possibilities. Together, they contributed to the medicalization and scientification of what was known as the "Jewish Question" in the official parlance of the empire.

As John Doyle Klier has shown, in these years the "Jewish Question" functioned as a middle ground where different representatives of the state, Russian imperial educated society, and Jewish actors met, interpreted, and misinterpreted each other, and variously engaged in dialogue or a war of words and policies.[92] The shift in the Jewish discourse on the "Jewish Question" in the 1870s and 1880s reflected changes in the imperial middle ground in general. The role of Russian Jews in facilitating this change was far from passive or even secondary. They were reacting to the arrival of mass politics, including genocidal pogroms and grassroots nationalisms, to the nationalization of the imperial state that was leaving less and less space for imperial particularism and hybridity, and to the spread of new epistemologies and methodologies in social and natural sciences that generated a new ontology of groupness. Such complex impulses prompted Jewish intellectuals and professionals to reinvent themselves for the future postimperial modernity, where "post" meant the scientifically predicted inevitable transformation of the current underrationalized and particularistic imperial order. In this sense, Russian Jews contributed to the racialization and medicalization of the "Jewish Question" at least as actively as other participants in this discursive middle ground, and perhaps even more so. As one Jewish commentator wrote in 1883: "Among different brochures and pamphlets on the unfortunate Jewish Question, brought about by recent aggravation of this eternal question, there was, I recall, a brochure titled 'The Pathology of the Jewish Question.' I do not know whether someone has published 'The Homeopathy of the Jewish Question,' but if no one has done it yet, this is very unfortunate."[93]

Obviously ironic, these words were hardly an exaggeration. The language of the participants in the debate about Jewish national deficiency was indeed overloaded with medical metaphors, diagnoses, and recommendations on how to treat the sick collective Jewish body. A typical diagnosis from *Voskhod* would read: "The Russian-Jewish body looks like something very remote from a unified whole that is full of life juices, youthful freshness and might; it does not resemble a body on which deep

wounds heal quickly and various losses recoup easily."[94] This pronounce-
ment was seconded by another typical diagnosis, this time from *Russkii
evrei*: "Currently Russian Jews are in the position of a critically ill [indi-
vidual] who has just recently survived a quite dangerous crisis, and who
after this crisis has been feeling as if not a single healthy spot was left in
his entire organism . . . as if, finally, his illness has produced a strong
anemia, which by itself is dangerous for his organism."[95]

Jews were deemed sick because of the persecutions and physical violence inflicted on them, but also, and even more so, because they were
not a modern nation—strong in its unity, self-conscious, living in harmony
with its inner and outer physical self, enjoying political rights and well-
developed social and cultural institutions. For some, the main cause of
illness was Jewish nonterritoriality; for others—their racial otherness.
These sick Jews inhabited a world populated by other sick and healthy,
normal and incomplete nations ("for quite a while we have spoken about
Turkey as a sick man; and even concluded that its illness cannot be
treated").[96] Jews struggled with these other nations for existence and resisted influences undermining their nation's vitality. The ultimate success
of the struggle depended on the success of the Jewish scientific self-
cognition that would allow treatment of the sick. Only modern science
could provide foundations for a new population politics targeting national
and social bodies. As a contributor to *Russkii evrei* put it in 1883, the political science of the future would be "a comparative biology of nations":

> Such a science would call to life other related sciences such as the embry-
> ology of nations, psychology of nations, psychiatry of nations, anatomy
> of nations, hygiene of nations, pathology of nations, and so on. The re-
> sults produced by these scientific disciplines could lay the foundations for
> a purely practical science—i.e., anthropo-valentologia, a theory of exact
> estimation of the value and merits of human individuals and nations.[97]

CHAPTER TWO

Samuel Abramovich Weissenberg

"Provincializing Europe" through Jewish Racial Self-Description

> At any rate, it will not be possible to offer a definitive answer to the question of the anthropological status of the Jews until research on the Jews in Western Europe, Asia, and Africa has been done. Such important research, undertaken following unified principles, will allow, it is to be hoped, a primordial type (*Urtypus*) of Jew to be discovered. . . . The gain [in knowledge] to be had from this investigation will be my greatest reward.
>
> —Samuel Weissenberg, 1895

SAMUEL ABRAMOVICH WEISSENBERG was one of Russia's provincial Jews who grew up bilingual, reading both the Jewish and the Russian press such as *Voskhod* (Dawn).[1] He embraced the mission of Jewish scientific self-cognition and the utopian goal of reconstructing the Jewish *Urtypus*—the anthropological standard of absolute primordial physical Jewishness. He thus adopted a view shared by many physical anthropologists of the day that contemporary human diversity developed from initial pure types. With it, he embraced the standard research procedure of deducing the primordial type from the presently existing diversity. The success of scientific reconstruction of the *Urtypus* depended on the availability of statistical data on all living Jewry world-wide. The existing imbalance between the anthropological data on European and non-European Jews prompted Weissenberg to take on a scientific mission on behalf of non-European Jewry. He became one of the best-known and most widely traveled Jewish race scientists of his time, described today in the *Encyclopedia Judaica* as "perhaps the most distinguished of that first generation of Jewish anthropologists" after Cesare Lombroso.[2]

However, regardless of his scholarly achievements, Weissenberg spent most of his life in provincial Elisavetgrad in the Pale of Settlement (today, Kropyvnytskyi in Ukraine). He never enjoyed the right to reside outside the Pale, never held any academic position or worked at an academic institution, and never formally participated in a political party. This initial liminality later made him almost invisible to historians of regular science or Jewish politics. John M. Efron rediscovered Weissenberg in the early 1990s and included him in his pioneering study *Defenders of the Race*. Recently, Amos Morris-Reich revisited Weissenberg's anthropology in a chapter that explores Jewish racial self-classifications.[3] To acknowledge Efron's precedence and avoid unnecessary confusion, I also use the now conventional German spelling of Samuel Weissenberg.[4] However, my interpretation of Weissenberg's Jewish anthropology diverges from Efron's and Morris-Reich's, who tend to see him exclusively as a protagonist on the German or rather German-language academic and ideological scene.[5] In this role, Weissenberg seems to be a familiar case—that is, a Jewish scholar responding to the rise of German scientific racism.[6] This view leaves behind a much more controversial and complex Weissenberg—a prolific contributor to the Russian-language scholarly press and participant in the imperial public sphere, as well as a Russian Jewish activist. This Weissenberg advanced Jewish self-racializing in an imperial situation where the discourse of race did not possess ultimate hegemony and was claimed simultaneously by various actors from above and below. My focus is on this other Weissenberg and his imperial situation.

Weissenberg's Intimate Empire

Elisavetgrad, where Weissenberg was born and raised, was his first window on the empire and the world. As everywhere else in the Pale, in the multicultural Elisavetgrad, where Jews composed just over one-third of the population, most of them spoke Yiddish, and many of Weissenberg's generation and social class were bilingual. Elisavetgrad was a county capital in Kherson province of New Russia (Novorossiia), a "no-man's-land" in the steppes north of the Black Sea. Long dominated by nomads and later by the Crimean Khanate, in the second half of the eighteenth century this territory became an object of the Russian Empire's active colonization efforts. Colonists were invited from all over Europe and recruited among Russia's subjects, including Jews.[7] New Russia attracted the most mobile and industrious among them, less bound by traditional restrictions and communal control. As a newly created community devoid of ingrained traditions that were left behind in the old shtetls of the former Polish lands, Elisavetgradian Jewish society exhibited more dynamism and readiness for integration.[8]

It is not easy to draw an ethnocultural map of the Elisavetgrad population in Weissenberg's time. Russian imperial statistics did not use categories such as ethnicity, nationality, or race, and heavily relied on categories such as mother tongue, religion, and the obsolete category of "legal estate." According to the first and only imperial census of 1897, conducted when Weissenberg turned thirty, the population of Elisavetgrad numbered 61,841.[9] Of these, 58 percent (35,868) spoke Russian, Belorussian, or Little Russian (Ukrainian) at home, and 56.8 percent (35,115) were Russian Orthodox Christians. These two groups largely overlapped, producing a fairly loose agglomerate of Orthodox Christian Eastern Slavs that together constituted close to two-thirds of the city population. Actually, the total number of Christians and Slavic speakers in Elisavetgrad was a few percent higher, but they were "different" Christians and Slavs. There were also 292 Tatar speakers (mostly Muslims). Of the Elisavetgradians, 37.6 percent (23,256) spoke Yiddish at home, and 38.8 percent (23,967) identified Judaism as their religion.[10] The discrepancy between these numbers means that, as in the case of Eastern Slavs, language and confession did not coincide neatly, and the effect of "nationality" produced by their constellation is more a statistical phenomenon than a characteristic of collective identity (a "category of practice"). A total of 711 Jews, who were assimilated into Russian, German, or Polish cultures, just as 753 Eastern Slavs who were not Russian Orthodox, demonstrate flexibility in combining different factors into one's cultural identity. This flexibility also suggests that even those who were Russian-speaking and Orthodox or Yiddish-speaking and Judaic did not necessarily experience the combination as a national identity in the modern sense. Therefore, the statistical portrait of Elisavetgrad as divided between the "Russian" and the "Jewish" majorities conceals the internal fluidity of each group and porousness of their boundaries, making political mobilization along national lines problematic. Using the names of ethno-confessional groups without additional qualification for the sake of brevity, it is important to keep in mind their approximated, context-sensitive boundaries.

Religiously or linguistically defined, the Jewish population almost tripled between 1861 and 1897, from 8,073 to 23,967.[11] The year 1861 is one of the most important turning points in Russian history: it marked the beginning of the peasant reform that put an end to serfdom and opened the floodgates of social mobility (hence the skyrocketing population growth after 1861). Another side effect of the reform was the rise of modern Russian nationalism: finally, the Russian nation could be conceived of as a popular and mass-scale phenomenon, rooted not in a tiny layer of the educated elite but in the "people"—no longer serfs but free peasants.[12] The year was fateful for Elisavetgrad for one more reason: since 1829, the town had been controlled by the Administration of Military Settlements (Upravlenie voennymi poseleniiami). In 1861, it finally regained its status as the regular capital of Elisavetgrad county, and a civic administration replaced the military. Elisavetgrad then developed as a regional trade and industrial center, with Jews playing prominent roles in commerce, trade, and crafts. (Weissenberg's father was registered in the second guild of the Elisavetgrad merchant legal estate, providing a comfortable living for the family of seven.) Jews also controlled three-fourths of Elisavetgrad's factories—gristmills, machine and tool factories, foundries, soap factories, brickyards, vinegar distilleries, and tobacco factories. The industrial labor employed by most of the factories was not Jewish (the exception being tobacco manufacture).[13]

Such an uneven pattern of industrialization and distribution of labor most probably explains why on April 15, 1881, Elisavetgrad became the site of the first Jewish pogrom in the infamous "pogrom wave" of 1881.[14] Weissenberg was fourteen at the time. The pogrom began with a dispute in a tavern provoked by the accusation that Jews used Christian blood for ritual purposes, and continued for two days, bringing much violence and destruction.[15] It left a lasting trace in Jewish memory and was incorporated into a liberal version of the history of Elisavetgrad, but not in the official narrative.[16] The official overview of Elisavetgrad's history published in 1897 did not mention the pogrom at all—not even in a chapter exclusively dedicated to disasters that had occurred in the town since its founding in 1752.[17] Later in life, Weissenberg and his fellow Elisavetgradian Jews witnessed several more pogroms: first during the Russian Revolution of 1905–1907 and then during the revolution of 1917 and the ensuing Civil War.

Still, the 1881 and 1905 pogroms did not seem to serve as decisive watersheds affecting the demographic dynamics or social landscape of the Jews. In 1895, when Weissenberg published his best-known anthropological work, *Die südrussischen Juden,* the structure of the Elisavetgrad population should have been close to that registered by the 1897 census, with 38.8 percent self-identifying with Judaism. In 1911, when his second most important study came out (*Das Wachstum des Menschen*), the City Sanitation Committee estimated the Jews (by religion) at 39 percent.[18] The remarkable stability of these numbers after fifteen turbulent years, during which the entire population of the city increased by 22 percent, implies that the rate of population growth was the same for Jews and gentiles. The growing number of Jews was accompanied by an evolving communal infrastructure. In 1913, Elisavetgrad had five synagogues and thirteen Jewish prayer houses, as well as Jewish hospitals, libraries, schools, colleges (not all of them religious), and mutual aid associations and cooperatives.[19]

The change came with World War I. By the time of Weissenberg's death in 1928, the Jews in his native town had dropped to 27.6 percent (18,500), mostly because of wartime dislocations and pogroms.[20] The population dynamics must have been experienced even more dramatically inside the city. First, the influx of wartime refugees from the front swelled the number of Jews in Elisavetgrad and fostered a sense of affinity within the Jewish community, brought together by the task of accommodating and supporting the newcomers.[21] Then, the pogroms psychologically demoralized and numerically decimated the Jewish community, while the end of hostilities and triumph of the Soviet regime in the early 1920s prompted the migration of thousands of the most entrepreneurial Jews to Moscow and Petrograd to escape overpopulation and unemployment.[22]

Weissenberg spent these difficult years in his native town and experienced all the painful transformations. During the span of his life the city changed its name twice. In 1918, its original eighteenth-century name, after the empress Elizabeth, was modified to reflect a modern pronunciation—Elizavetgrad. In 1924, the city was renamed Zinovyevsk, after the Bolshevik leader Grigory Zinovyev. A close associate of Vladimir Lenin and longtime head of the Communist International, Zinovyev was born in

Elisavetgrad as Hirsch Apfelbaum, sixteen years after Weissenberg. Thus, without moving to another city or emigrating, Weissenberg nevertheless died in a different city and country from where he had been born.[23]

The Jewish "Caucasians"

Weissenberg first left Elisavetgrad in 1884, after graduating from a local realschule, and headed directly for Germany. Russian universities admitted only graduates of classical high schools (gymnasia), so he decided not to take any chances and for two years studied mathematics and natural sciences at the Higher Technical School (Technische Hochschule) in Karlsruhe. He then transferred to the medical department of Heidelberg University, graduating in 1890.[24] As soon as he secured a medical diploma, guaranteeing him a stable income as a doctor, Weissenberg began pursuing his "greatest reward," as he put it—the reconstruction of the Jewish *Urtypus*. He spent a few years in Constantinople, practicing medicine and anthropology, before returning to Russia for good, establishing a permanent medical practice in Elisavetgrad.[25] Dr. Weissenberg became quite popular in the provincial town, which he gradually turned into one of the most anthropologically studied places in Russia, widely known to race scientists everywhere in the world. His last work, published posthumously in 1928 and exploring demographic patterns in Elisavetgrad from 1901 to 1925, became his final tribute to his town.[26]

When Weissenberg was rediscovered as a local celebrity by Elisavetgrad historians following 1991, after long years of having been completely forgotten, some interesting facts surfaced. During a cholera outbreak in 1893, Weissenberg treated poor patients free of charge in his private clinic, assisted by his wife, Dorothea, and a team of nurses. He even visited the gravely ill in their homes, also without any charge. The Elisavetgrad City Council honored Weissenberg's "diligent and selfless service to the poorest part of the town's population" in its official address.[27] He was one of the highest-paid gynecologists in Elisavetgrad, providing services to the city elite. While making a good living from the city's wealthy, he continued treating poor patients—working-class women, Jewish and gentile alike, free of charge. The reputation Weissenberg earned among the different strata of society worked for him: his clinic and apartment (in the same two-story building) were always protected during pogroms, including the genocidal Civil War–era pogroms by irregular partisan squads. Each time, a team of firefighters arrived to protect the doctor and his clinic from the pogromists.[28]

Weissenberg's reputation with Elisavetgradian Jews could partially explain the trademark feature of his anthropological work—he had practically unrestricted access to hundreds of local Jewish, and some gentile, bodies. He collected measurements on all these people in situ, without any backing from the state and with no access to regular loci of mass anthropological surveys, such as military barracks or school classrooms (which did not prevent him from accumulating the measurements of hundreds of Jewish schoolchildren). The systematic collection of data in one's own community was not common practice among race scientists. Today we would, perhaps, say that Weissenberg abused his patients' trust: all of them, poor and rich, men and women, trusted him as a doctor. They allowed him access to their bodies most probably without always knowing that he would use their data for anthropological research. On the other hand, his better-educated patients most certainly knew of their doctor's scientific fame and hence might have consciously participated in his scholarly project, being able to appreciate its usefulness for the common Jewish cause. One way or another, Weissenberg's method of collecting anthropometric data from his Elisavetgradian Jewish neighbors and patients exemplified self-exploration in a most direct sense.

At first, the measurements on Jewish men ages twenty-one to fifty, Jewish women ages eighteen to forty, and girls and boys ages one to five (1,300 altogether) were used for Weissenberg's groundbreaking work on the anthropology of Southern Russian Jews, originally published in German in 1895.[29] The work pursued a broad methodological and theoretical agenda, suggesting new methods of comparative anthropometry and evaluating the influence of racial and environmental factors on human development. Weissenberg assembled a substantial number of women's measurements, focusing on "the boundaries of gender roles and their relation to race," which Efron has reasonably interpreted as a reaction to the widespread stereotype of Jewish effeminacy.[30] It would be equally plausible to see here the influence of anthropological Darwinism: females, it was believed, better preserved the initial racial type (*Urtypus*) due to the working of the law of sexual selection (although exercised exclusively by men). In other words, those women who embodied the ideal of beauty accepted within a given group had more chances of finding a partner and hence procreating and reproducing their type.

Weissenberg separated characteristics he viewed as culturally determined (such as narrow chest or low height) from those he considered
to be features of race, including form and proportions of head, forehead, and face, long arms, short legs, and others. He concluded that
Eastern European Jews and their typical representatives, Southern Russian Jews, formed a more or less unified type that had little in common
with the Semitic type. Instead of long Semitic dolichocephalic skulls they
had short, brachycephalic ones; flat foreheads instead of slightly elevated
ones; relatively prominent cheekbones instead of the ideally oval faces;
rather wide mouths instead of small ones, and so on. This meant that
only the descendants of the ancient Jews, who in the course of their migration came in contact with the long-headed populations of the Mediterranean, maintained the primordial type.[31] Unlike them, the Southern
Russian (Eastern European) Jewish type resulted from the mixture that
occurred when Jews first migrated into the southern Russian region
prior to the destruction of the Second Temple. Their route traversed the
Caucasus, where they mixed with the local population; later, their descendants converted the Khazars to Judaism in the eighth century. There
was also another possible route from ancient Palestine to the northern
shores of the Black Sea and the Sea of Azov, which also involved racial
intermixture.[32]

Two very important implications followed from Weissenberg's reasoning. First, it rendered the European (including Eastern European)
Jews as "Caucasians." Because the majority of European Jews did not
resemble the Semitic type, the most logical explanation was that they must
have been descendants of the Jews whose racial features had been transformed during their sojourn in the Caucasus. While Western European
Jews, many of whom subscribed to general ideas about cultural and civilizational hierarchy between populations of Western and Eastern Europe,
treated their Eastern European brethren with a degree of disdain, Weissenberg's measurements of the Jewish type presented Russian Jews as a
crucial link in the chain of formation of Ashkenazic Jewry.[33] He literally
offered a Jewish version of the "Caucasian" race, replicating Johann
Friedrich Blumenbach's "discovery" of the Caucasus as the racial birth-
place of modern white Europeans (1795).[34] As Efron has summarized
Weissenberg's message: "It was not West European but East European
Jews, with their noble past and culturally vital present, who were the authentic creators and bearers of the European Jewish tradition."[35]

The second important implication was the claim of Jewish ancient presence on the territory of the contemporary Russian Empire, supported by objective scientific evidence. If Jews arrived in the region in the first centuries CE, they were not a "diaspora," an alien Semitic race living where they did not belong. On the contrary, Weissenberg's logic implied that the Jews were one of the mixed local races whose physical type resembled types of the surrounding population and who could claim for themselves aboriginal status, at least in southern territories of the empire.

The Jewish Mix

Die südrussischen Juden received international recognition for its unquestionable scientific merit and far-reaching analysis. At the same time, different audiences reacted to it somewhat differently. For German academics, the discussion of the *Urtypus* and the methodology of anthropological study of a living population held the most value. For Jewish race science and the Russian Jewish discourse of modern nationhood, proper scholarly aspects of the work were as important as the implicit ideological themes: the discussion of Russian Jews as "Caucasians"; the genealogical connection between modern European Jews and "traditional" non-European and Eastern European Jewry; and the archaization of the Jewish presence in the lands of the Russian Empire. And there was one more specific audience that was very important for Weissenberg—the Moscow-centered liberal network of physical anthropologists, who embraced Jewish scholars as equals and Jewish topics as legitimate. The network organization of this all-imperial community of race science enthusiasts suited Weissenberg's uneasy circumstances exceptionally well. He could remain in Elisavetgrad, outside institutionalized academia, and yet feel fully integrated into the modern scholarly community in the empire. Prior to receiving confirmation of his foreign medical degree in Russia in 1896, he did not have the right to apply for permission to reside outside the Pale of Settlement.[36] But even after 1896, Weissenberg remained in Elisavetgrad, visiting imperial capitals only on special occasions. At the same time, he regularly exchanged information with other members of the network and published in the network's *Russian Anthropological Journal* (*Russkii antropologicheskii zhurnal, RAZh*). One of Weissenberg's letters to the leader of the Moscow school of physical anthropology, Moscow University professor Dmitrii Anuchin, written in

1892, offers a glimpse at his situation as a Jewish scholar in the Russian Empire:

Most Honorable Professor,

First, please let me express my gratitude to you and the society for the regular delivery of the minutes [of the meetings of the Anthropological Division of the Society of Lovers of Natural Sciences, Anthropology and Ethnography (IOLEAE)[37]].

I am enclosing 8 rubles and asking you to register me as a member of the anthropological congress.[38] Unfortunately, I cannot come to Moscow personally: being a Jew who has received his higher education abroad, I don't have the right to reside in Moscow. As much as I want to get acquainted with some scholars, and with Moscow museums and exhibitions organized for the congress, I have absolutely no desire to get acquainted with the Moscow police.[39]

In another typical letter from Constantinople, Weissenberg asked Anuchin to send him, "in the interests of science," works by Nikolai Maliev on the anthropology of Bashkirs and by Nikolai Zograf on the anthropology of Meshcheriaks relevant for his own work on Russian Jews.[40] In return, Weissenberg promised to send information about his work in progress "to be published in the Diary of the Society."[41] In this network, Weissenberg was an equal member.

The Moscow school rejected the idea of pure races and advanced the concept of the "mixed physical type" as the main elementary racial unit in the empire. The interest in mixing partially resulted from a positivistic methodology embraced by the school's representatives. In most cases their statistics, whether collected at a provincial or county level, did not yield any sustainable pattern of racial purity. Yet an even more powerful impetus to explore racial hybridity came from the Moscow liberal school's conscious aspiration to account for the empire as a natural entity made up of complex elements, an organic wholeness not reducible to a mechanical sum of these elements. The relationships between the elements were therefore regarded as independent entities and problematized accordingly, stimulating the development of the scientific language of mixing, crossing, degrees of kinship, and so on.[42] Weissenberg, integrated in the network and allied with the school, could not escape being influenced by this methodology and approach. His insistence on the mixed nature of the Russian Jewish type and the spatial localization of the mixing that had produced the Jewish "Caucasians" within the existing imperial borders were not a coincidence. When presenting his study to a meeting of the Anthropological Division of the IOLEAE in Moscow in October 1895, Weissenberg specifically stressed the composite structure of Jews as a group and the mixed nature of the European-Jewish physical type. As he explained, despite the stereotypical belief in the physical likeness of all Jews, they were in fact split into a number of physical types that differed by eye color and hair color, skull index, nose shape, and other characteristics. Weissenberg's distinction between physical types and what he called facial types was not always clear. In *Die südrussischen Juden* he distinguished seven Jewish facial types, while in his talk at the IOLEAE meeting he identified eight: (1) Jewish (type) with rough facial features; (2) Jewish (type) with subtle facial features; (3) Slavic (Jewish type); (4) Northern-European (Jewish type); (5) Southern-European (Jewish type); (6) General Caucasian (Jewish type); (7) Mongol (Jewish type); and (8) Negro (Jewish type) (very rare).[43] One way or another, these types implied that Jews had historically mixed with local peoples and resembled local types. Moreover, the synthesis of all eight—modern national Jewishness—itself was a mixture of European and non-European racial mixes.

Russian liberal anthropology provided clear guidelines for establishing the mixed racial type as an elementary unit of the imperial demographics. The regional approach was one important requirement. It prompted a scientist working on one particular group to collect similar anthropometric statistics on all other groups from the same region (defined in accordance with the official imperial nomenclature, as an administrative unit that rarely coincided with any historical or culturally distinct territories). In addition, the network's members were expected to extend their comparison to all "physical types" in the empire and view the existing imperial borders as natural boundaries of their research field. Not unlike many of his colleagues, Weissenberg simultaneously followed these rules and violated them (for example, by arbitrarily choosing the objects of comparison). Remarkably, the leaders of the Moscow school—the custodians of its method—tended to close their eyes to the methodological transgressions by Jewish race scientists. In this case, the "affirmative action" philosemitic stance of the Russian liberal anthropological community was meeting a matter-of-fact perception of Jewishness as a transnational/transimperial phenomenon. Indeed, if one were to accept that Jews had originated in ancient Palestine and then spread almost all over the world, it seemed legitimate to study Jewish biology (especially when reconstructing their *Urtypus*) by comparing different groups of Jews across political borders.[44]

Weissenberg developed a distinct politics of comparison of his own, accepting the population diversity of Elisavetgrad as a norm and pursuing his ideological agenda of proving the aboriginal status of East European Jewry.

He compared Southern Russian Jews to "Great" and "Little" Russians of the region (nowadays ethnic Russians and Ukrainians), and to Galician Jews in the Habsburg Empire and other types of East European Jewry. Similarly, after having measured the entire Elisavetgrad Karaim community, all thirty members of which "without exception submitted themselves to [his] sliding compass," Weissenberg compared them to different Jewish groups and one of the Turkic peoples of the empire.[45] The most logical choice in this case would have been Tatars, in particular Crimean Tatars, who had controlled the steppes of Novorossiia for centuries prior to Russia's imperial expansion. However, Weissenberg chose Bashkirs, who lived over one thousand miles northeast, in the Ural Mountains region, and who had been considered a legal estate (like the Cossacks) rather than an ethnic group just several decades earlier.[46] Although unusual, this choice did not violate the Moscow school's orthodoxy and could be substantiated by the fact that, at the time, Bashkirs were among the most anthropologically studied groups in the empire. Weissenberg located Karaims between Eastern European Jews and Bashkirs. With the latter, they shared "extreme shortheadedness and general configuration of the face." The former had whiter skin and narrower faces. The overall conclusion read that Karaims presented a crossing (*pomes'*) of Jews and Tatars (a reversed substitution, based on the established racial kinship between Tatars and Bashkirs), with a possible participation of other, yet to be established, "elements."[47]

The Limits of Comparison

The analytical framework of mixed physical type effectively defused the nationalizing impulses of members in the liberal anthropological network. To be accepted as members of the profession, most practitioners of physical anthropology, often amateur scientists without any special training, tried to uphold the standards established by the authoritative scholars leading the school. They followed the instructions requiring them to compare the objects of their study to some reference groups, to think in terms of regional types, and to confine their explorations within the given imperial borders. The shared orthodoxy of mixed physical type lessened potential conflicts and overt politicization of anthropological discourse.

Some tensions, however, did arise, and Weissenberg's case reveals the limitations of the Moscow liberal discourse in imposing the mode of impartial and depoliticized scholarship.

In 1912, Weissenberg ran into a conflict with a fellow network member, Alexander Dzhavakhov (Dzhavakhishvili), a leading expert on Georgian anthropology of the Russian Caucasus. Their studies often appeared under one cover in the *RAZh*, the Moscow school's mouthpiece.[48] Just like his Jewish colleague, Dzhavakhov combined Western European education, profound Russian acculturation, and non-Russian (in his case, Georgian) ethnic origin and loyalty. Similar to Weissenberg's Jewish racial type, Dzhavakhov posited the existence of a special Georgian "anthropological group."[49] But unlike Weissenberg's composite and nonterritorial Jewish race, Dzhavakhov's Georgian race was territorial. He did not openly violate the orthodoxy of the Moscow school and took measurements on different population groups besides Georgians, calculating degrees of racial kinship between them and Armenians, Bashkirs, Jews, Ingush, Kumyks, and Great Russians. At the same time, in Dzhavakhov's estimation, this kinship always appeared to be quite remote, and the "Georgian group" numerically dominated the area coinciding with the imagined Georgia of modern Georgian nationalists. Rather unusually for a member of the Moscow liberal network, he tried to smuggle in an explicit anticolonial agenda by substituting the language of imperial territorial division into administrative units with authentic Georgian terms for historical territories. Thus, Dzhavakhov consistently used terms such as "Sakartvelo" and "Kartvelians," "sanctified by the national worldview," instead of "Georgia" and "Georgians."[50]

This Georgian anti-imperial nationalist responded aggressively to what he perceived as a nationalist claim by his Jewish colleague in the article "The Jews of the Caucasus in View of Anthropology," which discussed two distinct Jewish groups in the Russian Caucasus as being deeply integrated into the local texture of life:

> The first group [Mountain Jews] lives in the mountains and on the plain adjacent to the Caspian Sea. They speak a peculiar Tat language, which includes multiple [words] of Persian origin, and they practice mostly farming and gardening. The second group lives in the western part of the Caucasus adjacent to the Black Sea. They speak a pure Georgian language and are engaged almost exclusively in trade.[51]

Weissenberg collected anthropometric statistics for these two groups and compared them with similar data for non-Jewish groups of the region (assembled by Dzhavakhov and other imperial anthropologists). The centerpiece of his analysis, however, was the comparison of the Jews of the Caucasus to other Jewish types, in particular Yemenite Jews and Southern Russian Jews.[52] Here, he extended the so-called short scheme (basic measurements such as head and facial indexes and height plus color indicators) to include length and width of head, face and nose, and nasal index and form. The article claimed that the Mountain and Georgian Jews were anthropologically identical and hence had to be regarded as one racial type that "has assumed different ethnographic forms."[53] The Jews of the Russian Caucasus were short-headed and hence differed from the long-headed Yemenite Jews, whom Weissenberg regarded now as being closest to the hypothetical *Urtypus*. At the same time, eye and hair color indicators of the Mountain and Georgian Jews put them in a separate category from Southern Russian Jews. They thus presented another localized mixed Jewish type, whose short-headedness helped to trace the transformation of ancient Jews during their migration through the region, farther into Southern Russia and Eastern Europe.[54]

What infuriated Dzhavakhov about this study so much that he published a special article attacking Weissenberg?[55] Apparently, it was Weissenberg's first paragraph, which Dzhavakhov cited in full in his polemical response:

In his recent article in the *Russian Anthropological Journal* . . . S. A. Weissenberg . . . writes: "Having been for many centuries under the oppression of the surrounding population, whose Christian and Muslim parts hardly differed in their ignorance and fanaticism, and being not connected to their European brothers, Caucasian Jews, naturally, could not succeed. What is really perplexing is that even now, when schools have become accessible and culture has broadly penetrated the Caucasus, Jews still allow Georgians and Armenians to surpass them." We cannot agree with this characterization of the relationships between Caucasian Jews and the native population of the Caucasus.[56]

This was the only paragraph in the entire article in which Weissenberg allowed himself some emotional attitude and value judgments. Nevertheless, Dzhavakhov perceptively sensed that the brief introduction concealed a certain political agenda and worldview in which Georgians (and Armenians, who concerned him less) exemplified the underdevelopment and eternal stagnation of the Orient. Their "oppression" was the main reason for the deplorable state of Caucasian Jews. An extended version of the same paragraph, but from a different article Weissenberg penned a year later for the Jewish ethnographic journal *Jewish Antiquity* (*Evreiskaia starina*), proved Dzhavakhov right. Three pages of what was Weissenberg's report to the Russian Jewish audience on his expeditions to the Caucasus and Crimea covered the history of "religious wars and fanaticism" that formerly made the life of Jews in the Caucasus unbearable. The Jewish scholar-traveler vividly described deserted Jewish settlements (*auls*)—silent witnesses of the destruction of the once flourishing Jewish culture and numerous population. Themes of oppression, persistent underdevelopment, and hopelessness dominated the narrative: "Submissively stepping under the sliding compass, some [of the Jews] whispered naively: God knows, perhaps '*Geula* [Redemption] will come!' [Bog znaet, mozhet byt', "geula budet!"]. . . . Only this touching belief in the imminent redemption makes their sad life a bit better."[57]

Weissenberg put the blame not on the colonialism of the empire, but on the colonialism of the local Georgian population. Recently, he reported, in their quest for enlightenment, Caucasian Jewish communities had begun inviting better-educated rabbis and teachers from the Pale. Because these newcomers did not know the Georgian language but knew Russian, they unwillingly played the role of Russifiers and agents of empire in the eyes of local Georgians. Weissenberg admitted this but held Georgian nationalism responsible for growing anti-Semitism among the Georgian intelligentsia: "'Georgian Jews need not forget that Georgians regard them as one of their own,' a cultured Georgian noted to me with an unkind note in his voice, when we discussed how to elevate the cultural level of Georgian Jews. Once again, the old story repeats itself: Jews are treated as a thing, with which those who are stronger at a given moment do as they wish."[58]

Dzhavakhov, who was unaware of this article when working on his response, nevertheless correctly decoded Weissenberg's message from just one paragraph in the otherwise normative anthropological publication.[59] In his rejoinder, he reinstalled the true colonizer by insisting that Jews since time immemorial "have shared the historical destiny of the Georgian people," that is, the destiny of a colonized (by the Russian Empire) nation. Dzhavakhov firmly rejected Weissenberg's hypothesis that native peoples of the Caucasus in ancient times were assimilated into Judaism, Jewish culture and race, and that their mutual anthropological resemblance was due to this assimilation. On the contrary, wrote Dzhavakhov, these were Caucasian Jews "who bore a definite stamp of assimilation into the types of the aboriginal population of the Caucasus."[60] Their cultural status was not as low as Weissenberg believed, and their socioeconomic situation never differed in principle from that of the rest of the population. If this status had deteriorated recently, not Georgians but Jews themselves were to be blamed: stubbornly protecting their religious particularity, they gradually started drifting away from the "sociocultural organism of the Georgian people."[61]

In the end, Dzhavakhov claimed that Jews exhibited greater racial uniformity than Weissenberg had assumed, and that they in fact visibly differed from those whom he perceived as members of the Georgian nation, even if this difference was not directly reflected in the available anthropometric statistics. Regardless of how Georgian Jews diverged from the European Jews, he argued, they still "retain some distinctive features of organization, intrinsic to Jews in general." Dzhavakhov called these features "racial survivals," which presumably distinguished Jews "from other types even when they [Jews] exhibit diverse physical traits." Regrettably for Dzhavakhov, the "survivals" could not be measured by anthropological instruments, but the trained eye of an anthropologist, he insisted, would easily identify them in the "form of the eyeball, eyelid, nose, mouth, lips, ears, jaws, wrist, and nails."[62]

Nothing in Weissenberg's Jewish anthropology would support this view. He persistently rejected the idea of the universal recognizability of the so-called Jewish physiognomy.[63] Already in *Die südrussischen Juden* he made his position clear: the eye usually notices only traditional Jews marked by their peculiar habitus (costume, gestures, sidelocks). The more traditional the habitus, the more recognizable the type seems. An acculturated Jew was hardly distinguishable from his or her neighbor. That is why, Weissenberg concluded, the discussion of Jewish racial difference and uniformity should be based on exact anthropometric measurements, and only on them—as if measurements were immune to politicization and determined their "correct" interpretation.[64] The conflict with Dzhavakhov proved the illusionary nature of this belief.

Citizen Science: Redefining Coloniality

While it may not appear obvious in the light of their conflict, both Weissenberg and Dzhavakhov successfully operated within the confines of the hegemonic discourse of the Russian imperial anthropological community.

Their disagreements and even national competition did not impact their academic reputations within the network of Russian liberal race science. Both remained true to the orthodoxy of the mixed racial type (all subversive attempts to homogenize Georgian or Jewish types were rather inconsistent and camouflaged) and the correct politics of comparison. If they allowed ideological statements, these were still marginal in the context of their scientifically sound anthropological works and could be ignored as parascientific in nature. The Moscow anthropological school was capable of accommodating and disciplining both Weissenberg and Dzhavakhov and keeping their studies within the limits of acceptable consensus. Both were awarded the IOLEAE gold medals named after Moscow University professor A. P. Raztsvetov.[65] The preeminent Russian anthropologist and leader of the Moscow liberal school, Dmitrii Anuchin, specifically underscored the high scientific quality and hence objective neutrality of Weissenberg's work, in which

> for each category of measurements, the author offers a table with numbers, indicating a maximum, a minimum, and a percentage of different numbers. He also builds curves and diagrams, often comparing [his data] to the results of other scholars who studied Jews and other tribes and peoples. To determine the influence of income, the author compared data for the students of gymnasia and realschulen to that for the students of public primary schools [narodnaia shkola] . . . ; to determine the influence of occupation [he compared] data for coachmen, blacksmiths, carpenters, and so on to that for tailors, shoemakers, bookbinders.[66]

Anuchin concluded on a half-informal, friendly note: "For someone living in a provincial town (Elisavetgrad), such an award can be especially valuable."[67]

Fifteen years later, on September 15, 1912, another general meeting of the society recognized Weissenberg's cumulative contribution to Jewish anthropology and awarded him its highest anthropological prize, named after Grand Duke Sergei Alexandrovich, a famous anti-Semite who ordered the expulsion of Jews from Moscow in 1891. The liberal anthropological community thus asserted its independence from the political regime and confirmed the exemplary quality of Weissenberg's scientific contribution. At the award ceremony, another Russian Jewish anthropologist and leading member of the Anthropological Division, Arkadii El'kind, delivered a speech in Weissenberg's honor that was demonstratively "technical" in tone, allowing not a hint of the special nature of Jewish race science or suggesting a possible affinity between them as Jewish scholars researching Russian Jews and sharing, perhaps, some extra-academic agenda:

> He measured representatives of the autochthonous population of Palestine such as local Jews, Samaritans, and Fellahs; then, Yemenite Jews, Spanuoli, Jews of Central Asia and the Caucasus (Georgian and Mountain [Jews]); then Syrian, Persian, and Mesopotamian Jews and, finally, Moroccan and Egyptian Jews, and Egyptian Karaims. Prior to that he had studied a small group of Crimean Karaims. . . . Altogether Weissenberg measured 580 men (not counting German [Jews]) and 98 women. Measurements were taken in accordance with a short scheme. For the majority of individuals, he established height, arm span, horizontal head circumference, the most important diameters of head and face, form of nose and color of hair and eyes. For some other individuals, only height and head index were added to the hair and eye color. . . . If we further consider that this year Dr. S. A. Weissenberg undertook a new trip to Turkestan, the Caucasus, and Crimea with the goal of researching local Jews, and with that he has completed what he calls the cycle of observation of non-European Jews, the accomplishments of this tireless worker in the field of Jewish anthropology are undoubtedly remarkable and deserve full recognition and encouragement.[68]

El'kind's account of Weissenberg's work did not differ in tone from the report on his award in the *RAZh:*

> He measured in detail the following groups of Jews: Samarkandian—50 individuals, Bukharan—30, Marvanian—10, Geratian—10, Meshkelian (Jezids)—10, Persian—8, Kurdistani—30, Mountain—30, Georgian—17, Krymchak—30, and Karaims—30 individuals. In addition, according to a short scheme, 50 more Samarkandian, 50 Mountain, 50 Georgian and 20 Krymchak Jews were studied. Thus, altogether 255 Jews were studied in detail, and 170—briefly. . . . The main task consisted in explicating the relationship between Jewish types and types of the surrounding population.[69]

Obviously, the network collectively and its leaders in particular controlled the discourse by eliminating from it everything that seemed marginal and unimportant, awarding for the most normative contributions. However, the ability of the liberal anthropological network to do so was not backed by any real institutional or political power, and thus depended on its members' willing participation. Horizontally structured network science relied on intelligentsia self-organization, which in the Russian Empire was a traditional tactic of opponents of the political regime. The government had very limited control over public discourses but exercised tight political control, while the pan-imperial intelligentsia juxtaposed itself to the imperial political authority of the government and the state apparatus and enjoyed almost a monopoly on producing and disseminating public discourses. Their unchallenged discursive authority notwithstanding, the intelligentsia positioned themselves as subalterns of sorts (if only utterly eloquent), suffering from the oppression of the imperial government just as "regular" subalterns—nondominant ethno-confessional groups and lower classes. Thus, the entire repressive potential of Russia's empireness (both direct oppression and subtle coercion through cultural Russification) was associated with the government as a common foe.

Although the pan-imperial *obshchestvennost'*, including the Moscow liberal anthropology network, quite "imperially" imposed some compulsory views and norms—a fairly rigid cultural canon including the Russian language as its lingua franca—it still accepted and even encouraged the strategic "bilingualism" of its members. They were expected to cultivate loyalty to their local communities (of nationality or class) parallel to their participation in the pan-imperial and Russian-language public sphere. The benevolent imperialism of *obshchestvennost'* was not immune to colonialism (in the form of the *mission civilisatrice*) and promoted its own version of Russification. Yet, dissociated from any particular political structure or ethno-confessional group, it appeared as an incarnation of universal modernity and globalization, rather than a manifestation of any monolithic imperial hegemony. The hegemonic discourse of *obshchestvennost'* was abiding, but also dependent on the consensus of most participants in this public sphere. Although imperial in its scope and aspiration for universality, politically this public sphere was a republic (Republic of Letters), in which everyone had a say in forging and modifying the hegemonic public discourse cementing their unity. By joining the *obshchestvennost'* in any of its specialized branches (such as the Moscow anthropological network), a representative of a local community became empowered by the ability to influence the entire pan-imperial Russian public sphere. This was not a mere case of mimicry, but real empowerment through the discourse of global modernity and emancipation from the constraints of various forms of parochialism.

Weissenberg could not overcome his legal and political disenfranchisement as a Jew residing in the Pale and pursuing scholarship outside of normal academia, and all the responsibility for this discrimination was openly and eagerly accepted by the political regime of the Russian Empire. By embracing the discourse of Moscow liberal anthropology, he sided with the *obshchestvennost'* as the sworn enemy of the imperial regime. At the same time, Moscow's orthodoxy of the mixed physical type was in tune with Weissenberg's understanding of Jewishness, just as his own assumed Russian Jewish identity and his own vision of the future of the Jews as a national group correlated with the composite (two-tier) structure of the *obshchestvennost'*. The mixed type model reflected the everyday reality experienced by Weissenberg in Elisavetgrad, with its heterodox communities of various ethno-confessional groups. "Mixed type" was a useful analytical category normalizing hybridity (and hence modern Jewishness) as a fundamental social condition rather than a marginal phenomenon.

The Problem of the Method

Weissenberg's second major work, *Das Wachstum des Menschen nach Alter, Geschlecht und Rasse* (The growth of people by age, gender and race; 1911), did not mention Jews in the title; it formulated its task in the most general terms. Weissenberg set out to explore how the dynamics of human growth depended on age, gender, and race.[70] Eleven chapters covered prenatal growth; body proportions of children and adults; age variations of height, body length, length of arms, legs, hands, and feet; the changes of circumference of head and chest; arm span; body weight; and so on. The study considered relationships between age and height, age and gender, age and environment, and, finally, age and race. The established correlations, both positive and negative, became standard parts of anthropological and medical textbooks.[71] The broad scope of Weissenberg's study was its obvious advantage and reflected the ideal toward which his contemporaries aspired. Just as other similar projects utilized mass anthropometric statistics, this work combined Weissenberg's own data with the data collected by other scholars in different countries, who often used incongruent methods of measurements and calculations. However, one feature made the work special: Weissenberg's own and hence the pattern-setting anthropometric data were exclusively Jewish. In a treatise on general

61

problems of physical anthropology, this choice communicated a clear scholarly and political claim that a medicalized representation of Jewishness was just a variation of universal humanity.

No wonder Weissenberg's message resonated particularly well with readers sensitive to the agenda of Jewish activism. A reviewer for a journal published in Odessa, the *Jewish Medical Voice* (*Evreiskii meditsinskii golos*), singled out primarily one aspect of the work: "Dr. Weissenberg . . . took most of his height measurements from (Southern-Russian) Jews." He clearly interpreted Weissenberg's choice to treat Jewish statistics as a paradigmatic case as a manifestation of his antiracist stance vis-à-vis the Jews and all other races. The review approvingly reported on Weissenberg's discovery of the universal pattern of growth for all of them ("The established pattern of growth for white, yellow, and copper-red races most probably also applies to the black race") and of the generally limited influence of race on human growth.[72] According to the study, the correlation between race and growth revealed itself only during puberty. Weissenberg backed up this conclusion with comparative statistical data on Jews, Russians, and Englishmen. (Curiously, neither he nor his reviewer for the *Jewish Medical Voice* seemed to notice that the comparison involved two imperial nations.)[73] His politics of anthropological comparison undermined the idea of Jewish racial separateness and uniqueness, not to mention degeneracy.

Obviously, Weissenberg intended that his readers reach this conclusion. At the same time, he understood his work as truly scientific, objective, adhering to high standards of academic neutrality, and affording no place for an activist agenda. He would not tolerate a purposeful manipulation of data to explicate his public views, and, luckily, he did not need to compromise his scholarship in this way. Weissenberg's sociopolitical worldview, his perception of Jewish physicality as being hybrid and internally heterogeneous, correlated with his methodological approach, which, in turn, was recognized as legitimate by the Moscow liberal anthropological school and much of world physical anthropology of the time. The demand for equal civil and even national rights appeared to be methodologically harmonious with the monogenist stance in post-Darwinian science and the unrestricted politics of comparison. From this perspective, Jewish race science was just a branch of general physical anthropology, and the Jewish *Urtypus* project—a case of universal importance. And indeed, in physical anthropology of the turn of the century, Jews were broadly seen as a rare case of an ancient race that had preserved its distinctiveness and was easily available for comprehensive study of the dynamics of collective physical traits across time, staging an experiment through the ages. Obviously, this experiment could have been framed differently. Some sought to substantiate the reality of racial purity, and others wanted to understand the mechanisms of evolutionary transformation.

While adhering to the view of Jews as "the archetypal race," Weissenberg did not pursue the epistemological analysis of incongruent analytical categories of groupness (such as regional identity, religion, or even language) used in the *longue durée* study of Jews. Instead, he repeatedly reminded his readers about the universalist dimension of Jewishness (Jewish anthropology helped in solving problems such as "the influence of environment, migrations, mixing, isolation, and many others"[74]) and the equally universalist and apolitical nature of scientific racial exploration of Jewish "types." He viewed serious anthropologists who studied Jews as a race as having no political/ideological motivations. His discourse always tended to underline the pure academic logic of inquiry that guided colleagues in his field:

> Even the first researchers in this area noticed the fundamental differences between the two types, as well as the total transformation and distancing of the Jewish from the Semitic type. One attempt to explain this fact was to propose that within Jewry there actually existed two types, one next to the other: the first was the primordial, genuine, Semitic type, represented by the numerically quite substantial group of Southern European and African Jews—the Sephardim; the second group, meanwhile, was represented by a nongenuine, highly [racially] intermixed Eastern European Jewry (Vogt, Weisbach, Blechmann, Stieda, Hovelacque, Andree).
>
> However, the measurements done by [Joseph] Jacobs on several Sephardic Jews in England yielded the unexpected fact that they, too, were short-headed.
>
> Convinced of the invariability of the Jewish type from ancient times until the present, and thoroughly obsessed with the idea of Jewish short-headedness, a number of other authors (Alsberg, von Luschan, Jacques) on the other hand have tried to use these as proof of a strong intermixture of foreign blood even in ancient Palestine.[75]

The list of anthropologists cited here includes almost all the contemporary luminaries of race science: Karl Vogt, Augustin Weisbach, Ludwig Stieda, Richard Andree, Moritz Alsberg and Felix von Luschan, Bernard Blechmann, André Hovelacque, Victor Jacques, and Joseph Jacobs.

All of them had reputations as objective positivists. They represented, respectively, German, French, Belgian, and British academia. Two of them—Blechmann and Jacobs—were Jewish, but this factor (and their personal identification or nonidentification with Jewishness and any specifically Jewish cause) played no role in the academic discourse that demonstratively censored out any hint of the politicization of "real" science. What indeed equally characterized all these positivist race scientists, including Weissenberg, was their lack of critical reflection about the analytical apparatus they used. Positivist science fought over taxonomies, endlessly parceling larger entities into smaller groups that would better fit the rigid criteria upheld by scholars, rather than questioning the limits of applicability of those criteria, the "objective" nature of the accepted categories of analysis and the genealogical approach.

Weissenberg in particular did not question the historically evolving meaning of Jewishness or the idea of its unique longevity and cultural (not biological) wholeness. In his search for the Jewish *Urtypus*—a positivist utopia by itself—he once attempted the study of thirty-four Jewish individuals with the family names Kohan and twelve with the family name Levi, or close variants.[76] A Jewish reader would know, and a non-Jewish reader would learn from Weissenberg's succinct explanations, that the Kohanim (Cohen, Kohen) were descendants of the sons of Aaron who served as priests in the Temple in Jerusalem, and Levites were descendants of the tribe of Levi that performed particular religious duties for the Israelites. Their progeny were forbidden to marry outside the Jewish race and even outside their elite lineage. Taking this cultural and idiosyncratically Jewish grouping at face value, it was only natural for Weissenberg to see the shortest path to establishing the Jewish *Urtypus* in measuring representatives of the "pure cast," who must have preserved the original physical appearance of their ancestors. Alas, the result was disappointing: both Kohanim and Levites exhibited the same anthropological features as their regular Jewish neighbors.[77] Without contemplating the reasons for and implications of his experiment's results, Weissenberg concluded that the quest for the Jewish *Urtypus* had to be continued elsewhere—in Palestine, Africa, and the Caucasus. The failure did not inspire epistemological reflection or stimulate Weissenberg to question his genealogical approach and the legitimacy of relying on sociocultural categories in a study of physical anthropology.

The obvious logical flaw of this lame experiment reveals Weissenberg's confusion about coordinating the terms of cultural groupness in an anthropological study. His personal experience as a scholar and Elisavet-grad resident confirmed the fundamentality of hybridity and applicability of the mixed racial type approach to its analysis. However, the Jews of the past were invariably defined as homogeneous insulated groups, and the ancient *Urtypus* embodied purity by definition. This conflict must have raised a number of questions: How did something pure turn hybrid, and when did it happen along the way? Alternatively, how does one explain the survival of Jews as a distinct group throughout many centuries, if they have always been a mixed racial type? What criteria did Weissenberg him-self use to designate those whom he measured as Jews (their Judaism, their definition as Jews by the imperial state, their own self-identification, their family lineage)? In the end, only cultural arguments could help to overcome the tension implicit in the concept of mixed racial type and the goal of locating the initial pure type. Neither the German nor the Moscow mainstream physical anthropology schools encouraged this methodolog-ical "crossbreeding," so Weissenberg allowed himself this digression from the scientific canon mainly in the Russian Jewish professional periodicals, where they were validated by their implicit nationalizing Jewish agenda.

Imperial Jewishness: Reversing the Center-Periphery Divide

How are we to make sense of Weissenberg, a non–Western European Jewish scholar residing in a Jewish ghetto, who is nevertheless well inte-grated in the world of European, particularly German, scholarship? He is a scholar who researches the non-European origin of modern Jews in the most advanced scientific language of the time and often with support from German and Russian academic institutions and foundations, but who, depending on the venue, presents his findings as operational knowledge in modern Jewish politics. He is a self-conscious modern Jew who con-ceptualizes Jewish hybridity as an objective fact instead of providing a clear blueprint for Jewish national self-normalization and self-determination. And he is a scholar who simultaneously does all this in three distinct contexts: (1) German anthropology; (2) the Russian imperial situation, where he de jure occupies a weak subaltern position, yet de facto allies himself with Russian progressive science and the public (*obshchestven-nost'*) that endow his scholarship with unique symbolic political capital; (3) and Russian Jewish science and politics, where his position is complicated by his presence in the other two contexts.

Perhaps we should make a full circle and return to provincial Elisavet-grad, where multiple planes of Weissenberg's existence as a scholar and a public persona coalesced. Elisavetgrad was his primary research field; from there he sent articles on Jewish anthropology to Russian and German scholarly journals, and contributed to Russian Jewish and German Jewish periodicals on emigration, Jewish intelligentsia, the crisis of the traditional Jewish way of life, Jewish sanitation and demography, Jewish ethnography, and other topics. He participated in the work of the Moscow Anthropological Division of the IOLEAE as its Elisavetgrad member. He partook in the activities of Russian Jewish societies, most notably the Society for the Promotion of Culture among the Jews of Russia (Obshchestvo rasprostraneniia prosveshcheniia mezhdu evreiami v Rossii, OPE) and the Jewish Historical-Ethnographic Society (Evreiskoe istoriko-etnograficheskoe obshchestvo, EIEO). From time to time, Weissenberg visited St. Petersburg to present his work at the EIEO meetings, or Moscow to attend the meetings of the IOLEAE Anthropological Division. He also undertook longer research trips: in 1908, with support from the Virchow Foundation, he traveled to Syria, Palestine, and Egypt; in 1911–1912, with support from the EIEO in Petersburg, he went to Turkestan, Crimea, and the Caucasus. But he always returned to Elisavetgrad. His scholarship and his life revised and reversed center-periphery relationships, placing non-European Jews at the center of the conversation about the modern Jewish nation, and a town in the Pale at the center of a reconsideration of modern Jewishness as embracing not just the educated intelligentsia but also the people, simple folk, who for the most part resided in the Pale. "But one forgets," he wrote, "that aside from the few men of commerce, law, and medicine, there is a large Jewish mass which adheres to its traditions and ideals. They possess a life of emotions, and these traditions and ideals are rich, differing fundamentally from those of the surroundings, and are still to be researched."[78] By making a conscious choice to stay with the "Jewish masses" in the Pale and share their "life of emotions," Weissenberg firmly associated himself with the objects of his study. He thus undermined the ethnographic authority and hierarchy that normally separated Western European (including Jewish) scientists from the "non-European" Jewish objects of their inquiry, whose non-Europeanness was construed in cultural rather than geographical categories (modern versus traditional, civilized versus Oriental).

Weissenberg was not a pious and observant Jew. In a letter to S. Ansky (Shloyme Zanvil Rapoport), the writer, revolutionary populist, self-taught ethnographer, and organizer of the first Jewish ethnographic expedition into the Pale of Settlement (1912–1914), he described his "insufficient knowledge of ancient language and insufficient orthodoxy" as a "defect."[79] On the other hand, when in March 1912 An-sky invited Weissenberg to visit Petersburg for a planning gathering for the expedition, he first declined because the dates of the meeting coincided with Passover. As Weissenberg admitted to An-sky, he could not violate the tradition of family Passover celebration. "For me and for my family," he wrote in Russian, "not to celebrate Passover [he used a Russian word for Orthodox Easter—Paskha] together would be a great loss."[80] An-sky must have felt differently. He initiated the Jewish "going to the people" campaign with the purpose of collecting and saving for future generations Jewish ethnic traditions and the culture of popular Judaism that he perceived as the basis for the Jewish national revival. A nation in the making needed its own folklore, its ethnographic cultural canon, a tangible proof of the continuation of authentic Jewish life in Eastern Europe—the hegemonic center of modern Jewish history, to use Simon Dubnov's definition. Inspired by this vision, An-sky and his associates rediscovered the Pale as a "dark continent" to be explored. As we know from Nathaniel Deutsch, it was Simon Dubnov who in 1891 called the Pale the "dark continent," referring to Henry Morton Stanley's *Through the Dark Continent* (1878), an account of the English journalist's adventures in equatorial Africa:

> When Burton, Speke or Stanley undertook their bold expeditions into then uncharted Central Africa, they could not predetermine in advance exactly all the great discoveries—geographic, ethnographic, naturalistic—that would subsequently give us a completely new idea of that part of the world. . . . We future explorers of Russian-Jewish history find ourselves in just such a position. We also have before us our own kind of dark continent ("the dark continent")—as the English called the interior of Africa—that lies ahead to be explored and illuminated.[81]

Twenty-one years later, while designing the plan for the expedition, An-sky was still thinking and acting within the "dark continent" paradigm. But his colleague, Weissenberg, the real traveler to remote and exotic places, evidently did not find himself "in just such a position." He would hardly identify with the sentiment expressed by An-sky in a personal letter on the eve of the expedition: "I feel very nervous in the face of great uncertainty.

How will our project go? Will I be able to secure the trust of those poor and ignorant whose world used to be mine too, but from whom I have departed so far during the past years? Sometimes I even feel terrified."[82]

Weissenberg's writings about the Pale could be critical; they could expose sincere ethnographic curiosity and scientific rigor or convey simple human compassion—but never fear or distance.[83] Unlike An-sky, he did not have to invent an ethnographic persona and go native. Deutsch's reconstruction of An-sky's ethnographic self-fashioning presents a striking contrast to Weissenberg's identification with the Pale. "An-sky, the Socialist revolutionary and intellectual who had abandoned Vitebsk at the age of seventeen to escape the traditional Jewish milieu of his youth," writes Deutsch, "was now transformed into 'Reb Shlomo' who wore the long black coat of a Hasid, spent hours in Hasidic study houses listening to stories and songs, and behaved with *devekes,* a Hasidic term that typically refers to the ecstatic state of communication between a rebbe and his disciples or an individual and God."[84] Weissenberg did nothing of the kind even when briefly collaborating with An-sky in the field. He remained a true "mixed type": a German medical doctor living and working in provincial Elisavetgrad; someone who valued the traditional Jewish way of life yet was not particularly pious and preferred Russian to Yiddish, a Jew from the Pale rather than an ethnographer trying to go native.

This is not to say that Weissenberg disagreed with An-sky on the role of Jewish traditions and popular Judaism in Jewish national revival. On the contrary, he thought they cemented Jewishness more than race, serving as vehicles for transmitting authentic Jewish values and thus preventing assimilation and preserving the racially fragmented nation. During his last presentation at the general meeting of the Anthropological Division in Moscow on January 7, 1914, with the paper "On the Jewish Anthropological Type," Weissenberg specifically underlined his disagreements with Franz Boas, who posited the possibility of changing major race indicators in one generation under the influence of environment. He considered Boas's relativization of race irrelevant for understanding Jewish nationhood. The East European Jewish immigrants to North America could not be representatives of the pure Semitic type, and hence the "change" in their head index was not in fact a real change. Weissenberg explained "misunderstandings" such as Boas's theory of the extreme mutability of race (as proof of a Jewish propensity to assimilate) as a result of confusion of the notions of "race" and "nation." "Nation," Weissenberg declared, "is valued not on the basis of the form of a skull or nose, but on the basis of the ideals that the nation has elaborated, the directions it shows to humanity."[85] This statement, quite unusual for a nationalizing race scientist, revealed the role that culture played in his Jewish physical anthropology. It is, therefore, hard to agree with Morris-Reich that Weissenberg "had doubts about whether there even was such a thing as a Jewish Volk," and that "he was not invested in the creation of a Jewish nation, Zionist or otherwise. As a result, Rasse for Weissenberg was an independent category rather than one subordinated to Volk."[86]

Further complicating the image of Weissenberg the race scientist is his position on presumed Jewish degeneration. While popularizing in Russia the activities of the Berlin Bureau for Jewish Statistics and Demography and working himself with Jewish social statistics, Weissenberg advanced an explanation of Jewish national "decay" that differed from that of the Berlin Bureau's Zionists, such as Arthur Ruppin. Weissenberg blamed the "decay" on the decline of religious feelings, the crisis of the traditional way of life and family due to the penetration of "Western mores into the sphere of sexual communication." Freed of traditional restraints, early and informal contacts between males and females stimulated sexual desire and contributed to the spread of extramarital sex. "If we are to judge a nation's morality by the number of illegitimate children, then Jews should not be awarded last place [with regard to the growing number of such children]."[87] These views inspired his social conservatism and antifeminism, as well as general criticism of both Western Jews and the "educated part of contemporary Russian-Jewish youth" for not knowing "their own way of life" (he used the Russian *byt*).[88] At one point, Weissenberg considered publishing an ethnographic [*sic*] book with the title "To Be a Modern Russian Jew" (Byt' sovremennym russkim evreem).[89] Practical considerations of national self-preservation and rejuvenation motivated him to advocate the establishment of a Jewish ethnographic museum in Central Europe, for "as long as the Jews not only preserve their culture but also promote it . . . they will remain ineradicable."[90] Not the enigma of the "dark continent," but the concern for the preservation of Jewish traditions as the guarantee against physical and moral national degradation motivated Weissenberg to join An-sky's expedition.

The initial plan as detailed by An-sky for the potential sponsors envisioned an expedition team of five: the head of the expedition (An-sky himself); two deputies, "responsible for musical recording and anthropometry"; a stenographer; and a temporary member, an expert in one of the fields covered by the expedition. While the core members of the expedition had to work for seven months, temporary experts could stay with the team for brief periods of two to four weeks. Weissenberg's name featured in the earliest version of the plan alongside a few other names of volunteers for the expert position (Khayim Nahman Bialik, Sholem Asch, Leonid Pasternak).[91] Weissenberg also joined the committee entrusted with the task of developing the expedition's research program. As a correspondent member, he could not exercise any serious influence on the program's content, but he tried to talk An-sky out of the idea of including physical anthropology in it.[92] His reasoning was simple: Russian Jewish race scientists, himself included, had done a lot to explore the racial composition of Southern Russian Jews. Now the time had come for a more practically oriented scholarship focused on their cultural anthropology (which he saw as a more politically instrumental discipline than physical anthropology). He communicated his reservations in letters to An-sky:

> I am very glad that you included anthropology in the program, although I think that in this sphere in particular a lot has been done on Russian Jews, while I cannot say the same about their medical examination.

> I'll be home by the end of June and can be available for the whole of July. I think that doing anthropology would be excessive. Instead, I would emphasize ethnology (everyday life, forms of life, occupations, and so on).[93]

At the time of this exchange, Weissenberg traveled to the Caucasus. Along the way, he made ethnographic observations, which revealed a sense of cultural distance comparable to that separating An-sky from the Jews of the Pale.[94] While Weissenberg perceived and experienced the Pale as a living and changing universe, the Caucasus emerged in his reports and articles as a stagnating place in the classical Orientalist sense. In the past, it gave the Jewish nation its racial "Caucasians"; since then the region had lost its significance for the Jewish nation because local Jewish traditions and values had lost their authenticity under the degrading influence of colonialism of underdeveloped Caucasian natives—the "Judaized Caucasians," as he called them.[95] The Jews of Crimea were no less "exotic" and Oriental. However, the region was closer to the Pale and European Russia in general, and Jewish communities there had more intellectual leaders. Some of them assisted Weissenberg during his expedition, shared historical knowledge of the region, and showed their collections of archaeological artifacts and religious objects. Regarding one such collection in Simferopol, Weissenberg noted that although most of the objects were of comparatively recent origin, "they nevertheless testify to [the presence] of at least some aesthetic needs, which cannot be said about Bukharan or Caucasian Jews."[96] Ethnography in the Caucasus only nourished pity for its Jewish population. Just as it was with Greek classical heritage in the European tradition, Jewish ancient heritage in the Caucasus had to be acknowledged, studied, and saved from destruction—ideally, in museums. Weissenberg tried to educate Russian Jews about Jewish archaeological objects and monuments of the Jewish past, completely neglected by the culturally underdeveloped Caucasian Jewish communities: "As a popular rumor has it, there are numerous Jewish monuments scattered across the region; they should be studied before it is too late."[97] After all, "the Caucasus and Crimea most likely represent the gates through which Jews initially penetrated Russia."[98]

Weissenberg's veneration of the Caucasus as a place of racial origin of European Jews and his orientalization of the contemporary Caucasian Jews; his role in An-sky's expedition to the Pale and belief in the practical nature and political relevance of cultural anthropology; the combination of racial and cultural approaches in his construction of national Jewishness—all these add complexity to the already complex scientific persona of this race scientist. Ethnography as practiced by his contemporaries was still a particularistic discipline that sought to reconstruct unique traditions and cultures. Physical anthropology represented the opposing paradigm. "Race" as a concept and a nondescriptive language of numbers and indexes theoretically did not depend on local cultural, linguistic, or confessional peculiarities. But Weissenberg's Jewish race science was a combination of the two—anthropological universalism and ethnographic particularism—a combination very "imperial" in principle. The *Urtypus* stood for the universal myth of origin; Semites and European Jewish "Caucasians" were distinct racial types that had no direct equivalents at the level of ethnographic culture and could not be translated into national categories; local Jewish types, however, were ethnographically distinct. They embodied the idea of Jews not having one primordial "national" territory and suggested that Jews were biologically and socially embedded in societies and regions where they lived. The established fact of the non-European origin of the "Caucasian" type of modern-day European Jews pushed Europe to the backstage of Jewish conversation about national rejuvenation and decolonization through rediscovery of the Jewish authentic Self. While Zionists responded to this provincialization of Europe with the promotion of Palestine as the "soil" that could rejuvenate Jewish national authenticity,

Weissenberg implied that it persisted among the Jews of the Russian Empire, the original Jewish "Caucasians," in the form of their traditional values. The duty of activist experts consisted in collecting these values and traditions, rethinking them as an ethnographic culture and a cultural heritage, and thus making them usable as a basis for the national revival.

Several mutually exclusive logics and interpretative paradigms met in this vision of modern Jewishness, inviting conflicting readings of Weissenberg's works, Jewish complex subalternity, and Jewish contested nationality by different audiences. The multilayered, polyphonic, inconsistent nature of his scholarly discourse reflected a worldview that accepted the coexistence of multiple logics and understandings of groupness as a fact of life and nature, and also as a possible analytical framework and a form of personal and scholarly identification. In this, the leading Jewish race scientist of his time, Samuel Weissenberg, was a quintessential imperial intellectual.

CHAPTER THREE

Arkadii Danilovich El'kind

Constructing a Jewish "Cultured Nation"

> Being scattered between different countries, Jews nevertheless
> retain quite similar physical organization, which follows the
> same pace of development as that of other cultured peoples.
>
> —Arkadii El'kind, 1912

UNLIKE SAMUEL WEISSENBERG, whose work of decentering of
European Jewishness, imperial capitals, and disciplinary standards in
anthropology and ethnography was conducted from imperial borderlands, Arkadii (Aron-Girsh) Danilovich (Donov) El'kind identified with
the idea of Europe as the epitome of modernity and with Moscow as
the main locus of modern science in the Russian Empire. His paradigm
of modern Jewishness had been formulated at the very heart of the liberal anthropological network and in a conversation with primarily non-
Jewish members of this network. At first glance, it absolutely lacked the
deeper Jewish subcontexts that characterized Weissenberg's Jewish science. Whereas Weissenberg's alliance with the liberal Moscow school
implied a voluntary choice and allowed for a geographical and some
ideological distance and independence from the school's headquarters,
El'kind embraced the liberal school's approach, language, and general

vision as part and parcel of his disciplinary anthropological training. One may say that this Moscow-based Jewish intellectual inhaled them with the air as a medical student at Moscow University. He had formed as a scholar under the direct guidance of the leader of the liberal network, Moscow University professor Dmitrii Anuchin, and took his first steps as a researcher within the framework of the Anthropological Division of the Imperial Society of Lovers of Natural Sciences, Anthropology and Ethnography (Obshchestvo liubitelei estestvoznaniia, antropologii i etnografii, IOLEAE). El'kind always tended to ascribe to the liberal anthropological network political and institutional power that it never possessed, as if the model of "citizen science," which proved so suitable for Weissenberg, could not accommodate his scientific agenda and aspirations. The drive to normalize Jewish science informed El'kind's decision to defend a dissertation in physical anthropology—a rare achievement for a science semi-institutionalized everywhere, not only in Russia. But the title of his dissertation was quite daring for a scientist who wanted to build an academic career in the empire that discriminated against Jews: "The Jews: A Comparative Anthropological Study Based Primarily on Observations of Polish Jews."[1] His efforts paid off, and by 1914, El'kind had assumed the key positions in the liberal anthropological network: the secretary of the IOLEAE Anthropological Division and the editor of the *Russian Anthropological Journal* (*Russkii antropologicheskii zhurnal, RAZh*).

Regardless of how far he went on the path toward the normalization of Jewish race science and how Russified and integrated the Muscovite El'kind was, he made a conscious choice to retain his Judaic faith, and with it all the handicaps of his legal status as a Jew. Even more surprisingly, being located at the center of the liberal anthropological network, El'kind nevertheless did not faithfully subscribe to its "mixed racial type" orthodoxy. Unlike Weissenberg, who embraced the concept of the mixed type as his own and worked to localize Jewish types in European Russia, the Caucasus, and Central Asia, El'kind presented Jews as monotypical and racially distinct, sharing few traits with the surrounding population. Thus, his normalization of Jewish science had a side effect of a complete discursive separation of Jews as a race in the empire. This chapter explores the complex dynamics of inclusion and exclusion in El'kind's unusual career as a race scientist and in his controversial paradigm of modern Jewishness as defined from the imperial center.

Redefining Citizen Science

Unfortunately, little is known about El'kind's upbringing and family. The last volume of Brockhaus and Efron's *Jewish Encyclopedia,* published in 1913, when El'kind was at the peak of his career and prominence, gave only a Russified version of his name, Arkadii Danilovich, which El'kind used in everyday interactions and in all his publications, including in the *Jewish Medical Voice (Evreiskii meditsinskii golos)*—the only Russian Jewish periodical in which he ever published (only once!).[2] We find his Jewish name in archival documents only when they register El'kind's communication with the state—or rather the state's with El'kind. For example, the Imperial Moscow University graduation certificate, issued in 1895 and granting him the title of physician (*lekar'*), lists only Aron-Girsh Donov El'kind.[3] His birth certificate must have carried the same name (El'kind was born in 1869 in Mogilev, a provincial capital in the Pale). A biography penned after his death in 1921 by his younger colleague, the anthropologist Boris Vishnevskii, reveals that El'kind went to school (gymnasium) in Novgorod, which means that at some point the family had moved from Mogilev to Novgorod in inner Russia, far away from the Pale.[4] In the late nineteenth century, Novgorod was an insignificant provincial town that had neither well-developed industry nor important cultural institutions. Its population of 25,736 was one of the least diverse in the empire. The overwhelming majority spoke Great Russian at home and were Russian Orthodox (23,287). The imperial census of 1897 reports the number of Jews in Novgorod (by mother tongue—Yiddish, or religion—Judaism) as 893. The contrast with the ethnically, linguistically, and religiously diverse Mogilev (population 43,199) with its 21,453 Yiddish speakers and predominance of Catholics among local Christians is striking.[5]

The educated and commercially successful Jews legally residing in Novgorod must have been well acculturated into the dominant Russian milieu to effectively navigate it. El'kind excelled in his Novgorod Russian gymnasium, from which he graduated in 1888 with the highest distinction, the gold medal, and enrolled in the Medical Department of Moscow University, where the numerus clausus for Jews was 3 percent. While studying for his medical degree, El'kind also attended Anuchin's courses in physical anthropology and a practical seminar at Moscow University's Anthropology Museum.[6] After graduation in 1894, he started working as a physician in various Moscow hospitals and continued his participation in the activities of the IOLEAE Anthropological Division and the liberal anthropological network. This would be a normal path for an ambitious Jew in the late nineteenth-century Russian Empire who wanted to study a group with which he identified and partake in the pan-imperial intelligentsia community of modern knowledge.

El'kind, however, redefined the terms of his participation in citizen science by launching a career as an academic researcher who performed all the rituals of normal academia while still having zero chance of joining it (unless, of course, he converted). He negotiated his research topic with Anuchin personally. Having secured his support, in 1893 and 1895–1896 El'kind traveled to Warsaw to collect Jewish and related Polish anthropometric statistics for the project that would later become his dissertation.[7] As a typical graduate student would, El'kind presented the first results of these expeditions (the measurements of 240 Warsaw Jews) to his more experienced colleagues at a meeting of the Anthropological Division already in late October of 1895.[8] In 1897 he published his first study, based on the anthropological statistics of Warsaw Poles.[9] As was customary for regular graduate students in Russian universities ("doctoral scholarship holders"—*doktorskie stipendiaty*), El'kind spent the next two years, 1898–1899, in foreign academic institutions, perhaps with some financial support from the Anthropological Division. While in Europe, El'kind did not socialize in Jewish networks and did not study modern Jewish life. On the contrary, as was expected of a doctoral student in anthropology, he worked in anthropological institutions in Rome and Munich, perfecting general anthropological techniques.[10] In particular, he was fortunate to work in the Munich Anthropological Institute under the supervision of the eminent German physiologist and anthropologist Johannes Ranke. This leading world specialist on skull anthropology helped El'kind master, among other technics, the cranial morphology of the Italian scholar Giuseppe Sergi, a student of Cesare Lombroso and the recent founder of the Società romana di antropologia (1893).[11] It is not quite clear why someone interested in the anthropology of the living (Jewish) population needed such extensive training in craniology unless this someone perceived himself as a scientist preparing for an academic career in anthropology and, perhaps, university teaching.[12] Indeed, as did every academic anthropologist of his time, El'kind studied collections of crania in European museums, which was also very much in tune with the logic of Sergi's methodological innovation.[13] El'kind's report on the "studies abroad" years to the Anthropological Division resembled a typical academic report of a graduate student to the home department and stressed his mastering of a new methodological approach ("On Professor Sergi's Skull Types in Some Cranial Collections of the Munich Anthropological Institute").[14]

The only link connecting El'kind's anthropological craniology training during these two years with his Jewish research agenda was a study of Jewish skulls that he conducted in Italy. While exploring Jewish catacombs at via Appia in Rome, El'kind came across a few unattributed skulls.[15] He assumed they were Jewish, "received without any difficulty permission to take measurements," and tested his knowledge of cranial morphology on these objects.[16] Having examined four relatively well-preserved skulls and a single heavily fragmented one, El'kind established that the first four were brachycephalic and the last one—"judging by the part of the frontal bone adjoined to the parietal bone"—dolichocephalic; that according to their *norma verticalis* (a view of the skull as seen from above), the first skull, in Sergi's terminology, was *sphenoides rotundes,* and the rest *ooides* or *allipsoides,* and so on. The sophisticated analysis, however, culminated with a laconic conclusion: "There is good reason to believe that here we are dealing with two opposing skull types."[17] The abrupt ending indicated El'kind's reluctance to venture into speculations about Jewish racial origin, which would require him to subscribe to one of the existing racial genealogies of European Jewry. His reluctance to do this becomes especially apparent in the light of the obvious prototype of El'kind's study of five Jewish skulls from the Roman catacombs—the work of Lombroso, who had examined five Jewish skulls obtained from the catacomb of Saint Calixtus in Rome just a few years earlier (1894).

There are many reasons to believe that Lombroso's study directly inspired El'kind's catacomb project. Most probably, he learned from Lombroso about catacombs in Rome that could preserve Jewish skulls. The first sentence of El'kind's report mentions that the catacomb at via Appia was located near the Saint Calixtus catacomb. For specialists in anthropology, this was an unmistakable reference to Lombroso's study. Quite miraculously, El'kind also found exactly five Jewish skulls. It seems that he borrowed from Lombroso's study everything except the genealogical approach. Lombroso argued that the skulls belonged to the time before any considerable racial intermixture of the Jews with their neighbors could have occurred, and hence they could illuminate the origin and evolution of the Jewish race.[18] The five skulls turned out to be *not* dolichocephalic, which confirmed that ancient Jews were not a purely dolichocephalic race—long-headed Semites. On these grounds, Lombroso supported Felix von Luschan's theory that traced Jewish racial genealogy to the Hittites, a brachycephalic race that flourished in Syria and Asia Minor about 1500 BCE. Thus, five Jewish skulls from the catacomb of Saint Calixtus in Rome found their place in a particular genealogy of the Jewish primordial type and by extension a particular view of European Jews as a race.[19] El'kind made no attempt to contribute to this debate, instead presenting his study as a neutral, purely scientific exercise in craniological analysis. Working on Jewish skulls, he avoided being associated with any additional Jewish agenda and ideological motivations. His first study of the Jewish race thus projected a sense of the highest scholarly neutrality uncompromised by any possible external loyalty.

Defending the Undefendable

Once back in Moscow, El'kind made sure to carefully separate the three planes of his existence: a Jew living in Moscow and experiencing the humiliating aspects of his imperial situation; a regular Moscow physician practicing medicine for living; and a race scientist pursuing an alternative academic career in the IOLEAE Anthropological Division. Unlike Weissenberg, El'kind did not cultivate any intimate relationships with his Jewish community and did not treat his patients as objects of anthropometric research. This would have made little sense anyway, since Jews constituted a negligible minority among his Moscow patients.[20] The dissertation manuscript was ready by 1903, but El'kind's work was not only research and study, and it took him another nine years to prepare for the formal dissertation defense in 1912.[21] Now, he had to add a thirty-six-page appendix to account for new scholarship on Jews produced between 1903 and 1912.[22] Finally, in early December 1912, Moscow newspapers published advertisements registering another rare moment of El'kind's dealings with the official state: "The Imperial Moscow University announces that on December 12, in the [university's] Anatomy Theater, the physician Aron-Girsh El'kind will publicly defend his dissertation in pursuit of the degree of doctor of medicine, titled: (1) 'The Jews. Comparative Anthropological Study Based Primarily on Observations of Polish Jews. (2) Anthropological Study of Jews for the Last Ten Years.'"[23] Newspaper announcements of public defenses were a routine requirement. This time, however, the "public" reacted with unusual enthusiasm. By 2:00 p.m. on December 12, 1912, the university's anatomy theater auditorium was packed with listeners, who "welcomed El'kind with loud applause."[24] The public defense of a dissertation on Jewish race science was a highly unusual event in and of itself, and was very hard to insulate from current politics. It was not so long ago that the Russo-Japanese War (1904–1905) had fueled fears of the "yellow peril" and made "race" manifestly visible in the mass press.[25] Russia's humiliating loss contributed to the growing political crisis that culminated in the revolution of 1905–1907, which brought an unprecedented outburst of ethnic violence.

A new wave of Jewish pogroms followed the announcement of the October Manifesto of 1905, which granted basic civil rights to subjects of the Romanov dynasty. Pogroms unleashed the frustration and social disorientation of the part of society that feared radical change, but they also exhibited genocidal tendencies and were motivated by ideological considerations and racial hatred rather than traditional religious anti-Semitism.[26] The crisis of traditional authority and the deep societal frustration created the demand for a medicalized expert language of political analysis and diagnostics of the social body.[27] Debates in the newly introduced Russian parliament, the Duma, in turn contributed to modernizing and racializing the discourse of the "Jewish Question."[28] At the same time, the "black terror" against Jewish deputies from the liberal party of Constitutional Democrats in the First Duma, and the government's reluctance to persecute their killers, showcased the connection between Jews and the "threat" of liberalism as understood by conservative forces.[29] On the other hand, modern Jewish politics, especially competition between different versions (and parties and movements) of Jewish nationalism, socialism and liberalism, became a stable feature of the imperial political landscape. Finally, 1912 was marked by the police examination of the accusation of false ritual murder leveled against the Kiev Jew Menahem Mendel Beilis, which dragged on for the entire year and culminated in a notorious trial in 1913.[30] By the end of 1912, the public outcry provoked by the case was already a tangible factor in Russian political life. So intense was the non-academic context of the academic defense that it is hard to imagine how one could ignore it. However, El'kind assumed precisely such a position of complete political ignorance. He behaved as a normative doctoral candidate, whose motivations and research interests were purely academic. His official readers played along. The defense was a success, and the announcement of El'kind's doctor of medicine degree met with a long round of applause.[31] On December 29, 1912, he received his doctoral certificate, signed by the Moscow University rector, Matvei Liubavskii, and the dean of the Medical Department, Dmitrii Zernov.[32]

The defense that was conceived as a demonstratively apolitical event indeed received little political resonance in Russian Jewish circles. It did not alter El'kind's official position in academia either. A regular academic career still remained beyond the realm of possibility for him.[33] Nevertheless, there was one positive outcome of this existential struggle to normalize Jewish science and disentangle scientific knowledge about Jews from contaminating nonacademic contexts. Instead of a university chair, El'kind received the posts of secretary of the Anthropological Division and editor of the *RAZh,* which made him second only to Anuchin in the network's hierarchy. In his new role, El'kind spoke on behalf of members of the pan-imperial liberal anthropological network and could exercise authority over its members and their views. He used this power to diminish the elements of citizen science in the liberal anthropological project and stress its proximity to normal science. Resuming the *RAZh* in late 1915, after the break caused by the beginning of World War I, El'kind as newly appointed editor wrote:

> All areas of the science of man in which Prof. D. N. Anuchin has worked and which together comprise a broad series of anthropological sciences, will find response and representation in the pages of the journal: physical, zoological, anatomical, physiological, and biological anthropology, human paleontology and paleo-ethnology, *Volkskunde* [*narodovedenie*], ethnology, and anthropogeography. . . . Thus, serving the interests of the Anthropological Division—from the one side, and serving the interests of the science itself—from the other, we . . . define the journal's character and content.[34]

El'kind died soon after this high moment in his surrogate academic career, in early 1921, from typhus, in hungry and cold Civil War Moscow. By then, his legacy as a Jewish anthropologist was almost completely overshadowed by his role as the leading representative of prerevolutionary Russia's race science. Even the journal *American Anthropologist,* which belatedly (in 1924) and with some inaccuracies (El'kind's death date was given as November 1920) informed its readers about the fate of a few internationally known Russian physical anthropologists, portrayed El'kind only as "Editor of the *Russian Anthropological Journal* and Secretary of the Society of Natural History, Anthropology, and Ethnography in

Moscow."[35] His personal contribution to Jewish anthropology was not even mentioned and was soon completely forgotten. It remained largely unclaimed by Jewish race science and later by Russian and Soviet physical anthropology.

When Science Speaks for Itself

From 1900 on, the *RAZh* kept its readers updated on the progress of El'kind's research: "A. D. El'kind is preparing for publication his anthropological work on Polish Jews, whom he examined in 1895 and 1896 at Warsaw factories."[36] Two more years passed before El'kind fulfilled this promise: in 1902, he published the first part of his future dissertation as an article in the *RAZh*.[37] This first publication proved sufficient to earn him the prize named after Grand Duke Sergei Aleksandrovich, thus making El'kind the first Jewish scholar to receive the highest Russian award in anthropology.[38] Unlike Weissenberg, who by the time of his award in 1912 had published two major works and multiple articles (both enjoying wide international acclaim), El'kind's major monograph would come out only a year after the award (as a separate volume of the Anthropological Division's "News").[39] More importantly, Weissenberg's name was associated with major theoretical interventions: he posited his own original theory of European Jewish "Caucasians," tried to locate and conceptualize the *Urtypus,* and localized "aboriginal" Jewish racial types on the territory of the Russian Empire. Unlike him, El'kind did not explicitly advance any "theory." His Jewish race science was quantitative rather than qualitative. It prioritized assembling anthropometric statistics and cataloging views and interpretations advanced by others.[40] El'kind's colleagues in the liberal network praised his skills in determining "the dynamics of specific measurements and their correlations," for which he arranged anthropometric data "in rows, buil[t] curves, and in general applie[d] all methods available in anthropology."[41]

As did many other participants in the liberal network, El'kind venerated the school's leader, Anuchin—an encyclopedic scholar, head of the IOLEAE, a Moscow University professor of anthropology, and a key member of the editorial board of the oppositional liberal newspaper *Russkie vedomosti* (Russian news; 1863–1918). El'kind literally followed Anuchin's motto that the time had not yet come to make definite conclusions about human origin in general and racial types of the empire in particular.

Indeed, under Anuchin's guidance, the network of citizen science worked to accumulate scientific data and establish degrees of kinship and similarity of physical types in the empire. While successfully coordinating this gigantic positivistic enterprise and encouraging conceptual unanimity regarding racial mixing as the essential imperial condition, Anuchin himself avoided grand theorizing.[42]

Viktor Bunak, another dedicated student of Anuchin a generation younger than El'kind, once said about their teacher that he knew the "psychological secret of possessing great knowledge without using any theory."[43] After Anuchin's death in 1923, Bunak personally prepared his brain and found in it physical evidence of "the dominance of objective concrete memory over abstract synthesizing thinking." This was Bunak's explanation of the combination in Anuchin of astonishingly broad scientific erudition with a disdain, or even inability, for theoretical thinking. Anuchin most valued the "factual basis of the phenomena that he studied," Bunak wrote. "He believed that our attempts at theoretical interpretations were often premature, and that the main task of science in general and of ethnology in Russia in particular consisted in the accumulation of factual material."[44] Only from the vantage point of this positivist episteme, nothing less than brain preparation could be required to substantiate something that all Anuchin's students and followers knew anyway from their observation of Anuchin's academic style. El'kind, like Anuchin, absolutely prioritized the accumulation of anthropometric data as a self-sufficient exercise that would eventually result in a major theoretical breakthrough, but almost without any human agency (and hence any subjective interference). Still, Anuchin saw empire as a natural anthropological field and encouraged thinking about diversity as a natural phenomenon, which could be viewed as his major theoretical contribution. El'kind ignored empire and diversity precisely for this reason—as implying a choice, a theoretical position he desperately wanted to eliminate as a subjective element that could contaminate and compromise real science.

His first publication on the anthropology of Poles of the Vistula region—the statistical by-product of his study of the Warsaw Jews—lacked any theoretical ambition.[45] Almost four hundred pages of detailed anthropometric measurements predictably confirmed the mixed nature of the Polish type. El'kind carefully shared some observations regarding the possible sources and historical circumstances of this mixing, yet generally remained true to his precise technical approach. In his concluding statement, he endorsed the model of anthropological scholarship as a collective enter-

prise of data accumulation and processing and relegated the task of conclusion making to the next generations of scholars: "We should hope that
further anthropological observations and historical studies will better
elucidate the issue of the anthropological type of Vistula Poles."[46] A similar fear of conclusions and generalizations, the reluctance to assume a
definite position in a debate and thus expose his personal stake in it, was
evident in all El'kind's works. Not the scientist with his biases and loyalties but the science itself had to speak in the objective and authoritative
voice of numbers, indexes, and diagrams. This was El'kind's strongest
possible argument in support of the claim that normal Jewish science
was possible in the Russian Empire, regardless of the current configuration of the "Jewish Question," the growing pressure to nationalize,
and Jews' own request for usable scientific knowledge that would redeem
them as moderns.

Normative or Dissident? The Dissertation

No wonder El'kind's magnum opus, his dissertation "The Jews," is best
described as a collection of anthropometric statistics composed for 200
men and 125 women at several Warsaw factories. It also featured abun-
dant data assembled by other students of Jewish race worldwide. First,
El'kind analyzed descriptive characteristics of the Warsaw Jews, such as
color and texture of hair, color of iris, shape of nose, and facial features.
Then he considered their anthropometric measurements: height, head size,
skull index, facial measurements, body proportions, and chest circumference. El'kind found very few blond- or light-brown-haired individuals
among his Jewish group, which deviated from findings by other students
of Eastern European Jewry, including Weissenberg, Julian Tal'ko-
Hryntsevich, or Józef Mayer and Izydor Kopernicki.[47] The established
ratio of dark-haired individuals (30 percent) allowed El'kind to locate his
Vistula Jews between the Jews of the Caucasus as observed by Ivan Pantiukhov (84.1 percent), Southern Russian Jews (Weissenberg, 54 percent),
and Italian Jews (Lombroso, 52.9 percent) on the one hand, and Lithuanian, Ukrainian, Galician, and Baltic Jews (4–15 percent) on the other.[48]
Vistula Jews thus presented a transitional type, with predominantly dark
brown hair. El'kind also established that 39.5 percent of them had gray
and light blue irises, and 49.5 percent had deep brown and black irises
(the rest had irises of some mixed tone).[49] The most frequent combination

of color indicators was dark hair with dark eyes. The dissertation included corresponding data for non-Jews, but El'kind did not pursue a direct comparison because he was inclined to see Jews as an isolated type. The only part of the work where he did recognize the obvious parallelism and resemblance was the chapter on height. The average height of Vistula Jews was 1.610 millimeters (5.3 feet), and the number for Poles was not much different. On the basis of his and others' calculations, El'kind claimed that "height, chest circumference, and weight are in reverse progression to the years of work at a factory."[50]

Another major part of the dissertation dealt with cephalic index, which El'kind found to be 81.89 for Polish Jews. The number closely correlated with measurements produced by other students of European Jews (from 80.1 for the Jews of Spain to 83.5 for Galician Jews), and diverged from the same index for the hyper-brachycephalic Caucasian Jews (85.2). El'kind suggested that due to Jewish social isolation in the Diaspora their brachycephaly could not result from any racial mixing. He therefore tended to agree with race scientists who assumed that the primordial Jewish racial type had formed through mixing in ancient Palestine, before Jews spread into other regions. Suppressing further theoretical or historical speculations, El'kind let the calculations speak for him. First, he showed that everywhere in Europe Jews were less brachycephalic than their brachycephalic Christian neighbors. Then he analyzed the Jewish head index "in its reduction to the skull," that is, "the result of subtracting two units from [the head index] . . . for living individuals."[51] After this reduction, the number of "real brachycephalic" individuals among Jews had decreased twofold, making evident the tendency toward dolichocephaly of the Jewish type and increasing the distance separating European Jews from their hyper-brachycephalic coreligionists in the Caucasus. All these calculations remained unconnected by any overarching narrative that would embrace all the multiple measurements, comparisons, and established patterns. Against the richness of the assembled anthropometric statistics, the dissertation's overall conclusion was ridiculously modest:

> In terms of their anatomic composition, contemporary European Jews make up quite a uniform ethnological group, which in its main characteristics (hair and eye color, head index, and height) is everywhere equally different from the indigenous Christian population. On the other hand, however, there are reasons to believe that compared to Jews of other coun-

tries, Polish Jews have better-preserved anthropological features typical of this nation.[52]

Only the claim that Polish Jews were closer to the primordial Jewish type than the rest of the Jews signaled some originality. But even this observation could be read as a reiteration of the familiar trope of "traditionalism" of Eastern European Jews, isolated from the outside world.

In 1902, when members of the Anthropological Division nominated El'kind for their highest award, they correctly interpreted his lack of theoretical ambition as a reflection of his positivistic aspiration toward absolute objectivity:

> The author remains true exclusively to factual data in his possession, and avoids making bold conclusions and hypothetical assumptions. In this regard, some earlier scholars had been more outspoken, but this does not mean that their conclusions indeed resolved the issues at hand. . . . The careful nature of conclusions in this case should be seen more as an advantage than a weakness.[53]

By the time of El'kind's dissertation defense in 1912, even Anuchin, himself skeptical of grand theories, found it necessary to reproach his student for excessive empiricism: "The author criticizes different approaches, but himself does not make any definite conclusion. Instead, he expresses a hope that future studies will contribute to the final resolution of the problem. Of course, although this is a sign of his careful attitude, it would be desirable to know which of the opinions he is inclined to support."[54] The most interesting aspect of these reactions to El'kind's dissertation was their silent acceptance of his most consistent and seemingly revisionist—in relation to the mixed type orthodoxy—conclusion about the uniformity of the Jewish race. No one seemed to view it as a major theoretical intervention or be concerned about the interpretation of European Jews, as represented by the Warsaw Jews, as a homogeneous racial group that did not mix with the surrounding population. Obviously, this picture contradicted Weissenberg's theory of mixed and localized Jewishness, but more fundamentally it was seemingly at odds with the general philosophy of the school with which El'kind identified. As the reactions suggest, El'kind's "revisionist" approach in fact looked familiar to members of the liberal network. What they found lacking in the dissertation was El'kind's own voice, not a justification for something that conflicted with their own scientific understanding of Jewishness.

Unconventional Subalternity

El'kind began his dissertation by pledging allegiance to Anuchin:

> The idea of undertaking an anthropological study of Jews was suggested
> to me by Prof. D. N. Anuchin. Thanks to his support, I enjoyed quite fa-
> vorable conditions during my field research. In the course of my work on
> the project, he gave me valuable instructions and advice. . . . It is my most
> pleasant duty to express here my intense and sincere gratitude to Prof.
> D. N. Anuchin and A. A. Ivanovskii.[55]

These respects were standard in German-like "Mandarin" academic culture (although not in citizen science). What was unusual about El'kind's
praise for Anuchin was the presentation of his *Doktorvater* as a leading
authority on the topic.[56] Hitherto, nobody had called Anuchin an expert
on Jewish race. His only published work on the Jewish topic was the article "Jews" in the *Brockhaus and Efron Encyclopedic Dictionary,* mostly
based on his reading of others' works.[57] Neither Weissenberg nor other
specialists on Jewish race could later alter Anuchin's views once expressed
in 1893: "As one of the subdivisions of the Semitic type, the Jewish type
had parted already in deep antiquity. The images of Jews on Egyptian
and Assyrian monuments corroborate this, for they [the Jews] are easily
distinguishable from the images of [representatives] of other Asian and
African tribes."[58]

We have minutes of at least one meeting of the Anthropological Division where Weissenberg presented his findings and Anuchin criticized his
approach. He reminded the presenter about the "striking resemblance between the types from ancient Egyptian monuments and characteristic
types of contemporary Jews." Anuchin also doubted Weissenberg's "Caucasian theory," pointing to the probability of earlier intermixing with
"African blood" in Egypt and to "significant mixing with ancient peoples
of the Armenian plateau." Finally, Anuchin rejected Weissenberg's view of
Jews as a predominantly brachycephalic race. He referred to Joseph Jacobs's study of 363 English Jews, 33 percent of whom had dolichocephalic
skulls.[59] As we know, these disagreements neither undermined Weissenberg's standing within the Moscow-centered anthropological network nor
forced him to modify his conclusions. Weissenberg embraced the school's
general philosophy of imperial hybridity and did not think that Jewish
race science had to be anything special and different.

El'kind's dissertation, on the contrary, closely followed Anuchin's views, which construed Jews as a special case in the empire of interrelated and mixed racial types. In particular, El'kind supported Anuchin's claim that "even if Jews did engage in some blood mixing with other peoples, the admixture was not substantial enough, at least in the recent times, to significantly alter the Jewish type."[60] However, where Anuchin used words, El'kind relied entirely on numbers and indexes.[61] He also preferred to have the eloquent Anuchin as a participant in the truly international specialized scholarly conversation about the Jewish race. In the text of El'kind's dissertation, Anuchin was grouped together with Jacobs and Felix von Luschan, who viewed Jews as one racial type, against the mostly German anthropologists (Karl Vogt, Augustine Weisbach, Bernard Blechmann, and others) who split Jews either into two distinct racial types, Ashkenazim and Semites, or, like Constantine Ikow, into three, with only the Semitic type recognized as the original and the other two as products of the Diaspora.[62] References to Anuchin in this context were overly artificial and made sense only to members of his liberal anthropological network. By making Anuchin speak on behalf of Russian race science about specific problems of Jewish anthropology, El'kind achieved at least one important goal—he endowed his own work with additional legitimacy and relegated his own speech to the background.

> Prof. D. N. Anuchin has expressed a similar view. He sees the Jewish prototype in their [the Jews] depictions on Egyptian and Assyrian monuments, where Jews are easily distinguishable from the pictures featuring [representatives] of other Asiatic or African tribes. Prof. D. N. Anuchin points to the efforts of Jews to avoid blood mixing with the surrounding peoples, and the weakness among them of the spirit of proselytizing. Therefore, he is critical of Ikow's conclusions and is more inclined to think that the Jewish type had formed much earlier.[63]

The design of El'kind's research project conformed to the strictest requirements of the liberal anthropological model as defined by Anuchin and other leading participants in the network. The Polish Jews in the dissertation were *Russian* Polish Jews—the Jews of the Vistula region. Unlike Weissenberg, El'kind did not personally cross the imperial borders to measure foreign Jews, although like Weissenberg he broadly used comparative anthropometric statistics. Even his interest in Warsaw Poles was only a tribute to the regional approach advanced by the Moscow school.

Never again did he pursue Polish anthropology. On the other hand, El'kind also never studied Jewish ethnography and never partook in Jewish cultural and educational societies and movements that could compromise his academic neutrality. Speaking in Anuchin's voice and perceiving it as the voice of science itself, El'kind indirectly confirmed Jewish separateness and uniformity as an ancient and modern nation.

This implication (rather than conclusion) of the dissertation could be regarded as "neutral" only within the philosemitic context of the Moscow liberal anthropological network. Indeed, if Jews as a distinct Semitic race had been formed in ancient Palestine and historically did not mix with the racial types of the present-day Russian Empire, their place in the future regional political arrangement looked rather uncertain. The exclusion of Jews from the imperial mix as a model of the future "transnational nation" smelled of scientific anti-Semitism, similarly relying on the idea of Jewish absolute racial otherness. Members of the network often preferred to ignore this similarity. In the classification of racial types of the Russian Empire developed by El'kind's and Anuchin's colleague at Moscow University and the Anthropological Division, Alexei Ivanovskii, Jews were placed in a separate category of their own. Meanwhile, Ivanovskii's entire classification advanced the idea of universal racial kinship. It was published under the title *On the Anthropological Composition of Russia's Population* in 1904, a year after El'kind's "The Jews," and embodied the very spirit of the liberal school—its central concern with hybridity and mixing. Ivanovskii called on Russian anthropologists to study "meticization of our population," because this process "provides a rare opportunity to examine the very mechanism of the formation of new anthropological types."[64] He seemed to make an exception only for El'kind or rather for Jewish race science. As El'kind wrote in his dissertation, "A. Ivanovskii treated my work with constant interest and always readily supplied me with literature. He offered his invaluable, sincere, comradely support when, having been himself overwhelmed with multiple tasks, took it upon himself to read the whole manuscript and then the complete galleys. Here and there he made important corrections and instructions."[65]

Evidently, Ivanovskii's "comradely support" did not motivate El'kind to explore "meticization" and the "mechanism of the formation of new anthropological [Jewish] types" through racial crossing on the territories within the contemporary Russian Empire (the way this was done by Weissenberg). El'kind did not embrace Ivanovskii's philosophy of the mixed type. He only used his method of evaluating racial elements that made up the type to underline Jewish separateness.

For his classification, Ivanovskii selected those physical indicators that featured more frequently in the studies of Russian anthropologists (otherwise poorly compatible in scope and research design). His classification was based on the following indicators: (1) hair and eye color, (2) height, (3) cephalic index, (4) height-longitudinal skull index, (5) facial index (a ratio of the maximum width of the face to its length), (6) nasal index (a ratio of the maximum nasal width to nasal length), (7) body length, (8) chest circumference, (9) length of arms, and (10) length of legs. El'kind considered all of them. Ivanovskii assigned each classificatory group a letter symbol, for example: A—light type, B—mixed type, and C—dark type. This basic code was developed further to designate smaller subgroups: a—"real blonds," a1—"blonds of the light type," b1—"sharply mixed type," and so on. Using these symbols, Ivanovskii coded each population group measured in the Russian Empire and thus made it possible to compare them. Identical letters indicated similarities of racial makeup. If in relation to the same racial indicator one group belonged to type A and another to type B, they differed from each other by one "unit of difference." If these were A and C, the "unit of difference" was two. On the level of subdivisions coded with small letters, the "unit of difference" had to be divided by the number of subdivisions (if there were two subdivisions, the "unit of difference" equaled half). Using this mathematically "precise" scheme, Ivanovskii calculated "units of difference" among all peoples of the Russian Empire, but then classified them according to three degrees of "physical similarity." The highest degree of similarity had a ratio of differences between the indicators of less than one; the second degree was less than two; and the third was no more than three.

The resulting classificatory narrative told the story of interconnectedness through variative differences and relative intermixture. This was the story of hybridity conceptualized as a phenomenon in its own right and, moreover, as a fundamental condition transcending the isolationism of individual groups but at the same time not necessarily leading to their eventual amalgamation. To avoid building in any implicit teleological scheme, the narration of this imperial hybridity was organized simply in alphabetical order starting with Afghanis and followed by Aisors, Armenians, Bashkirs, Buriats, Belorussians, Great Russians, and down to Yakuts. But even the neat structure of this ABC of imperial diversity was constantly interrupted by parallel entries for each category. For example, instead of a single comprehensive article for Buriats, there were several alternative sets of measurements: "Buriats/[as measured by] Tal'ko-Hrynstevich/"; "Buriats/Portnov/"; "Buriats—integrated indicators" (because each set was produced by a different anthropologist). The same pluralism could result from measurements taken in different localities: "Great Russians of Tver' province," "Great Russians of Kursk province," "Great Russians of Tula province," "Great Russians of Yerevan province," and "Great Russians—integrated." "Russians" (without predicates, such as "Great," "Little," "White," or "of a given district") were demonstratively absent from this classification. Ivanovskii did not think this political concept could be validated as a physical type by the available anthropometric statistics.

Moreover, his Slavic group (to which different "Russians" belonged) was not a unified type either. Ivanovskii's "Slavonic anthropological group" included Great Russians, Little Russians (Ukrainians), Belorussians, Poles, and Lithuanians, as well as Kazan Tatars, Bashkirs, and Kalmyks, yet excluded Little Russians of Kiev province and the Kuban Cossacks. To complicate matters even more, those Little Russians, Great Russians, and Belorussians who composed the backbone of the group were themselves characterized by an extremely broad "scope of fluctuation of anthropological traits."[66] Whatever Ivanovskii's primary motivation, the absence of Russians and Slavs as single types and their representation as an array of regional composite groups related to various non-Russians profoundly undermined imperial hierarchies of power, while his classification in general naturalized the empire as a horizontally organized space of interconnectedness and hybridity. That is why Weissenberg's mixed and localized Jewish types theoretically fitted in this model so well. In practice, however, this was Anuchin's view of Jews as a homogeneous race that Ivanovskii endorsed. In his all-inclusive classification, there was only one racial type set apart from the family of imperial racial relatives, the Jewish racial type. Below is the entry for Jews in Ivanovskii's classification:

JEWS OF WARSAW PROVINCE [GUBERNIIA]
   I. Jews
  II. Jews of Mogilev and Odessa
 III. Jews of Kovno [Kaunas] and Kurland

JEWS OF MOGILEV PROVINCE

  I. Jews
  II. Jews of Warsaw and Odessa
  III. Jews of Kurland

JEWS OF KOVNO PROVINCE

  I. Jews of Kurland
  II. Jews of Odessa, Jews
  III. Jews of Warsaw and Vitebsk

JEWS OF VITEBSK PROVINCE

  III. Jews of Kovno

JEWS OF LIFLAND PROVINCE

  III. Great Russians of Yerevan province [in Armenia]

JEWS OF KURLAND PROVINCE

  II. Jews of Kovno province

JEWS OF ODESSA

  I. Jews
  II. Jews of Warsaw, Mogilev, and Kovno

JEWS

  I. Jews of Warsaw, Mogilev, and Odessa
  II. Jews of Kurland.[67]

This coded entry meant that Jews of the Russian Empire were related only to other Jews of the Russian Empire. Among themselves, they exhibited all three degrees of sameness, but they were equally alien to all their non-Jewish neighbors. Only a negligible number of Jews of Lifland province (Estonia), eleven people, showed the lowest degree of kinship with Great Russians of Yerevan province (Armenia) in the Caucasus (apparently, the exiled Eastern Orthodox sectarians). Statistically this fact was insignificant, and Ivanovskii concluded with confidence that the "accumulated Jews" represented "a single and quite isolated anthropological group that is not connected with any other people."[68]

Ivanovskii was not prejudiced against Russian Jews. On the contrary, he was known for his philosemitism.[69] However, both he and Anuchin retained the view that Jews had originated in Palestine, and hence could not participate in the ancient mixing that had occurred in the spaces of northern Eurasia. They selectively used Weissenberg's anthropometric statistics, welcomed his search for the Jewish *Urtypus*, accommodated him in the network, but evidently did not find his theory of the mixed nature of the local Jewish types convincing. And the leaders of the Moscow network were not unique in such an inconsistent anthropological and political interpretation of Jewishness. Ivanovskii in particular was inspired by the famous anthropological survey of 6.7 million German schoolchildren initiated and administered in the 1870s by the anthropologist and head of the German Society for Anthropology, Ethnology and Prehistory, Rudolf Virchow, whom both Anuchin and Ivanovskii greatly admired as a scientist and liberal politician.[70] In Virchow's project, the teachers who conducted the survey following the instructions of professional anthropologists had to separate Jewish pupils from all the rest and collect their racial statistics separately.[71] The analytical segregation of Jews was justified by strictly scientific considerations: if in the ongoing scholarly debate about one or many, monotypical or mixed Jewish racial type(s) one saw Jews as a Semitic race, it seemed only logical to assume their non-Germanic origin. Against the backdrop of rising anti-Semitism in the recently united Germany, no one used this survey to accuse Virchow of trying to exclude Jews from the German nation.[72] On the contrary, as a liberal candidate consistently opposing anti-Semitism, whether racial or confessional, Virchow defeated anti-Semite Adolf Stoecker in the Reichstag elections of 1881 and 1884. At the same time, as Andrew Zimmerman has pointed out, Virchow "pursued the racial difference between 'Germans' and 'Jews' more systematically than any anti-Semitic ideologue. . . . Perhaps following professional imperatives of anthropology, [he] also gave (pseudo-)scientific credibility and basis to racial conceptions of German nationalism."[73]

The same was true of Anuchin and Ivanovskii, who acted out of the best intentions and were sincere opponents of anti-Semitism. Moreover, the liberal network's opposition to the oppressive colonial policies of the imperial regime and general philosemitism of its members generated subconscious support for Jewish claims to national status and national wholeness. In El'kind's apolitical, carefully crafted, statistical discourse, these implicit claims did not suggest any territorial separatism and hence did not bear a threat of disintegration of the future (democratized and scientifically reorganized) empire of knowledge. This could not be said about Weissenberg's localized Jewish "aborigines," who theoretically could claim a national territory and hence contribute to the highly undesirable territorial disintegration of the empire along national lines. While both Anuchin and Ivanovskii denied Russian, Georgian, or Ukrainian national projects' scientific racial and territorial grounds, they thought differently about Jewish national identity. The profound tension between the analytical segregation of Jews as a race and their complete political inclusion (in the nation of unified Germany or of the democratized Russian Empire) was not obvious to liberals and believers in the objective and neutral nature of real science, such as Virchow, Ivanovskii, and Anuchin. However, one would expect El'kind, who experienced this duality of exclusion-inclusion personally, to be more sensitive to this controversy. And indeed, he was.

Jewishness: Speaking between the Lines

Some of El'kind's choices of words and some telling omissions suggest that in fact he was not completely comfortable with the model of Jewishness recognized as objective and scientific by the school of anthropology with which he identified. While Weissenberg viewed the anthropology of Caucasian and Central Asian Jews as essential for his scientific study of Jewish "Caucasians," El'kind chose to focus exclusively on the most geographically European group of Russian Jews. And unlike Weissenberg, who measured Jews in their traditional habitat—shtetls, Jewish quarters of ancient cities, or Caucasian auls—El'kind studied Jewish factory workers in Warsaw, most of whom were migrants to the city. Their constructed collective Jewish body exhibited no ethnographic peculiarity. It was uprooted, quintessentially modern, suffering from the negative influences of urbanization and industrialization, and easily imaginable in any other industrial center of the Russian Empire, Europe, or the United States. The racial anthropology of Russian Jews as a dissertation topic did not necessarily require the selection of such a distinctively modern and urban group, so the choice of Warsaw factory Jews could provide us a glimpse of El'kind's own understanding of modern Jewishness.

El'kind described the conditions of his work at Warsaw factories in his early publication on the anthropology of the Vistula region Poles:

In my observations I followed the scheme of measurements supplied to me by the Anthropological Museum of Moscow University, thanks to the assistance of Prof. D. N. Anuchin and A. A. Ivanovskii. Unfortunately, working conditions did not allow us to accomplish the program in its entirety. The main obstacle that we encountered was the unwelcoming attitude of the workers. Only the help of the Factory Inspection, factory and plant owners and factory physicians, who often tried to influence workers in our favor, enabled us to fulfill our task. Another unfavorable moment was that because we had to make our observations directly in the factories and during working hours, workers . . . pressed us to let them go [to continue their work] and hence we had to shorten our program.[74]

This description of the work "in the field" differs strikingly from Weissenberg's sentimental recollections depicting naive Central Asian Jews, who agreed to submit themselves to his caliper, or enthusiastic and supportive local Jewish intelligentsia in Crimea and the Caucasus, or his native Elisavetgradians, with whom he shared a sense of home. El'kind depicted modern Jews, whose life was regulated by mechanized industrial time and unfolded against a faceless industrial background. Instead of enthusiastic volunteers from the local intelligentsia, he collaborated with factory owners and modern experts who most probably supported his study of the industrial workforce as a precondition for efficient management. While Weissenberg took measurements of Jews of all ages, from babies to seniors, El'kind's sample included mostly men and women ages fifteen to their late forties, who together made up a fairly young collective body. Obviously, left behind were the oldest and weakest, those who could not sustain the hard life of an industrial worker and "had to abandon the factory and clear their places for younger and healthier elements."[75] El'kind's Jews were members of the modern mass society, mechanical "elements" of the depersonalized capitalist machine.

From the outset, El'kind introduced them as "European," while for the Jews of Russian Central Asia and the Caucasus he used a rare word— "allotypical." Simply put, these Jews differed so much from the generally uniform Jewish type that their anthropology was largely irrelevant for understanding "European" Jews (the majority could even have been locals converted to Judaism in the past). Thus, El'kind's study confirmed the possibility of complete separation of the two groups of Jews of the Russian Empire. Like many other Jewish anthropologists, he insisted on cultural and social, and hence temporary and curable, causes of deficiencies of the collective European Jewish body, such as a narrow chest or feeble constitution, as well as characteristically "Jewish" physical illnesses.

"Nowadays, one can hardly doubt that most of the somatic traits of the Jewish people are the result of the negative influence of their environment," wrote El'kind, characterizing "European" Jews.[76] At the same time, the otherness of the "allotypical" Jews of Central Asia and the Caucasus was of a different nature. If Weissenberg blamed the colonialism of the underdeveloped nations for their cultural degradation and called on the Jews of European Russia to regenerate themselves and support their poor oriental brethren—the living Caucasian "ancestors" of modern Jewry—El'kind saw no need to integrate these anthropological Others into the future body of the nation. By sacrificing discursively one "allotypical" part of the imagined Jewish national collective, he elevated the other, most modern and European, part that was already living a life in common with the rest of the modern world. This replacement was never clearly spelled out in his dissertation, but El'kind's most sensitive and interested readers could read between the lines.

El'kind's findings suggested that some Jewish anthropometric characteristics and color indicators were racially specific, yet the general picture of Jewish physical development revealed no racial degeneration or deviation from the developmental path of other modern nations. His Jews could be described as assimilated, but not into some alien or even colonial culture. They were assimilated directly into modernity, with all its dark and light sides. By itself, this view was not original. Moreover, in the context of the Jewish politics of the day it signified a clear political stance. In anthropology proper, among its most prominent advocates was Maurice Fishberg, who in 1889 emigrated from the Russian Empire to the United States, where he established himself as the chief medical expert of United Hebrew Charities, a professor of medicine and race scientist specializing in Jewish physical anthropology. Fishberg was known in Russia mostly due to El'kind's enthusiastic reviews of his works.[77] Fishberg's anthropological statistics exposed the great transformative power of American society, economy, and political regime, which presumably attracted the healthiest and strongest Eastern European Jews. Naturally, Fishberg polemicized with Zionist race science, which posited an intimate connection between Palestinian soil and the Jewish national body. One year before El'kind's defense, Fishberg summarized the results of his own long-term studies in *The Jews*:

> While speaking of the changes which the contemporary Jews have been undergoing within the last fifty years, we have avoided taking the position

of a partisan or advocate, and have treated the subject objectively. . . . While pointing at the process of assimilation of the Jews, we by no means advocate their absorption by the surrounding people. . . . We do not find it important for the welfare of the remnants of Israel, or of those around them, that Jewry should commit race suicide. . . . The fact that the differences between Jews and Christians are not everywhere racial, due to anatomical or physiological peculiarities, but are solely the result of the social and political environment, explains our optimism as regards the ultimate obliteration of all distinctions between Jews and Christians in Europe and America.[78]

Visible Jewish otherness, Fishberg explained, had predominantly social causes. This did not conflict with his other claim that "Jews present a homogenous type of head-form without any indication of racial intermixture" (at the time, head form was still regarded as one of the most stable racial markers).[79] In other words, Fishberg recognized that Jewishness had a racial basis, but race played no role in the process of assimilation into modernity, which he identified with the United States. He understood assimilation as a positive process of eliminating visible Jewish cultural and physiological peculiarities, social in nature, and acquiring new traits through integration into some advanced modern culture. Jews constituted a (normal) race and a religious community, argued Fishberg, but not a nation "with a distinct history, traditions, aspirations, ideals, and so forth. . . . The only thing they have in common is their religion. It is the consensus of opinion of all modern statesmen as well as ethnographers that religion alone cannot be considered a basis of the very existence of the social group."[80]

Hence, Jewish assimilation into the life of a modern nation was historically inevitable and progressive, provided that "nation" did not refer to an ethnolinguistic community of common origin. Fishberg's *The Jews* conveyed this message not only in the language of numbers and indexes but also through a well-articulated narrative that clearly stated his political stakes and disagreements with Zionism. In addition, Fishberg's monograph included a powerful visual narrative. A portrait of a skinny "Polish Jew in Jerusalem," with beard and sidelocks, wearing a traditional Jewish caftan, demonstrated that the Holy Land had not improved his physical condition. At the same time, American Jews were represented by men and women in fancy dresses and suits, looking particularly healthy, urban, and "modern." All the men were shaven and bareheaded, and all the women displayed fashionable hairstyles. Only a subtle imprint on their faces, expressed through hardly verifiable manifestations, exposed them as Jews. Having been assimilated into modern American society, they became better versions of *Jews* and successful *Americans*.

El'kind shared many of Fishberg's ideas on Jewish assimilation into modernity. At the same time, neither the Russian Empire of his time nor the future democratized empire of knowledge in which Jews were "objectively" destined or privileged to remain isolated as a nation resembled an American "melting pot," obviously idealized by Fishberg. In addition, strong nationalizing tendencies in both the Russian and the Habsburg Empires left not much hope for the triumph of any universal modernity in the region, while assimilation or integration into ethnic/national modernities would result in the complete disappearance of Jews from modern European history.[81] Such an outcome could equally be the result of a voluntary decision to assimilate or of the logic of exclusion. In any case, a corresponding intuitive assessment of the difference between the US nation and the Central and Eastern European imperial and/or nationalizing future explains El'kind's particular alternations of the model of Jewish assimilation into modernity: his greater stress on the role of race in stabilizing Jewishness; a more consistent advocacy of the view of Jews as a single unified race; and more persistent references to Jewish difference from the surrounding population. To prove the last point, El'kind even turned to cultural and historical arguments—"their lives are completely separate [from those of their non-Jewish neighbors], a situation arising out of myriad distinct or peculiar historical conditions."[82]

One of El'kind's most interesting and typically muted discussions with Fishberg concerned the possibility of Jewish racial mixing in Eastern Europe. In his review of one of Fishberg's works for a 1907 issue of the *RAZh*, El'kind lingered on a seemingly marginal observation by his colleague about pogrom violence, rape in particular, as a possible source of racial mixing ("the author takes covert, forced mixing [*heimliche Vermischung*] as a source of the origin of blonds among contemporary Jews"[83]). El'kind responded with a confidence and eloquence unusual for him:

> Fishberg draws parallels between the pogroms of the Haidamacks and today's pogroms, which makes his conclusion even more questionable. As contemporaries and even witnesses of these pogroms, we clearly see that they destroy the last Jewish possessions, devastate Jews physically, but have absolutely no influence on the change of their anthropological type, regardless of the scale of violence and abuse involved.[84]

Assuming a position of double authority as an objective scholar and a "participant observer," El'kind resolutely rejected any possibility of mixing in both positive and negative contexts. He wanted to preserve the Jewish anthropological type unaffected, pure and ready for national existence either in the nationalized European modernity of the future or in the modernized Russian Empire.

In any case, finding a balance between the model of "Jewish assimilation into universal modernity" and the concept of Jewish separateness and racial purity was not an easy task, especially for someone like El'kind, who wanted objective science to speak through him. But it was even more challenging to reconcile the two concepts in the visual narrative of the dissertation. Fishberg's *The Jews* as an obvious reference point for El'kind's *The Jews* set a high bar: Fishberg's visual images effectively reinforced his ideological message. Many of these images represented "types" rather than actual Jews whose statistics Fishberg used. Even when borrowing this approach, El'kind's task was more challenging—to show Jewish "types" as simultaneously participating in universal industrial modernity and remaining isolated and recognizably Jewish. The de-ethnicized industrial bodies (and often faces) of Jews from the Warsaw factories could have been modern enough to illustrate the universalizing dark power of capitalism, but they were simply not sufficiently Jewish to convey the idea of a separate Jewish presence in this historical drama of inevitable progress. Therefore, El'kind asked his younger brother, then a Warsaw University student, to photograph the Jews of the traditional small town of Kholm (Chełm; Yiddish Khelem) of Lublin province—a prominent target of Jewish humorist folklore. Alongside illustrations from the Moscow University Museum and the Moscow Rumiantsev Museum, El'kind included in the dissertation these pictures of Kholm Jews, seventy portraits altogether. Jewish faces from a traditional Jewish town were to complete the anthropology of modern Jewish bodies, compensating for the ethnographic neutrality of the "European" Jewish physical type that El'kind constructed. The visual narrative represented eternal Jewishness, as imprinted on Abyssinian and Egyptian monuments, on the faces of Rembrandt's rabbis, and on the twentieth-century photographs of Kholm Jews.

Even Anuchin was perplexed: What was the meaning of this rich visual narrative? How did it relate to the narrative of the dissertation? During the dissertation defense, he wondered why El'kind did not analyze his visual material and thus missed the chance to uncover the enigma of "Jewish physiognomy": "The very fact of assembling this great number of Jewish portraits in the dissertation invites such an analysis."[85] Indeed, El'kind did not scientifically objectify the existence of "Jewish physiognomy." His analysis of Jewish facial features in the dissertation did not yield any race-based explanation of Jewish distinctiveness. For example, El'kind examined the "Jewish nose" using the general classification that distinguished straight, snub, hooked, and flat noses. He showed that straight noses were equally prevalent among both Christian and Jewish peoples of the Vistula region. The difference lay in the proportion of hooked and snub noses. Contrary to anti-Semitic stereotypes, Jews had more snub noses, their Polish neighbors more hooked noses. Such statistical patterns contributed little to the discourse of "Jewish physiognomy," at least for El'kind. He obviously knew that the well-known British anthropologist Joseph Jacobs viewed the nose as the main element of "Jewish physiognomy" or, as he called it, the "Jewish expression." As Jacobs wrote, "The nose does contribute much toward producing the Jewish expression, but it is not so much the shape of its profile as the accentuation and flexibility of the nostrils." Jacobs used an italicized 6 to chart the form of the Jewish nose characterized by "nostrility" and claimed that it made the face look Jewish.[86]

El'kind silently ignored Jacobs's 6 as statistically unsubstantiated. But his helplessness in this matter seemed to bother Anuchin. During the defense, Anuchin suggested that Jews exhibited "some distinctiveness in how their upper and lower lips met . . . evident in many Jewish physiognomies." Rembrandt had caught this physiognomic feature especially well, Anuchin thought. He suggested to El'kind that the same "Jewish lips" could be seen on the faces of Kholm Jews.[87] It is hard to say what exactly El'kind saw on their faces, for he appeared to be least interested in measuring either their "nostrility" or the way their upper and lower lips met. In general, he said very little about the portraits in the dissertation, as if protecting their metaphorical vagueness and the ability to produce an impression of lasting and eternal Jewishness, even under modern conditions. All El'kind could articulate as justification for his visual narrative was his trivial conclusion about "contemporary European Jews" as comprising a "uniform ethnological group."[88]

This time, Anuchin challenged El'kind's use of the term "ethnological": "Ethnos can be applied to a people only if we understand it in a spiritual and cultural sense, but not as a physical type."[89] Naturally, El'kind knew this basic distinction from Anuchin's own university lectures. Nevertheless, he made a few such "mistakes" in the text. For example, in another place he wrote: "Old views used to ascribe to Jews a physical organization that differed from that of other *cultured peoples*."[90] Anuchin's sensitive ear once again registered a dissonance:

> The author claims that the Jews' physical organization follows the same developmental pace as that of the "other cultured peoples," and that "some of their somatic peculiarities" have developed under the influence of "environment." The expression "other cultured peoples" is not quite clear; first, we have not yet studied the developmental pace of all cultured peoples equally well; second, what does the level of cultural development have to do with this anyway?[91]

A careful analysis of these rare verbal slips suggests they occurred when El'kind needed a word to convey the sense of being modern, industrially developed, urban, and civilized. Late nineteenth-century German anthropology, which El'kind studied during his two years abroad, was known precisely for such use of the "cultural people" trope (*Kulturvölker*). Unlike the Moscow school of liberal anthropology that based its legitimacy on the embrace of Darwinian evolutionism, the pre-turn-of-the-century German tradition rested on a fundamental distinction between nature and culture.[92]

> For German anthropologists, culture was not a universal human property, but rather the exclusive possession of Europeans as the "historical peoples" or "cultural peoples" (*Kulturvölker*). Africa, the Pacific Islands, and much of the rest of the world was a static realm of natural peoples (*Naturvölker*) and natural resources. Natural peoples were natural both because they did not change historically and because they could not transform nature with science and technology. The division between what we would call race and culture did not apply to the *Naturvölker*, whose language, customs, and artifacts existed in the same natural register as their flesh and bones.[93]

It might well be that the isolated place of Jews in the classification of the interconnected races of the Russian Empire and the rhetoric of Jewish unchangeability from ancient times (expressed in Anuchin's ability to recognize contemporary Jews on ancient Egyptian and Assyrian monuments), as well as the need to present Jews as a "historical people" partaking in industrial, scientific, and cultural modernity, prompted El'kind to use the Russian version of *Kulturvölker* as a euphemism for civilization status.

The semantics of "cultured peoples" in El'kind's dissertation becomes more apparent when one traces this trope in El'kind's other publications, especially his only article in a Jewish periodical (*Jewish Medical Voice*), "On the Social Struggle against Degeneration."[94] In this article, El'kind almost completely ignored the discourse of degeneration as a Jewish problem constructed by anti-Semitic science, and approached it as the key problem of the French, English, and North American "cultured" societies. El'kind claimed that "the higher forms of social and cultural life" always developed in a "fatal combination" with the processes of degeneration, such as growing social inequality; urbanization and the growth of industrial labor; peasants' pauperization; the refusal of bourgeois women to breastfeed and the impossibility for working-class women to breastfeed due to life and work conditions; decreasing birthrate and rising child mortality; and so on. The evils El'kind summarized under the rubric "degeneration" were commonplace in the contemporary critique of existing capitalism. However, the majority of Eastern European Jews—petty traders, craftsmen, estate managers living in small townships—did not yet experience the full power of *this* kind of capitalist degeneration. On the other hand, the degeneration that had been ascribed to Jews (and embraced by many of them as a legitimate argument) had more to do with their racial makeup and physical and moral fitness than with their participation in capitalist modernity. El'kind used the first meaning of degeneration and ignored the second, thus shifting the conversation from Eastern European Jews as victims of the degeneration discourse to degeneration as a common problem of "cultured" industrial nations. The solution to the problem was also common to all "cultured" European peoples. Specifically, El'kind called for a focus on "the problem of labor in all its aspects; labor hygiene in the broadest sense; the issues pertaining to school hygiene and, finally, a comprehensive struggle with pauperism of all types."[95] The "allotypical" Jews of the Caucasus and Central Asia were protagonists of a very different civilizational story. El'kind could not imagine them as industrial workers, urban masses, schoolchildren in standardized schools, or working-class women who could not breastfeed their babies. In these, they were simply not Jewish—both physically and in terms of their civilizational habitus.

El'kind's Jewish anthropology resolved the problem of the analytical separation of Jews as a race and their potential political alienation based on this "scientific fact" by shifting the focus from the theme of Jewish difference from the surrounding "cultured" and "natural" peoples to the theme of Jews' mutual physical, or even more ephemeral—physiognomic—sameness. The Jewish text of his dissertation described normative European Jewish modernity as a typical habitat of contemporary Jews. It talked about a "cultured people" integrated into a universal context of nonterritorial modern industrial civilization, but still remaining recognizably Jewish. The only semantics he was ready to ascribe to "culture" was capitalist modernity, released of any ethnic connotations. Any other interpretation of culture, and Jewish culture in particular, could be a dangerous source of potential contamination of the discourse of a scientist. This left El'kind with physical bodies and "physiognomies" as the only space for Jewishness to reveal itself and persist into the future.

Revisiting El'kind's Paradox

El'kind's dissertation did not address any specific Jewish audience, which may explain why its Jewish subtexts remained largely undisclosed. He did not look for specifically Jewish venues where he would be able to discuss his views on the future Jewish nation with Russian Jewish scientists and activists. After the defense, he continued pursuing his successful surrogate academic career, now at the top of the network's hierarchy. It seemed as if all he wanted and was able to say within the given paradigm of objective science had already been said in his dissertation. His only publication after the defense and before the complete disintegration of the old life in the revolutionary turmoil of 1917 was an article that appeared in late 1912 in the *RAZh*, "On the Anthropology of Negroes: Dahomeans."[96] It presented the measurements of twenty-two men and three women from a group of traveling artists performing in the Moscow zoological garden. El'kind was not the first race scientist who expressed interest in these black performers—before their visit to Moscow the cast had been examined by anthropologists in Berlin, London, and Zurich. Since Dahomeans were "natural people" by definition, completely defined by their natural environment, El'kind collected extensive information on Dahomey's geography, climate, nature, population, and history from the sources available in Moscow. He did not modify the program of measurements compared to his Jewish project and honestly recorded the results showing that Dahomeans' body proportions "were in many ways close to body proportions of Europeans." He even directly compared Dahomeans to Jews: "Thus, while exhibiting no substantial deviations in terms of their general bodily organization, Dahomeans nevertheless leave one wanting more in terms of their physical wellness. Their chest circumference is too small, much smaller than that of the Jews—the most narrow-chested people in Europe."[97] This poor physical state of Dahomeans resulted from the impact of European colonialism, El'kind explained. And yet, however "benevolent" his account of Dahomeans was, he refused to consider the performers, who had left their native land decades earlier, as part of the European urban crowd.

There was nothing in the article on Dahomeans that El'kind would not be able to articulate openly without undermining the greater objectivity and neutrality of his science; there was no place for language slips and no need to invoke the authority of leaders of the liberal school such as Anuchin; and there was no reason to camouflage his embrace of capitalist industrial modernity as a model for the future. The Dahomean project had released El'kind's anthropological discourse from so many restrictions and thus revealed the degree of his identification with Jewishness in the dissertation that he strove to neutralize and redefine all his life. Retrospectively, it seems that the posthumanist universalist science with which El'kind allied himself failed him. More exactly, it failed to accommodate the complex imperial situation of modern and acculturated individuals, such as El'kind, who did not prioritize ethnic Jewishness as their identity but were defined as Jews by the state, society, and the very science they embraced, and who had to think about the future in nationalizing terms and develop some acceptable and scientifically convincing (including to themselves) form of collective Jewishness of the future.

CHAPTER FOUR

Lev Iakovlevich Shternberg

The Political Science of Jewish Race

Culture can also function like a nature, and it can in particular function as a way of locking individuals and groups a
priori into a genealogy, into a determination that is immutable
and intangible in origin.

—Etienne Balibar, "Is There a 'Neo-Racism'?"

I took them all for pure-blooded aristocrats.

—Lev Shternberg, 1890s

L EV (BORN KHAIM LEIB) SHTERNBERG (1861–1927) represents yet
another, radically different, type of Jewish scholar promoting a distinctive version of Jewish race science parallel to his main specialization
in cultural anthropology (ethnography). Unlike Samuel Weissenberg
and Arkadii El'kind, both physicians by training, Shternberg started as a
law student at the Imperial Novorossiisk University in Odessa.[1] His arrest for revolutionary activities in 1886 prevented him from graduating.
Unlike El'kind and Weissenberg, Shternberg did not hold a postgraduate
degree. Regardless, his was the most successful academic career of all the
protagonists of this book. His exile-built reputation as an ethnographer
of Sakhalin islanders encouraged members of the Imperial Academy of

Sciences to appeal in 1900 on his behalf to the St. Petersburg police department for a resident permit. In 1901, Shternberg accepted a job as a curator at the Museum of Anthropology and Ethnography (Muzei antropologii i etnografii, MAE), but his residence permit had to be renewed every three months. Finally, in 1902 he took university exams and graduated, thus having normalized his legal status in the capital. Only then was Shternberg officially hired as a full-time MAE curator; in 1904, he was promoted to senior curator, and some years later—to MAE deputy director.[2] For a Jew, a former political criminal, a recent exile, and someone with no special training in ethnography, this was an amazing success story. El'kind could not even dream of such a degree of acceptance and normalization in official academia.

Shternberg's subsequent career as a leading imperial and then Soviet ethnographer further enhanced his standing in academic science. He represented Russian and then Soviet ethnography at many international anthropological and ethnographic congresses, presided over the Russian chapter of the International Committee for the Study of Asia, and together with two other former exiles and populists turned ethnographers, Vladimir Bogoraz and Vladimir Jokhelson, participated in the publication project of Franz Boas's Jesup North Pacific Expedition. During the years of World War I, the revolutions of 1917, and the Civil War, Shternberg spread himself thinly, "organizing commissions for the Russian Geographic Society and Committee for Study of Tribal Composition of the USSR [KIPS], traveling to the war front on behalf of the Committee to Aid Jewish Refugees, and teaching in a number of institutes and universities around the city rechristened as Petrograd."[3] In 1918, Shternberg assumed a professorship in ethnography in the newly opened Geography Institute in Petrograd. In 1924, he was elected a corresponding member of the Russian Academy of Sciences. A year later, the Geography Institute became a department of Leningrad State University, and Shternberg—a university professor.[4] He was highly regarded as the founder of the Leningrad school of ethnography, defined by a stadial view of human evolutionary development, the comparative-historical method, and an emphasis on field research and the mastery of aboriginal languages.[5]

Shternberg's spectacular success was not completely accidental. In turn-of-the-century Russia, people like Shternberg—political exiles turned scholars—formed the most capable cohort of ethnographers. They were well-educated intellectuals whose shared populist bias made them keenly interested in "autochthonous peoples," whether Russian peasants or non-Russian "primitives." They tended to think about these peoples theoretically, through the lens of populist or later Marxist sociology and evolutionism, stressing forms of social organization and pursuing ethnography as a version of social science. Finally, they had to spend long years in exile, that is, in the "field," and thus could practice what would later become known as the stationary method of ethnographic research. It is no wonder, therefore, that modern ethnography in the Russian Empire was developed primarily by political exiles of the populist generation, such as Bogoraz, Jokhelson, Dmitrii Klements, Bronisław Piłsudski, and Shternberg. Many of them were Jewish and some also contributed to the emergence of Russian Jewish ethnography, bringing into it their entangled perspectives as revolutionary populists, imperial intellectuals, modern ethnographers, and Jewish activists. All these identifications coexisted in the minds of individuals like Shternberg and informed the dialectic of cultural and racial in their Jewish science. The inquiry into this peculiar dialectic frames my analysis in this chapter, in which I trace the transformation of the cultural ethnographic idiom into an essentializing language of Jewish national groupness.

An Imperial Intermediary as Ethnographer

Unlike Weissenberg and El'kind, the young Shternberg's main passion was not scholarship but politics. Already in high school he joined the most radical wing of the revolutionary underground, the People's Will (Narodnaia volia) movement, which promoted terrorism as a political strategy for instigating popular uprisings. Besides the promise of radical sociopolitical change, participation in the revolutionary underground offered a chance to transcend one's Jewishness and provincialism through integration into the revolutionary nation. Shternberg was arrested in April 1886 and spent three years in prison (two and a half in solitary confinement), followed by eight years in exile on the island of Sakhalin, known for its harsh climate and difficult living conditions.[6] Fortunately for Shternberg, his populist-inspired interest in Sakhalin aboriginals appeared consonant with the imperial civilizing mission and the actual needs of governance on the island, which by the time of his arrival had been under Russian control for only fourteen years. Becoming an intermediary on behalf of the Russian government was not his calculated choice, but it proved to be fateful.

In 1891 Sakhalin's top official, General Vladimir Kononovich, a well-educated and progressive administrator, asked Shternberg to conduct a census of the Northern Gilyaks (Nivkhs), the least-known inhabitants of the entire island. A few months later he requested that Shternberg undertake another expedition to the Gilyaks and Uil'ta on the eastern part of the island. In August 1893, the authorities sent him to conduct a census of the Gilyak and Evenk inhabitants of the western coast of Sakhalin. In late June 1895, the Vladivostok-based Society for the Study of the Amur Region, together with the Khabarovsk Branch of the Russian Geographic Society, asked the Sakhalin administration to send Shternberg from Sakhalin to the mainland to study the Amur River Gilyaks. In late April 1896, Shternberg undertook another expedition to the Oroch of the Udsk area. Finally, in November 1896 the authorities requested that he return to Sakhalin to take part in the upcoming pan-imperial census of 1897.[7]

As one would expect from a populist, Shternberg valorized the pure nature, egalitarian lifestyle, and institutions of the peoples he studied and wanted to protect them and their just social formations from distracting colonial influences. He learned the Gilyak language and befriended some of them. At the same time, the Sakhalin and Siberian natives cooperated with Shternberg not least because they saw him as a government official. Sergei Kan, his biographer, explains that in each new settlement Shternberg would first visit the dwelling of the Russian-appointed native headman and present a government document, instructing local elders and chiefs to collaborate and assist. The impossibility of leaving his place of exile, the absence of other jobs, and the proximity of the natives prompted Shternberg to discover the field research method well before it would become the main approach of ethnographic study.[8] Shternberg's ambivalent experience as an involuntary imperial intermediary eventually yielded him a regular position at the empire's main anthropological museum.

Today, Shternberg is known primarily as a specialist on Sakhalin ethnography, a museum anthropologist, and a Russian collaborator of Boas.[9] In Soviet scholarship, he was hailed as a Russian revolutionary populist and a founder of the Leningrad ethnographic school.[10] In 2009, Kan brought these multiple personas together in his exhaustive and innovative intellectual biography of Shternberg, tellingly titled *Lev Shternberg: Anthropologist, Russian Socialist, Jewish Activist*. Kan approaches Shternberg's life as "a window on an important period of his discipline's history," thus confirming the paradigmatic nature of his personal and scholarly biography that is equally important for this study. As a pioneer of this approach, Kan also added Jewishness to the mix of Shternberg's populism and general ethnographic erudition:

> In fact, one cannot fully understand his scholarship without examining his Populist ideology and strongly philosemitic views. These two commitments not only heavily influenced his views but also contradicted and undermined them. His Populist admiration for the social organization of precapitalist societies and his firm belief in the uniqueness of Judaism as a system of moral philosophy clashed with his classic nineteenth-century evolutionism.[11]

Kan argues that Shternberg's religious Jewish education and upbringing in Zhitomir, the capital of Volyn' province—a major center of Hasidism and a provincial center of the Pale of Settlement—left a strong imprint on his understanding of moral and ethical matters.[12] In heder, Shternberg learned to appreciate Judaism as a unique symbolic system that coded and expressed the essence of Jewish "nationality." Kan finds in Shternberg's letters and prison notes evidence of his fondness for the messianic and justice-oriented religion of the Hebrew prophets.[13] The emerging picture seems to confirm a type well established in historiography—the Jewish revolutionary whose radical choices are at least partially inspired by Jewish messianism and a sense of philanthropic responsibility and social justice.[14] However, Shternberg's early life, as reconstructed by Kan, suggests a rather limited applicability of the "Jewish upbringing" trope. For one thing, Shternberg had become a committed *narodnik* (populist) quite early, while still in high school. Until much later in life, when he became a Jewish liberal activist criticizing all Jewish "renegades," Shternberg had never publicly condemned his fellow Jewish populists for their assimilationist views, or the People's Will Party for its official toleration of popular anti-Semitism as an elementary expression of the anticapitalist spirit of the Russian common folk. It might be more accurate to say that, like many other Jews of his generation, Shternberg did not interiorize any coherent tradition or identity. Rather, he responded to various challenges by selectively borrowing from familiar cultural repertoires—whether of experiential and learned Jewishness, or the imperial intermediary scenario, or the Russian intelligentsia subculture, and so on. To understand the significance of Jewishness for people like Shternberg one needs to inquire into the immediate contexts of their actions and the specific challenges they faced.

In Shternberg's case in particular, such an inquiry can help to elucidate the difference between his ethnography of primitive *inorodtsy* in which race had no part and his Jewish ethnography in which the category of race played quite a prominent role.

Jewish Science—Political Science

Shternberg's return to European Russia and acceptance of the position at the MAE in St. Petersburg landed him practically at the center of institutionalized academia and at the same time put his Jewishness on display as a legally compromising identity that could undermine his newly acquired authority over Russian ethnography. These were the years when the rising significance of modern science as a means of social diagnostics mobilized intellectuals like Shternberg to take on scholarly projects with explicit public agendas.[15] Politically, Shternberg now adhered to moderate Populist views represented by certain factions within the Party of Socialist Revolutionaries. As a scholar, however, he shared with Russian liberals the ideal of the empire of knowledge, that is, a democratized supranational state where modern knowledge directly informed political choices and decisions. He therefore collaborated with liberal politicians, especially in seeking a resolution to the "Jewish Question."

Shternberg powerfully entered the scene of imperial Jewish politics in 1905 with the publication of the pamphlet "The Tragedy of the Six Million People."[16] Soon he became a founding member of the Jewish People's Group (Evreiskaia narodnaia gruppa, ENG), officially registered in January 1907 and functioning as an expert club for the parliamentary caucus of the liberal Constitutional-Democratic Party (Konstitutsionno-Demokraticheskaia Partiia, KD). Although the group was officially disbanded after the dissolution of the Second Duma by the emperor on July 3 1907, it continued its semilegal collaboration with the KD Duma deputies until the beginning of World War I, and Shternberg remained one of the ENG's chief ideologists and experts and an active contributor to the group's newspaper, *Svoboda i ravenstvo* (Freedom and equality).[17] The ENG stood on an anti-Zionist platform, offering a synthesis of autonomism with some elements of Bundism and the left-liberal ideology of the KD Party.[18] It combined a slogan of civic integration for individual Jews with a demand for the recognition of Jewish collective national and cultural rights.[19] These goals were to be achieved through a general reconstitution of the empire on democratic principles and its administrative decentralization, but only "under the condition of retention of Russia's integrity."[20] In the ENG political discourse, the pragmatism of small deeds, such as reforms of Jewish self-taxation and communal organization, coexisted with a global theory-backed expert ethos informed by modern science.

Shternberg masterfully balanced these two levels of political thinking. On the one hand, in his speech at the ENG founding congress in St. Petersburg in February 1907, he outlined two concrete strategies of Jewish liberal politics: lobbying Jewish interests in the parliament and developing the tactics of national self-help. The latter implied cultural and social activities that not only would help to attain economic self-sufficiency and civic and national rights but would also create a firm basis for modern Jewish national self-consciousness.[21] On the other hand, Shternberg's seminal "The Tragedy of the Six Million People," reprinted in the ENG collection *On the Eve of Awakening* (*Nakanune probuzhdeniia*; 1906), contributed to the collective theoretical message of the ENG activists who embraced modern Jewish science and science about Jews as the ultimate framework for thinking about Jews as a (politically empowered) nation. At the peak of political turmoil in the country, the authors of the collection still found it necessary to lecture their readers about Jews as representing "one of the most characteristic and distinct anthropological types that has preserved its particularity for centuries despite differing geographic and other conditions." The "contemporary formulation of the Jewish Question" in the collection rested on a number of basic scientific facts: Jews belonged to the Semitic race; Jewish racial type had been formed in antiquity and today could be seen in the "images of Jews on Egyptian and Assyrian monuments, where they are easily distinguishable from the images of other Asiatic and African tribes"; the stability of the Jewish type was reinforced by isolation, endogamy, and a weakly developed spirit of proselytism; and the distinct Jewish psychological personality developed on this biological and social basis.[22] Nationality was, therefore, a higher cultural formation resting on racial foundations. For the ENG's theoreticians, this conclusion signified "the Alpha and Omega of normal human existence." It served as objective proof that the postrevolutionary progressive empire of knowledge would inevitably embrace the principle of nationality, including the Jewish nation. Then "the common motherland of all these nationalities will impress the world with the richness of its spiritual content and with the splendid blossoming of its spiritual culture."[23] In 1906, Shternberg still perceived the unfolding revolutionary events as the beginning of this new Russia, where all individuals and all nations would rise together for freedom and human and national dignity, and create a new union in which "all the peoples of Russia will finally attain their rights."[24]

With the passing of this turbulent yet hopeful moment, Shternberg and many like-minded Jewish intellectuals concentrated on the "self-help" part of their program, including the politics of knowledge that was to prepare the expected future transformation of the old empire into a new empire of nations. The ENG's newspaper (*Svoboda i ravenstvo*, February 1, 1907) instructed that the meaning of nationality was never firmly fixed; it was a combination of a "shared language, shared blood, shared historical past, and shared way of civic life [*gosudarstvennyi byt*]," but all peoples arranged these elements differently in different historical periods and under varying circumstances and constraints. "The content of national self-consciousness . . . develops together with the development of national culture and the changing conditions of national existence."[25] Shternberg returned to this point time and again. He reiterated it in 1911 in the pages of *Novyi voskhod* (New sunrise), a newspaper that mirrored the political platform of the ENG. Shternberg's regular column, "Conversations with Readers," reminded readers that "territory, race, language, and religion" were all variables of national self-consciousness, and their specific arrangement was historically conditioned.[26] Moreover, echoing Ernest Renan's concept of "nation as a daily referendum," Shternberg insisted on the conscious—intellectual and affective—participation of individuals in a national groupness. He did not consider "objective events" by themselves as the source of "national self-consciousness." It rather resulted from "emotions produced by these events," and, more specifically, from emotions generated by the shared experience of struggle, cooperation, creativity, and scholarly self-exploration.[27] The latter was particularly significant as a source of emotional education.

A similar philosophy inspired another project to which Shternberg had committed from the moment of its inception in 1908—the Jewish Historical-Ethnographic Society (Evreiskoe istoriko-etnograficheskoe obshchestvo, EIEO). Unlike the German Wissenschaft des Judentums, "which imagined history as a science divorced from contemporary politics and external influences," the Russian EIEO advocated a different politics of knowledge, most famously personified by Simon Dubnov, who at the time had just founded his Folkspartei (1907).[28] Dubnov's Jewish history was a collective project of all conscious participants in the nation, and a crucial element of the politics of Jewish diaspora nationalism.[29] Both Shternberg and Dubnov carefully differentiated between the political nature of Jewish science as an instrument of collective self-cognition and the petty politicking of their opponents. Shternberg, for example, criticized Kiev University professor Ivan Sikorskii, a renowned neurologist and race scientist who, as an expert at the Beilis trial, supported the accusation against Jews as a degenerate race that practiced ritual human sacrifices.[30] "Besides the innate stupidity" Shternberg ascribed to Sikorskii, he called him "a not quite selfless petty politician [*politikan*], who transferred into his 'academic works' the racial philosophy of the learned men from *Novoe vremia* [a monarchist and Russian nationalist newspaper] and of their authoritative sources such as [Houston Stewart] Chamberlain."[31] Sikorskii's indulgence of political partisanship in his scholarship was inspired by national and racial hatred and thus had little to do with the politics of knowledge as a form of national self-cognition, cooperation, and participation. To Shternberg, Sikorskii's racist science about the Other differed in principle from racializing science about the Self, even when both shared the same epistemology (of race or culture) and methodology. The difference here was political: Shternberg advanced a critique of the epistemological power of the hegemonic discourse in a political situation defined by uneven access to political resources. While Shternberg's Jewish knowledge was the authentic and self-liberating knowledge of the oppressed, Sikorskii's science supported the hegemonic political force of Russian nationalism with its policies of exclusion. It is worth noting that Shternberg advanced his openly politicized paradigm of knowledge only as a Jewish scholar, while remaining a theory-driven and politically neutral, albeit sympathetic, student of indigenous "primitives" of Siberia and Sakhalin in the context of regular academia.

Jewishness—the Blind Spot of Imperial Ethnography

How typical was such a personality split in an ethnographer? Turn-of-the-century academic ethnography in Russia was spread across a few university departments that had specialists qualified to teach the subject. Local and central academic societies, with the Moscow Imperial Society of Lovers of Natural Sciences, Anthropology and Ethnography (Obshchestvo liubitelei estestvoznaniia, antropologii i etnografii, IOLEAE) and the Imperial Russian Geographic Society (Imperatorskoe Russkoe Geograficheskoe Obshchestvo, IRGO) in St. Petersburg, occupied the leading positions.[32] Their heralds and proceedings, as well as the intelligentsia's "thick journals," published articles on ethnography that set standards in the field. In 1889, the IOLEAE had founded the first professional ethnographic periodical in the empire, the *Ethnographic Review* (*Etnograficheskoe obozrenie*). In 1891, the IRGO's Ethnography Division launched its own journal, *Living Antiquity* (*Zhivaia starina*). Shternberg contributed to these journals only as a specialist on the Gilyaks or as a writer of obituaries for his colleagues.[33] He never published an article on Jewish ethnography in either journal, which was not at all surprising. From the moment of *Ethnographic Review*'s foundation in 1889 until 1916 when it was discontinued, it did not feature a single article on Jewish ethnography.[34] The same was true of its St. Petersburg counterpart. The reason for ignoring Jews was not anti-Semitism—at least, this was not the main reason. Jews, it was believed, did not have "the people"—the common folk, usually associated with peasants—who would be the natural object of ethnographic exploration of traditional culture, folklore, and forms of life. Before S. An-sky's expedition had identified the Pale as a place inhabited by the Jewish "people," Jewish ethnography dwelled more comfortably in the literary works of writers such as Sholem Yankev Abramovich (Medele Moykher-Sforim) and Yitskhok Leybush Peretz than in scholarly publications.[35]

The absence of Jews on the imagined map of legitimate objects of imperial ethnography, which otherwise embraced almost all imperial peoples, including Russians, complicated the position of a Jewish scholar. In Siberia, Shternberg and other Jewish or Polish exiles turned ethnographers could study indigenous populations and be perceived as enlightened representatives of the empire. Once back in European Russia, the same Jews or Poles encountered a more complex and politicized hierarchy. Even Shternberg, the most successful among his peers, experienced problems with acceptance by the imperial community of ethnographers. One controversy in particular revealed the compromising nature of his Jewish identity. In a letter to Boas, Shternberg called this story his "l'affaire Dreyfus." In Russia it was known as the Zhuravskii affair after the man who leveled false accusations at Shternberg in 1908–1910. In Kan's detailed reconstruction of the scandal, it emerges not as a personal conflict but as

113

a struggle over the national identity of Russian ethnography and ethnographers. Andrei Zhuravskii, a monarchist, Russian nationalist, and self-taught ethnographer, assembled for the MAE valuable collections of artifacts of the Russian Old Settlers, the Nenets, and the Komi. It so happened that the MAE sold some of these objects to German museums via a wealthy Jewish businessman who in return procured for the MAE artifacts the Russian museum lacked. Despite such exchanges being standard practice at the time, enabling academic museums with limited funding to build and diversify their collections, Zhuravskii treated it as a case of abuse of Russian science by Jews and Germans, who monopolized administrative positions in the museum. In his complaint he blamed specifically the Jew "Khaim Leib" (instead of Lev) Shternberg and the head of the MAE, the German-born academician Vasilii Radlov (Wilhelm Radloff). The scandal was formally resolved only in early 1911 by the Academy of Science's court of arbitration, which decided that the accusations were unmerited.[36] Still, the controversy revealed that in the eyes of many ethnographers and students of folklore, Shternberg's right to act on behalf of Russia's ethnography was questionable. Imperial ethnographers as a consolidated community never publicly condemned Zhuravskii's clearly nationalist accusation of the "foreigners" in their midst.

Just when the academy's court of arbitration put an end to the Zhuravskii affair, the IRGO's Ethnographic Division rejected the request of Edward Piekarski, who demanded an official denunciation of blood libel speculations. A former political exile turned ethnographer of the Yakuts and Tungus and a Pole, Piekarski wanted his expert community to protect not only Jews but modern science itself. The Ethnographic Division's chair disagreed: "Currently, this question is being raised not from a scholarly but from a political point of view, which is beyond the division's competency." Scholarly literature, he said, has already exhausted this question, so there was no need to take special actions and meddle in politics. Members of the division voted against Piekarski's proposal and thus refused to challenge political anti-Semitism with their expert knowledge. This took place in December 1911, when the police investigation of the Beilis case had been developing full speed, and when the consolidated expert opinion of the imperial ethnographers could have become a weighty argument.[37]

In this respect, the situation in ethnography—a humanistic discipline focused on local uniqueness and hence more susceptible to nationalist narratives—differed from the situation in race science that embraced the model of universalist and transnational natural science. To be sure, there were enough individual race scientists upholding explicitly nationalist views. However, as we have seen, the Moscow liberal school and network effectively controlled interpretations and applications of "race" (at least in academic discourse) and, unlike ethnography, relativized differences between various groups of the imperial population, fusing them in the concept of the mixed racial type. Moreover, the Moscow liberal network openly dissociated itself from Russian nationalism and opposed racism and anti-Semitism, which had direct political implications for Jewish scholars. That said, the case of El'kind's anthropology revealed that even for the Moscow school the Jews represented a unique case of simultaneous political inclusion and epistemological exclusion, a curious exception to the general orthodoxy of mixed racial types. Together, the invisibility of Jews in the field of imperial ethnography and their segregation in the liberal anthropology reinforced the propensity of Jewish scholars-activists to develop modern Jewish science as a separate anticolonial project of producing authentic Jewish knowledge.

On Ethnographic Self-Positioning: Between Sakhalin and the Pale

The two modern anthropologists, Bruce Grant and Sergei Kan, who have written about Shternberg the ethnographer, place him at the vanguard of world anthropology of the 1890s. Not only was his fieldwork model ahead of its time, so was his strong topical focus on kinship and social organization. Unlike the mainstream descriptive ethnography of the day, especially in the Russian Empire, Shternberg's studies were theory-driven. American scholar Lewis Henry Morgan, whose pathbreaking work set in motion debates on forms of kinship and group marriage, exercised a major theoretical influence on him, both directly and via the writings of Friedrich Engels. The emerging exiled ethnographer perceived his ethnographic data through these specific theoretical lenses, and hence, as Grant insists, it is "in the context of both Engels and late 19th century theories of group marriage that many of Shternberg's observations on Gilyak life can be understood."[38] Shternberg clearly experienced the effect of recognizing the familiar in a culture that looked so unfamiliar and strange, but this was a learned, theoretical familiarity that did not correlate with any of his previous experiences. As he wrote "from the field" in a letter to a friend on May 19, 1891:

> My main accomplishment has been the study of their [the Gilyaks'] social organization and marriage system. I discovered among them a system of kinship nomenclature and a system of family and clan law [*semeino-rodovoe pravo*], which are identical to those [that] exist among the Iroquois and to the famous Punulua family (in the Sandwich Islands). In other words, I found the remnants of that form of marriage, upon which Morgan had built his theory and which serves as the starting point of a brochure *Ursprung der Familie* [*Origin of the Family*].[39]

From Engels, Shternberg interiorized the association of marriage forms with historical formations. Thus, group marriage corresponded to the period of savagery, a loose pairing arrangement between husband and wife—of barbarism and monogamy—to the civilized stage of historical development. According to this evolutionary sequence, the Gilyaks were transitioning from savagery to barbarism. Their current kinship terminology, Shternberg believed, still reflected actual marital practices, that is, every Gilyak had a right to have sex with his brothers' wives and his wife's sisters. He thus established a "survival of group marriage" on northern Sakhalin. When in 1892 Engels came upon a summary of Shternberg's first report on Gilyak group marriage for the Moscow IOLEAE, which was published in the newspaper *Russkie vedomosti* (Russian news), he found it congenial, and immediately had it translated and published in Germany in *Die Neue Zeit* (New era).[40]

Kan suggests that over time, Shternberg's analysis of the clan system became less structuralist and more functionalist. Shternberg started paying more attention to the functioning of aboriginals' folklore and religious beliefs and to the specifics of their language, and he acknowledged that the Gilyak clan system was less fixed than he had presented in the early reports.[41] Nevertheless, forms of social organization, as they were distributed chronologically in the Morgan–Engels scheme, always remained at the center of his ethnography of Sakhalin and Siberian natives. This prevented Shternberg from a more consistent revision of the initial evolutionary account through incorporating in it the fact of the widespread Gilyak practice of marrying outside their clan and even their own society or the difference that existed between Gilyak sexual relations, structured by the nuclear family, and the extended network of support and cooperation as reflected in their kinship terminology.[42]

Under Radlov and Shternberg, the MAE exposition communicated a straightforward evolutionist narrative, which overshadowed the fact that the kinship idiom, so central to Shternberg's ethnography, resonated with the nationalizing moment in the empire and an emerging concern with the nature of modern groupness.[43] Obviously, the themes of blood ties, clan endogamy, racial purity, and extended family as a prototype of the nation were far from new; they were already central to romantic nationalism. However, positivist ethnography and sociological evolutionism had successfully readjusted the old imagery to new scientific requirements and public sensibilities. The kinship idiom became the only theme Shternberg borrowed from the Gilyak ethnography and applied in his explicitly national Jewish project. There was nothing else to "borrow" from the ethnography of a people living at a prenational developmental stage and not having a separate cultural sphere, distinct from the practicalities of social life. Kan attributes to Shternberg the view "that only ethnic groups that are professedly religious, literate, and have their own intelligentsia can develop an ethnic consciousness."[44] Jews were such a group, but Gilyaks were not. Lacking ethnic consciousness, they could not experience the feelings of belonging and mutual responsibility that were required for conscious participation in a nation as a spiritual and political community.

A very different mindset emerges from Shternberg's Jewish ethnography. First, it was a political project formulated in the context of the struggle for personal and national rights and survival in modernity. As if directly echoing his colleague in the EIEO, An-sky, who wrote that "at the present moment, only one militant slogan remains . . . nationalism," Shternberg admitted in 1911 that he and An-sky were developing their Jewish science "under the sign of militant nationalism [*pod znakom boevogo natsionalizma*], . . . it seems that there is no Jewish group for which the problem of nationalism would not be at the forefront."[45] The announcement of An-sky's ethnographic expedition in *Novyi voskhod* (June 7, 1913), with which Shternberg politically identified as a collaborator and an editor, offered an effective summary of this shared discourse of a "militant" science of national self-cognition:

> Each people aspiring toward normal national life, first of all, faces the task
> of self-cognition. . . . Without sparing themselves, scholars of all cultured
> countries diligently collect works of popular poetry and study physical and
> spiritual traits of their peoples. . . . Moreover, civilized nations compete
> among themselves for the exploration of anthropological traits, the

everyday life [*byt*] and creativity of underdeveloped, savage and semi-savage tribes and peoples. . . . The only exception to this rule, however strange and unbelievable this may seem, is the Jewish people. . . . In the West and the United States, at least something has been done supporting the study of the spiritual production and anthropological characteristics of the Jewish masses ("Vereine für Jüdische Volkskunde" with their periodicals; expeditions to study Jews of Arabia and Africa; Jewish museums; collecting, specifically in the United States, of the mass anthropometric measurements of Jews, and so on). In the Russian Empire, where the main mass of the Jewish people is concentrated and the basic foundations of the old Jewish way of life are better preserved, almost nothing has been done in this regard (not counting two or three collections of popular songs and proverbs and a few journal articles on Jewish ethnography).[46]

Reducing the richness of existing anthropological and ethnographic literatures to a few titles, the announcement demonstratively ignored publications by Russian anthropologists, including Weissenberg and El'kind, in academic venues such as the *Russian Anthropological Journal* (*Russkii antropologicheskii zhurnal*) or the proceedings of the Anthropological Division of the IOLEAE or of the Petersburg-based Russian Anthropological Society. Apparently, the problem was the non-Jewish nature of these academic venues, which thus did not contribute to Jewish national self-cognition. The EIEO's own journal, *Jewish Antiquity* (*Evreiskaia starina*), could partially improve the situation by providing a Jewish venue for academic Jewish research and targeting simultaneously Jewish producers of modern knowledge and a general Russian Jewish audience. Indeed, the archival file, "Correspondence of the Journal *Jewish Antiquity* with Subscribers," includes all possible categories of Jewish readers, from Jewish intelligentsia residing in and outside the Pale to Jewish industrial workers, and even other Jewish periodicals (such as, for example, the journal *Karaim Life*).[47] It is to these academic and nonacademic audiences that Dubnov explained the meaning of historical sources and instructed how to collect them, thus enabling everyone to participate in the process of national self-cognition.

Shternberg did not do quite the same with regard to sharing the basics of his "stationary" method and theoretical approaches to primitive cultures. But this was because he did not perceive Jewish culture as "primitive." In addition, he believed that educated yet nationally conscious Jews were already natural ethnographers. First, many Russian Jews de facto lived "in the field" or had departed from it fairly recently. Second, they a priori possessed an authentic knowledge of Jewish life and religion and a sense of psychological identification with other Jews. Without these essential conditions, that is, basically, without being Jewish, one could not understand and satisfactorily interpret Jewish culture and Jewish forms of social organization. Shternberg saw his task as discussing with these naturally prepared readers the concept of Jewish national (ethnographic) culture. He understood it as being ethnographically specific and fundamentally universal at the same time, and hence fully compatible with modernity:

> Only *messianic Jewish ideas* articulated the higher monism of humanity. . . . The great concept of a common origin of the whole human genus that had been created by the sole God and had then fallen as the result of a free choice, split and disunited, but was supposed to be saved . . . to form a united herd under a sole shepherd—this great concept that could be born only at the highest stage of monistic religious thinking was a first formulation of the higher idea of humanity that laid the foundation for all ideas and ideals . . . of all later periods of human intellectual development.[48]

After the 1905–1907 revolution, Shternberg's discourse of the Jewish culture had become more political. Judaism remained central to it, but now Shternberg claimed that it produced a unique kind of nationalism that differed from "zoological" nationalisms of other peoples.[49] Jews were thus pushed even farther from the "primitive" objects of Shternberg's Sakhalin ethnography. Already as a leading Soviet ethnographer in the 1920s, Shternberg continued to elaborate on the "the lack of 'primitiveness'" in Jews. Like El'kind, he characterized them as a "cultured people," explaining that "even that which could be called 'primitive' about them is a bookish provenance, or else borrowed from the surrounding peoples."[50] Shternberg clearly distanced himself from the fin-de-siècle discourse of primitivism, which, according to Gabriela Safran, allowed Jewish ethnographers–former exiles (such as Jokhelson and Bogoraz) to unproblematically lump together Siberian and Sakhalin *inorodtsy* with Jews as traditional peoples, unspoiled by capitalism and preserving communal forms of life. As Safran writes, for these people around An-sky, "Siberia stood for the questions in the heart of the *Dybbuk*'s [An-sky's famous play] plot: marriage, family structure, and the tension between old and new ways of life."[51] While such parallels were indeed possible in the context of Russian populism and the fin-de-siècle aestheticization of the primitive instinct, they seemed less obvious from the point of view of academic evolutionary ethnography.[52] For Shternberg, "primitivism" as a concept excluded developed cultural and spiritual spheres and hence could not be applied to Jews. The only connection between his Siberian and Jewish ethnographies remained the theme of "family structure," that is, kinship and endogamy.

More unexpectedly, Shternberg embraced race science as a key part of his Jewish ethnographic project. Race had never seriously interested him as an imperial ethnographer, and the turn to race did not automatically follow from the interpretation of Judaism as a superior ethical, cultural, and national worldview. And yet the title of his talk at the first planning meeting for the An-sky ethnographic expedition (March 24–25, 1912) was "On Collecting Jewish Anthropological Data." He presented physical anthropology as "a national cause" and insisted that the question of Jewish race was of principal significance for Jewish history. Some of the participants enthusiastically supported Shternberg. The physician and student of Jewish social hygiene from Moscow, Solomon Vermel, emphasized the medical relevance of Jewish racial statistics, which was "no less important for the salvation of the people." Shternberg's friend and colleague, and also a former exiled populist turned ethnographer, Jokhelson prioritized the collection of anthropometric statistics over Jewish folklore because the latter was "a survival of the past" that could not "guide our youth in the direction of progress."[53] In general, Shternberg's advocacy of Jewish race science as a "national cause" did not raise many eyebrows at the meeting. We know that An-sky's understanding of the expedition as being almost exclusively about Jewish folk culture had prevailed.[54] In the words of Nathaniel Deutsch, for An-sky,

> Jewishness itself was a kind of performance consisting of certain types of music, rituals, customs, and other traditional practices, that together constituted Jewish folk culture. Ethnography, in turn, was the best means for recording and documenting these traditional cultural performances, thereby producing the raw material for new Jewish cultural forms, such as theater (including An-sky's play *The Dybbuk*), museums, and so on.[55]

Shternberg, as is evident from his engagement with the race concept, understood Jewish ancient—but also very modern in its humanitarian monism—culture as rooted in something much more tangible than performative folklore. He continued using the EIEO and its public network to advance this emerging racialized understanding of Jewishness.

The originator of the field research method did no field research among the Jews. Shternberg's academic imperial ethnography was about the Other, and hence it presumed distance between a scholar and his object, which had to be overcome through long-term observation, careful research, mastery of an aboriginal language, and even personal immersion. This universalist, evolutionary ethnography was free of racialism. Shternberg's Jewish ethnography was about the Self—individual and collective as well as physical and spiritual. As a form of national self-cognition, it did not necessarily require any special going to the field, and as an anticolonial project it could promote exclusivity, use racialized idioms, and openly advance a political agenda. The only theme that remained important for Shternberg in his Jewish project was kinship or tribal endogamy, which became a connective tissue between his cultural and racial idioms of Jewishness.

Jewish Race and "Universal Races"

As Shternberg was recovering after his "l'affaire Dreyfus," an extraordinary event took place in London—the Universal Races Congress (July 1911). It was initiated by Felix Adler, Gustav Spiller, and other scholars and activists of the transatlantic pacifist, humanitarian, and ethical movement as a step toward promoting a new world order based on racial harmony, peaceful coexistence, and cultural dialogue. The participants came from Africa, Asia, the Americas, and Europe, and included anticolonial, national, and feminist activists (such as John Tengo Jabavu, Mojola Agbebi, W. E. B. Du Bois, Duse Mohamed Ali, and Charlotte Despard), on the one hand, and renowned scholars (such as Felix von Luschan, Giuseppe Sergi, and Franz Boas), on the other. Some of them arrived in person, and others sent their lectures to be published in the *Papers on Inter-racial Problems* that came out on the eve of the congress.[56] Their common goal was "to discuss, in the light of science and the modern conscience, the general relations subsisting between the peoples of the West and those of the East, between the so-called white and so-called colored peoples, with a view to encouraging between them a fuller understanding, the most friendly feelings, and a heartier cooperation."[57]

The Russian Empire was represented at the congress by the chairman of the Third Duma (1907–1910), Nikolai Khomiakov, and professors Dmitrii Anuchin, Otto Eikhelman (international law, Kiev University),

Vladimir Grabar (international law, Dorpat University), and a few others. Political émigrés, especially activists of the Georgian national movement, were listed as "Russian" along with others such as Jacques Novicow, a sociologist from Odessa who was famous for advancing a psychological critique of social Darwinism, primarily wrote in French, and built his scholarly career outside of Russia. Another member of the Russian group was the St. Petersburg University professor of international law Boris Nolde, who theorized the Russian Empire through the political concept of Greater Britain. There was also Ludwik Lejzer Zamenhof from Warsaw (then part of the Russian Empire), the creator of Esperanto. The law professor Alexander Yastchenko submitted a paper for the congress's collection, "The Role of Russia in the Mutual Approach of the West and the East," in which he employed the familiar trope of Russia's "twofold nature, its profound dualism" and hence its "grave duties and great mission."[58] Regardless of such active participation of individual politicians and scholars, the congress received a rather modest resonance in Russia. The association of colonialism with racism and the tendency to see race as a problem of skin color did not quite reflect the Russian imperial situation (where race did not necessarily imply the "color line," and academic discourse on race often outpaced its political instrumentalization by the state). The same consideration could explain the relatively marginal role allocated to Russia in the congress's deliberations.

Shternberg was among a very few Russian commentators who showed a keen interest in the congress and published on it. Amazingly, he managed to completely misrepresent the event. His long column in *Novyi voskhod* reported:

> Regardless of how one sees the Congress—as a parliament of the nations that came together to unite in one universal human family, or as a scholarly Areopagus that discusses the theoretical substance of racial and national collectives—the Jewish Question appears to be the most significant, the most central. From the point of view of interracial justice, there has never been and does not exist on Earth another example of a more persecuted people. And from a theoretical viewpoint, there is a no more complex, controversial, and hence interesting question than the question about the Jewish race and the Jewish nationality.[59]

Shternberg literally suggested that the "Jewish Question" stood at the fore of the congress's discussions. In reality this was not the case. The delegates who did mention Jews in papers and presentations, as a rule, referred to them in passing and together with Christians and Muslims, or when discussing the phenomenon of race in general terms, as did Luschan in his paper "The Anthropological View of Race" or the French philosopher Alfred Fouillée in "Race from a Sociological Standpoint." In the talk of Oxford University professor of Arabic David Samuel Margoliouth, "Language as a Consolidating and Separating Influence," Jews were mentioned together with other groups speaking a Semitic language. A similar role as an "example" was allocated to Jews in the talk "Turkey" by the philosopher, poet, and "Young Turk" politician, Dr. Riza Tevfik, "a deputy of Adrianople in the Imperial Ottoman Parliament" as he was introduced at the congress.[60] Overall, only one out of sixty-four papers submitted to the congress specifically dealt with Jews—"The Jewish Race" by Israel Zangwill, then president of the International Jewish Territorial Organization.[61] In his column, Shternberg focused exclusively on this single paper, as if it indeed reflected all the themes and ideas that brought the congress's participants together. Such an idiosyncratic choice exposed a view in which "race" acquired *real* scholarly and political significance only in connection to Jewishness. Racism and racial inequality—these shameful results of ignorance and colonialism—were destined to disappear with the advancement of science and progressive forms of political life. But the Jewish case was different and indeed deserving of concerted scholarly intervention, due not only to the seemingly timeless power of anti-Semitism but also to the very nature of Jewish otherness.

Shternberg claimed that Zangwill had failed to present this complexity of the Jewish problem because he cared "less about Jewry and more about territorialism and Zionism."[62] Again, this was not an entirely objective characterization of Zangwill's position. A journalist and writer, he was known in England as "the Dickens of the Ghetto," mostly for his novel *Children of the Ghetto: A Study of a Peculiar People* (1892). His play *The Melting Pot* enjoyed great success in the United States in 1909–1910, and significantly enhanced the currency and popularity of this metaphor as a description of the American model of immigrants' absorption. Zangwill's literary and journalistic works reflected and reacted to Jewish experiences in different national settings and societal strata, while politically he came to Territorialism from Zionism, which he quit in 1905.[63] Zangwill was one of those committed Europeans and westernizers who, in the words of John M. Efron, "before bidding farewell to Europe . . . sought to demonstrate the invaluable contribution Jews had made to the development of European civilization."[64]

In his paper for the congress, Zangwill distinguished between the two "Jewish problems." The first one was the general "Jewish problem" of Jewish persecution everywhere in the world. This problem could be resolved through civic and political emancipation. The second was the "Jewish 'Jewish problem,'" and its "comedy and tragedy" derived from the same political emancipation. "As Shylock pointed out, his race cannot eat or drink with the Gentile"; as Heinrich Heine pointed out, "Unless Judaism is reformed it is . . . a misfortune, and if it is reformed, it cannot logically confine its teaching to the Hebrew race, which, lacking the normal protection of a territory, must be swallowed up by its proselytes"; and, finally, as Zangwill himself pointed out, "Social intercourse would lead to intermarriage" and with the disappearance of the structural ghetto Jews would have to contribute to a non-Jewish culture.[65] Racism played its role in the general "Jewish problem," but was seemingly irrelevant for the "Jewish 'Jewish problem.'" Zangwill personally held the view that "every people is a hotch-potch of races," and that Jews were also mixed and hence typical. He argued for the unity of humanity, but at the same time believed that under the existing conditions the emancipation of Jews would result in their national disappearance (if granted without a territorial solution). On top of this he added an observation about anti-Semitism as an almost natural reaction to Jewish presence in gentile societies. As a kind of racial othering, it manifested itself in the "law of dislike for the unlike." "In the diaspora," Zangwill reasoned, "anti-Semitism will always be the shadow of Semitism."[66]

Shternberg attacked Zangwill from the position of a scientist who better understood the objective nature of Jewishness and the laws of human evolution. He exposed the contradiction implicit in the claim that all human races were mixed and mutually akin but, at the same time, subject to the "law of dislike for the unlike." He ridiculed Zangwill's misuse of Darwinian concepts such as "protective coloration," which Zangwill evoked to account for cases of crypto-Judaism and other forms of assimilation. Shternberg criticized Zangwill for approaching Jews as a religious congregation rather than a nationality, and for other mistakes. The only conclusion that he partially accepted was the predicament of unreformed Judaism, but he rejected the territorial state solution. Instead, Shternberg asserted that Jews in Eastern Europe and some other parts of the world were already living as a territorial nation in their consolidated minority communities. This was as apparent to him as the fact that neither religion nor any uninterrupted tradition kept Jews together in places like New York, where they were free to integrate and intermarry. What force did keep Jews together, then? The only plausible explanation seemed to be their fundamental biological unity. This unity expressed itself in Jewish kinship and prompted Jews to retain communal endogamy.

Shternberg would return to these themes in all his subsequent writings on the "Jewish Question." In the original context of the early 1910s, these themes and the entire misreading of the Universal Races Congress reflected his degree of preoccupation with the problem of Jewish nationality and the depth of his involvement in Jewish politics, where his main opponents were Zionists and Territorialists. A few months before the congress Shternberg wrote:

> Out of all these attributes [race, national territory, language, religion—as listed in the paragraph directly preceding the quotation] only one stands for Jews without any doubt—the anthropological, racial, attribute. From the point of view of this attribute, Jews, thanks to the purity of their blood, are of course a nation par excellence; they are unrivaled. It is remarkable, however, how nationalists completely ignore this most obvious and exceptional attribute when they discuss our nationality.[67]

By "nationalists" Shternberg meant Zionists and Territorialists, between whom he did not differentiate as both failed to understand that only the race factor ensured that the Jewish nation could survive and blossom without a national state. From this position, Shternberg criticized both the Territorialist Zangwill and the Austrian Zionist race scientist and physician Ignaz Zollschan, whose book *Das Rassenproblem unter besonderer Beruecksichtigung der theoretischen Grundlagen der juedischen Rassenfrage* (The race problem especially considered on the theoretical basis of the Jewish racial question) came out in 1910. Zollschan offered anthropological evidence in favor of the reality of Jews as a homogeneous race, but pessimistically predicted their disappearance under the conditions of nonterritoriality and cultural colonialism. Like Zangwill, Zollschan argued that legal emancipation intensified racial mixing and made Jews incapable of manifesting any independent historical subjectivity.[68] Shternberg challenged this assessment with the unorthodox understanding of Jewish territoriality, which he then reproduced in his polemic with Zangwill. Austrian Galicia, Russia's western borderlands, and New York City—these were "only ghettos according to Zollschan and [Max] Nordau, just as these were Golus [outside the land of Israel]—for the Orthodox," Shternberg wrote. They could not see that Galicia, western borderlands, and New York were territories where "Jews lived in a continuous mass," preserved their endogamy, and shared "similar conditions of environment and existence, material interests, and sociopolitical demands and needs." From here, two conclusions followed: one was political—the Jews were already a territorial nation and they had to fight for individual and national rights on these territories; the other was scientific—Jews were a single, homogeneous race, and this biological fact defined their cultural and national specificity even when they formed a compact minority in a non-Jewish state.[69]

Remarkably, Shternberg did not question Zollschan's Zionist tendentiousness in constructing Jews as a pure race. As scholars, they agreed that the Jewish race did not intermix with European races, that Jews constituted a *Kulturvolk* and had contributed key moral and ethical principles to universal human culture, that theories of racial superiority had no scientific basis, and that endogamy was a positive and desired choice, whereas intermarriages could lead to the disappearance of Jews as a race and nation. In addition, as committed Darwinists, both believed in the power of intraracial transformism to produce modifications within the existing "type."[70] The real disagreements between Shternberg, Zangwill, and Zollschan (and other Zionists or Territorialists) were political; they concerned the prospects of modern Jewish nationhood without a national territory, of the Jewish struggle for human and national rights—here and now.

Mastering the Science of Race

As a specialist in cultural anthropology, Shternberg had to educate himself about race science. His voluminous archive in the St. Petersburg Branch of the Archive of the Russian Academy of Sciences includes the collections "L. Ia. Shternberg's Notes and Extracts on the Problem of Race" (234 pages) and "Materials on the 'Jewish Question'" (792 pages!). The second collection contains many small cards, covered with Shternberg's almost illegible handwriting, and organized under the common rubric "Jews. Heredity." One of his notes reads:

> National psychology expresses itself through the stability of certain spiritual predispositions, mental abilities, [unreadable] . . . character traits, and so on. These peculiarities are formed during a period of the most extreme

isolation; therefore, they acquire an especially strong durability and are most capable of lasting eternally. Just as one's constitution [unreadable] . . . is formed and stabilized for good [unreadable] . . . as a result of consanguineous marriages between close relatives, some acquired psychological traits intensify under the conditions of endogamy. The Jewish racial psyche had been formed [unreadable] . . . when the way of life was still nomadic; when one random child-genius, through intergroup marriages, could transmit her traits—thanks to tribal endogamy—to the whole tribe. There are no doubts [unreadable] . . . endogamy did exist among the Jews. Sacral marriage to a sister's daughter. It is true that Jews also took wives from outside, however, according to [unreadable] . . . , racial types can return [to initial types], and these recessive genes are bearers of racial psyche. . . . The question of the purity of the Jewish race is irrelevant here, because genes return. Moreover, regardless of how substantial the admixture of alien blood in Jews was, Jews have been endogamous longer than other nations.[71]

These and similar notes on separate cards Shternberg titled "national psychology," "Jewish national psychology," "Jewish soul," and "racial morale" reflect extensive reading about race, heredity, and psyche. The paragraph above grew into a full-scale article on the Jewish racial psyche published in 1924, but the ideas that informed it originated in the 1910s.[72] The very sheet of paper Shternberg used to record the text above also features the handwritten word "Chicago" and a few other words in English. "Chicago" could refer to a library in Chicago where the note was taken. Indeed, back in 1905, Boas invited Shternberg to the United States as a guest of the American Museum of Natural History in New York, to examine and arrange its collection from the Amur River region. In addition, Boas commissioned his Russian colleague to write a monograph on the indigenous population of the Amur region for the Jesup North Pacific Expedition series. Shternberg arrived in early April and, according to Kan's account, did not spend much time working on the collection. He socialized with Boas and they became lifelong friends. He also observed with great curiosity the life of Jewish immigrants from the Russian Empire. Kan quotes from Shternberg's letter to his wife sent on May 30, 1905: "Generally speaking, how touching it is to see crowds of Jews who feel that they are masters here, who feel strong and self-confident."[73] The impressions must have been very strong, for they even interfered with Shternberg's memories of exile. In 1912, when Shternberg published an account of his involuntary journey from Odessa to Sakhalin, the Jewish-related aspects, previously absent in his writings, came to the fore, but were re-configured in light of his observations of American Jews: "When I finally encountered Siberian Jews, I could ascertain that what seemed to be natural peculiarities of Jewish physiognomy and speech were quickly dis-appearing in the course of a long life in a new environment. I observed the same phenomenon in the United States, where the first generation of emigrant Jews had already turned into typical Yankees."[74]

In late June 1905, Shternberg traveled to Chicago to examine the Amur-Sakhalin collection in the Field Museum and, as his column in *Novyi voskhod* reported, to attend the convention of the radical Indus-trial Workers of the World.[75] It seems that between specific museum-related scholarly assignments, which had nothing to do with his Jewish interests and remained largely unfinished, and political socialization, which had more to do with his real concerns at the moment, he found time to read on the topic that in 1905 absorbed him more than the Amur-Sakhalin ethnography—the phenomenon of Jewishness as a race, a nation, and a spiritual community. While Shternberg was educating himself about race in a country where race was deeply ingrained in political and academic thinking as well as social and cultural practices, in his own country revo-lutionary urban riots were progressively taking the form of Jewish pogroms. In April 1905, he learned both from American newspapers and from his wife's and sister's letters about a major pogrom in his hometown of Zhit-omir. Shternberg's family resided in a more upscale part of the city and remained physically untouched, but his mother suffered a nervous break-down and died as Shternberg, who had to cut short his American visit, was still on his way back to Russia.[76]

Thus, Shternberg's trip to the United States and his inquiries about the problem of race coincided with the beginning of the first Russian Revo-lution and his radical turn toward Jewish politics and Jewish science. As we know, Shternberg did not take the seemingly self-evident and direct path to the desired end; that is, he did not turn to works by Anuchin or Russian Jewish experts on Jewish race such as El'kind and Weissenberg. Apparently, he was discouraged by the philosophy of the mixed racial type and of the empire as a natural field for racial mixing, as well as by the accentuated academism and apolitical stance of the liberal Moscow an-thropological school. In addition, unlike Russian liberal race scientists, who believed that culture, including human psyche, should be considered sep-arately from race, Shternberg was fascinated by racial psychology as a field bridging cultural and physical anthropology. The files from his archive suggest that he remained true to this initial understanding of Jewish race science until the very end of his life. Some cards and separate sheets in the collection "L. Ia. Shternberg's Notes and Extracts on the Problem of Race" can easily be dated to the mid-1920s, when Shternberg taught regular university ethnography courses and used the cards in his classroom. One of his students in the 1920s, who was later interviewed by Grant, recalled: "As a lecturer he was very difficult. His lectures were always substantive and very deep, but he read very, very slowly. He would bring entire card catalogues with him to the lecture hall and read out long passages from various cards that he would take from his files."[77]

Another student from the same years, more sympathetic to Shternberg's teaching style, recollected the "cards" rather sentimentally:

> A thin old man, who seemed to have been charred by some internal burning, spread a pile of cards with notes on the podium and raised his eyes to the audience. For a whole minute his dark, burning eyes intently looked at us through the glasses. Then he began to speak. . . . He would periodically bend down to his cards in order to read a citation supporting his thought.[78]

These cards, together with selectively recorded questions from his students and attendees of public talks, give a pretty good idea of how Shternberg discussed race and how this topic resonated with the audience:

> [On a scrap of paper:] Professor, please explain why Negroes produce bad odor. Perhaps, their coloring produces it? [signed, but signature is illegible].[79]
>
> [Also on a scrap of paper:] Lev Iakovlevich! You say that the coloring of human skin changed under the influence of geographic conditions and that, perhaps, the same race under different geographic conditions could have yielded two different color branches. However, according to the theory of heredity, acquired characteristics cannot be inherited. If this is true, then a child born to a Negro couple should be white and become darker under the influence of geographic conditions. Is this so?[80]
>
> [Another minuscule scrap of paper;] What are the results of the examination of the skull of V. I. Lenin, who was undoubtedly a genius? What are the results of the study of his brain?[81]

The reference to Lenin's brain clearly dates the note to after January 1924. But there are other notes, mostly by Shternberg himself, which most probably belong to the period before 1917. Some of them could have been written even before 1905 for encyclopedia articles, such as notebook pages covered with information on the "fertility of hybrids" or on the physical unity of humankind.[82] A special card that refers to no particular source features just one line: "Nature does not know [unreadable] . . . a truly hybrid race."[83] Others resemble sketches on racial classification: series of notes on the black, "Caucasian or white" races and on coloring and pigmentation; classificatory trees, organized along the three main branches—the white, yellow, and black races, with the Semitic race placed under the white race branch.[84] In one classificatory tree the yellow race has the caption "a variety of the white."[85] Regardless of such nuances, everywhere the criteria of classification remain the same: coloring, hairiness (*volosistost'*), and face shape. From these notes, Shternberg as a racial classifier emerges as much less sophisticated than his contemporary Russian physical anthropologists, who use the term "race" cautiously and in whose classificatory discourse visual indicators played the least important role. Unlike most of them, Shternberg was not a physician or natural scientist. Among his many notes, one thin folder, "The Problem of Race," contains materials on craniometry and brain anthropology, but these were not his strongest fields.[86] Shternberg was much more at home with something that did not require special medical training and allowed him to combine race and culture—that is, racial psyche.

Eventually, the intellectual work that had left such a voluminous archival trail culminated in a coherently articulated narrative of Jewish ethnography and anthropology, which Shternberg advanced after Russian Jews had acquired full political emancipation in 1917. However, before arriving at this synthesis, he had to fight a few more political and epistemological battles.

Boasian Revolution

At the EIEO meeting on April 23, 1912, Shternberg presented a paper titled "The Race Question in the Newest Works on Jewish Anthropology (Zollschan, Fishberg, Boas)."[87] Ten days later, a very detailed and long anonymous summary, adapted for the interested lay audience, appeared in *Novyi voskhod*.[88] Subsequently, the EIEO journal *Jewish Antiquity* published the full text of the presentation.[89] This was enormous publicity for a strictly academic paper. Its focus on the race question reflected a broader trend that marked the EIEO activities in 1912, when three out of five of the society's scholarly meetings considered papers involving Jewish blood or race. At the meeting immediately preceding Shternberg's presentation, Moisei Trivus talked about "ritual murder trials in the prereform Russian courts (the Saratov affair and others)." The meeting following Shternberg's talk featured Weissenberg's lecture "Jews in Turkestan, Caucasus, and Crimea: Report on a Scholarly Expedition."[90]

This particular sequence of presentations makes Shternberg's omission of Weissenberg's works on Jewish anthropology even harder to explain. We know that Shternberg and Weissenberg collaborated in the EIEO on projects pertaining to Jewish ethnography, corresponded occasionally, and never criticized each other publicly.[91] At the same time, they tended to ignore each other's publications on Jewish physical anthropology. Although Weissenberg had every reason to reject them as amateurish, Shternberg definitely could not accept his colleague's view of the Jewish race as mixed. Perhaps Weissenberg's close affiliation with the liberal Moscow school of imperial anthropology bothered him as well. It is telling that Shternberg selected only foreign Jewish scientists for his paper. Their political orientation differed, and as Shternberg himself admitted, they were not equal in terms of "scholarly competency . . . and objectivity." At the same time, each of the selected works offered a solution to the "currently militant problem of Jewish race" (*boevoi po nyneshnemu vopros o evreiskoi rase*), and each scientist was ascribed by Shternberg a clear political position.[92] Zollschan's *Das Rassenproblem,* for example, was "not only a treatise on Jewish anthropology, but a coffee-table anti-Chamberlain [book] [*nastol'nyi anti-Chemberlen*], the best pamphlet against quasi-scientific anti-Semitism. For Zionists, Zollschan's book is a genuine ideology of Zionism based on anthropology."[93] Besides *Das Rassenproblem,* the works that satisfied Shternberg's criteria included Maurice Fishberg's *The Jews: A Study of Race and Environment* (1911) and Boas's *Changes in Bodily Form of Descendants of Immigrants (Final Report)* (1911), conducted for the US Immigration Committee.[94]

Shternberg's paper opened with a brief overview of scholarly debates about the Jewish race: For many decades, anthropologists had accepted the stability of the cephalic index, color indicators, and facial form and assumed that Jews were endogamous for most of their history. Therefore, they regarded the Jewish race as an archetypal example of racial purity and stability. Newer studies revealed that Jews exhibited variations of pigmentation, and this gave rise to theories of Jewish racial mixing, either in biblical times or in the Diaspora. "However, no anthropologist has yet raised the question about the possibility of light types emerging within a dark race independently from mixing." Another unresolved problem concerned Jewish head index: Were Jews a dolichocephalic race that had evolved into a brachycephalic race, or vice versa? As Shternberg preferred to put it suggestively, was "the contemporary Jewish short-headedness the result of some other factors, apart from mixing?" Finally, academic anthropologists debated whether the Jewish race was uniform, or the Sephardic and Ashkenazi Jews belonged to two different races.[95] Having established this general problem field, Shternberg moved to factors that, "having nothing to do with science," had influenced scholarly debates, pushing them in the wrong direction (for example, the Aryan theory, the association of language and race, anti-Semitism). As a result, the problem of Jewish race "has been transformed from a purely scholarly one into a militant question full of burning contemporaneity [*polnyi zhguchei sovremennosti*]."[96]

It is in this context of contemporary politics that Shternberg criticized Zollschan, with whom he otherwise agreed on scholarly matters. Most important, Shternberg subscribed to the view that Jews constituted a single, homogeneous race, and that if any racial mixing had ever occurred, it had happened in ancient Palestine. He prized Zollschan's acceptance of the malleability of all major racial indicators (except for the proverbial "Jewish physiognomy"), including cephalic index and pigmentation, which, as the Austrian scholar explained, could change under the influence of "mechanical-functional factors, somatic correlation, type of food, way of life, and, finally, geographic environment." While questioning specific interpretations of racial changeability, and especially Zollschan's explanation of Jewish brachycephaly based on their "intellectualism" that had developed as a strategy of resistance and survival, Shternberg still confirmed with satisfaction that Zollschan's approach was akin to Darwinian transformism.[97] This was particularly evident in Zollschan's discussion of light-colored individuals among Jews, whose existence he explained as mutation—a weak form of albinism resulting *not* from racial crossing but from intraracial transformism. Finally, Shternberg supported Zollschan against his opponent in the Zionist anthropological camp, Ignacy Maurycy Judt, on the problem of the Jewish place in the classification of races.[98] Following Luschan, Judt ascribed the predominantly short-headed Jews (descendants of ancient Hittites) to the Mediterranean and Alpine races. Zollschan, on the contrary, claimed that Jews were Semites

belonging to one of the two subgroups of the white race. Their particular subgroup "had given humanity the Sumerian-Akkadian culture,
Babylonians and Assyrians, Persians, Hittites, Phoenicians, Jews, Egyp-
tians, Greeks and Romans, Arabs and, to a great degree, the culture of
the Renaissance."[99]

Here, however, their scholarly agreements ended and differences began,
spearheaded by politics. Today, from a distance of a hundred years,
Zollschan's position can be interpreted as an example of "how it was possible to create an anthropology of resistance without betraying . . . alle-
giances to Enlightenment values" of human adaptability and equality.[100]
Yet, from Shternberg's "burning" contemporary perspective, the two
Zollschans—that of a Darwinist and a man of Enlightenment values and
that of a Zionist politician—were clearly differentiated. Zollschan's Zionism negated the value of his anthropology: in addition to embracing a
wrong political ideal, he aspired to "defeat Aryanists with their own
weapon": "They assigned Jews to lower races mostly on the basis of their
short-headedness. Now it turns out that the short-headedness is a sign of
higher intelligence, while long-headedness becomes a stamp of mental
underdevelopment."[101]

Shternberg found that Zollschan's version of relationships between
race and nation mirrored the view of scientific racism. Judging by the
notes in Shternberg's archive, in 1912 he himself was intensely searching
for an answer to the dilemma of conflation of race and nation, which he
would later find in the national psyche.

Although Fishberg did not identify race with nation and was not a Zionist, his work—the "real monograph" that touched on a broad range of
important topics, from the "history of proselytism and miscegenation, demography, and pathology" to criminality, Zionism, and anti-Semitism—
received even less endorsement from Shternberg. The reason was Fishberg's "extreme assimilationist" position and the denial of the reality of
the Jewish race.[102] The rich anthropometric data assembled in *The Jews:
A Study of Race and Environment* demonstrated that Jews were racially
mixed, and moreover that their racial mixing had continued for the entire duration of Jewish history. Shternberg did not even consider this view
worthy of serious scrutiny. As he wrote with obvious disdain: "To say
something like that about the people who for twenty centuries have observed the strictest endogamy is complete absurdity, which does not require any refutation."[103] Still, Fishberg's main crime consisted in denying
the Jews "any individuality at all":

He denied Jews biological attributes such as propensity toward acclimatization, high capability of survival, and specific national traits of their intellectual character. He talks with irony about the alleged Jewish thirst for knowledge, challenges [Werner] Sombart on the role of Jews in the development of capitalism and on their commercial talents. One cannot even talk about a national culture: he thinks that Jews do not possess even folklore of their own.[104]

Because Fishberg and Boas both studied American Jews in New York, Shternberg tended to see them as immediate rivals. Moreover, Fishberg's continuous reliance on head index after Boas had "undermined the whole teaching regarding the stability of the head index" only reinforced this perception.[105] But while treating the two anthropologists as academic rivals, Shternberg, operating within the paradigm of the "militant" science, also ascribed to them opposing political views. As a result, Boas's report emerged in his paper as a normative example of Jewish race science.

Boas conducted his study at the request of the US Immigration Commission (known as the Dillingham Commission) in the politically charged atmosphere of Progressive Era America tackling the issue of growing Eastern and Southern European immigration.[106] US political elites, on the one hand, and leaders of the organized labor movement, on the other, believed that these new immigrants, among whom Jews from Eastern and Central Europe composed a visible majority, were bringing with them feeble physical constitution and wrong values, which prevented their assimilation. The combination of popular eugenic views and political and economic concerns, specific to this moment in US history, produced a consensus in society that the new immigrants endangered the existing racial balance between whiteness and nonwhiteness and also that they were unable to comprehend American political processes and practices of social organization.[107] Under such circumstances, Boas—himself a Jew—selected Eastern European Jews as the main test group for his study of the transformation of generic immigrants' physical characteristics. Southern Italians made up his second test group, but they were studied less extensively and afforded a lesser role in the overall research design. First published in its entirety as a 573-page volume by the Dillingham Commission and then as an article in a 1912 issue of the *American Anthropologist*, the study presented anthropometric statistics and calculations proving the plasticity of racial traits, including the very cephalic or head index that was previously seen as the main and most stable anthropological indicator. Boas's research showed that the cephalic index in fact "undergoes far-reaching changes coincident with the transfer of the people from European to American soil. For instance, the east European Hebrew, who has a very round head, becomes more long-headed; the south Italian, who in Italy has an exceedingly long head, becomes more short-headed; so that in this country both approach a uniform type, as far as the roundness of the head is concerned."[108]

From here it followed that even those characteristics of a race "which have proved to be most permanent in their old home" could change in new surroundings. "We are compelled to conclude that when these features of the body change, the whole bodily and mental make-up of the immigrants may change."[109] Specifically, Boas's study revealed that American-born children of Eastern European Jews in New York differed from their siblings born in Eastern Europe, as well as from their parents. They resembled the surrounding population both physically and in terms of their mental makeup. The less complete data for Sicilians supported the same pattern. On these grounds, Boas insisted that "the adaptability of the immigrant seems to be very much greater than we had a right to suppose before our investigations were instituted."[110]

Shternberg did not comment on the political context of Boas's research; most probably, he had no understanding of its complexity. Besides communicating a sense of personal respect and admiration, Shternberg's introduction of Boas projected the picture of an ideal scholar working under ideal circumstances. Boas did not know the tension that could exist between one's formal academic affiliation and self-identification with Jewish science, and he had no reason to feel any contradiction between working for the state and being useful to the Jews. Shternberg's stay in the United States in 1905 notwithstanding, or, perhaps, because of this experience, he idealized the situation of American Jews, which he observed from the vantage point of a Russian Jew with limited political rights and protections (in fact, a Jew whose hometown was just recently devastated by a pogrom). Similarly, he interpreted Boas's academic and political self-positioning from the vantage point of his own uneasy situation as an academic ethnographer unable to pursue Jewish science in official academia. In Shternberg's words, Boas was "currently one of the most authoritative anthropologists not only in America, but in the world . . . , the Dean of the Department of Anthropology at Columbia University in New York," and "the chair of the scholarly division of the U.S. Congress's Special

Committee for the Study of Immigration," which was not true.[111] Shtern-berg told almost a parable about a democratic state that hired a respectful scientist to investigate changes in the bodily form of first-generation im-migrants and their children, with the goal of perfecting state policies of their adaptation and integration. The Jewishness of the scientist did not matter. "Boas decided to conduct his study not on immigrants in general but divided them into national-territorial groups; he decided that his first sample would be limited to the New York Jews—natives of Eastern Europe (Russia, Romania, Galicia)." (Obsessed with the image of an un-constrained scholar-demiurge, Shternberg greatly overused the verb "de-cided."[112]) The assumption was clearly that Boas was politically neutral and scientifically objective, and the division of immigrants into national and territorial units, including the homogenization of Jews from dif-ferent regions of East and Central Europe into one group, was natural. The ensuing research project was of such high quality and impeccable objectivity that its results came as a surprise to Boas himself: "Under the influence of American conditions, not only the general physical con-stitution of the descendants [of immigrants] changes, but so does their anthropological type, including the most stable racial index—head index." Boas thus proved the "elasticity of human types" and the im-provement of the entire "physical habitus" of Jewish immigrants to the United States.[113]

Overall, Shternberg's discussion of Boas's study was remarkable in its unintentional distortion of the original political and academic context in which it had been done. Shternberg's Russian Jewish listeners and readers would never guess that Boas was not Fishberg's rival, and that they shared the same goal of rebuffing the racialization of Jews in the Progressive Era anti-immigration discourse. "I also greatly profited by conferences with Dr. Maurice Fishberg, whose investigations among the New York Jews indicated the practicability of the present investigation," Boas wrote in his report.[114] Shternberg's selective blindness in this case had to discredit Fishberg's view of Jews as a racially mixed population. Opposing the two anthropologists, Shternberg insisted: "Boas shows that this differentiation is not necessarily connected to the process of mixing, but can be caused by factors that have nothing to do with meticization."[115] Shternberg un-derstood this as a proof of the existence of the pure Jewish race, easily adaptable to modern conditions (such as in New York) but without be-traying national Jewishness. Since Boas's discourse on this matter was vague and he never specified the exact causes of this found intraracial transformism of Jewish immigrants, Shternberg speculated that "the very unexpected nature of changes points to psychological influences on the parents."[116] He took it as a major task of militant Jewish science to explain the mechanism of Jewish transformation in democratic, inclusive, and economically advanced modern societies. Shternberg offered the Boasian model of anthropological study of immigrants to modernity—the result of his creative misinterpretation—as a foundational framework for Russian Jewish science:

> In this regard, Jewish anthropologists in Russia enjoy rich opportunities. . . . We should admit that the anthropology of Russia's Jewish population has not been sufficiently studied. Until now, anthropologists operated with hundreds of cases, but applied trivial methods [of analysis] to them. Boas's discovery has given Jewish anthropologists a completely new task. In addition to increasing the number of observations, they must carry them out among different classes and in different localities, under differing climate, environmental, and social conditions. It is necessary to study the anthropology of Jews residing outside the Pale, in Siberia, in big intellectual centers such as Petersburg, Moscow, Kharkov, and so on. A special questionnaire for Jewish student youth is needed. In the Pale, the Jewish population should be examined in different regions and according to occupation. . . . This collective work of Jewish intelligentsia would yield priceless material for resolution of the problem that the Jew-anthropologist [Boas] has formulated so brilliantly.[117]

Only after having reinterpreted the project of the world-renowned American scholar, Boas, in the context of the Russian Jewish political struggle for general and national rights, Shternberg felt confident to call him a "Jew-anthropologist." The price for this was a complete silencing of the assimilationist implications of Boas's research, so apparent to Shternberg in the case of Fishberg.

A German-born Jew, raised in an assimilated family identifying with the liberal ideals of the 1848 revolution rather than with national Jewishness, Boas emigrated to the United States in 1887 in search of complete acceptance in academia. Unlike Shternberg and his fellow Russian Jewish intellectuals, he did not think that "being Jewish might in itself operate as a formative element in a social environment."[118] As George W. Stocking Jr. has suggested: "The experience of Jews in Germany provided him the archetype of an ostensibly racial group that was in fact biologically heterogeneous, which had assimilated itself almost completely to

German national culture and which in multitudinous ways had enriched the general cultural life of modern civilization."[119]

Having examined the rhetoric of Boas's speeches on American white racism and anti-Semitism, Ezra Mendelsohn selected two words that were most central to his discourse: "forget" and "disappear." "It seems . . . that Boas would have no regrets were the Jews as a separate group to peacefully disappear," Mendelsohn concluded.[120] Later in life, Boas clearly articulated his preference for Jewish assimilation, which, not unlike Fishberg, he justified using scientific arguments.[121] Boas's explicit statements and life choices inspired Mendelsohn to characterize him as "a living advertisement for assimilation."[122] If Boas's anthropology could qualify as a "Jewish science," it was a very different "Jewish science" from the one envisioned by his Russian Jewish friend and admirer. This Jewish science would rather conform to a pattern described by the sociologist John Murray Cuddihy as an "ordeal of civility." Cuddihy thus referred to the newly emancipated Western European Jews, who, in the nineteenth century, turned an acute understanding of the ambiguity of their social position into rigorous intellectual reflection about the nature of groupness.[123] From this perspective, a major shift in anthropology from biological constants (races) to physical and cultural contacts and mutual borrowings, associated with Boas, can be partially explained by his personal "ordeal of civility." Among the American students of this founder of the country's first department of anthropology at Columbia University (1896) were quite a few "men of German or 'hyphenate' origin, several of them Jewish, several of them political radicals"—the latter quality often being associated with Jewishness as well.[124] At the same time, Boas's students and the school of anthropology that he had founded did not study Jews for their own sake (only as "immigrants"), which was another consequence of the ambivalent situation of Jewish internal belonging/nonbelonging and external acceptance/nonacceptance.[125]

Shternberg not accidently missed all these nuances of Boas's own Jewish condition. He needed a scholar—a humanist and a Jew—in the role of a producer of useful knowledge for a modern democratic supranational state with which Jews could identify. This was a model to which Shternberg himself aspired. Boasian anthropology gave him and other Russian Jewish modernizers scientific proof that the Jews of the Russian Pale could be transformed into a modern, integrated, strong, and self-conscious nation. As we have seen, in Shternberg's mind, both New York and the Russian Pale were spaces of Jewish territoriality where endogamy protected the homogeneity of the race and helped to further cultivate kinship ties, while Judaism could persist as a set of moral beliefs and humanitarian principles (creating an affective participatory sphere for the nation). However, the success of the model of the "Boasian revolution" ultimately depended on the success of the Russian imperial revolution—a global change of the political environment in the empire.

Biologizing Jewish "National Psyche"

Andrew Sloin has characterized the Bolshevik Revolution of 1917 to the 1920s as "the last great project for Jewish emancipation in modern European history," which offered Jews "a qualitatively new model of Jewish emancipation that would transcend the mere 'negative' logic of liberal emancipation by engaging in an activist, positive program for the total economic, social, and cultural transformation and integration of Soviet Jewry within a postcapitalist order devoted to social and national equality."[126]

Shternberg's reflections on the revolution in one of his last articles, "The Problem of Jewish Ethnography," published posthumously in 1928, advanced a similar "qualitatively new model of the Jewish emancipation" paradigm. Soviet Jews in his account were experiencing no less than a total revolution. They had been deprived of the "old economic base" and were acquiring new economic functions; their material world was evolving, together with old traditions, toward new forms; Jewish youth were rejecting religion and actively embracing civic culture; and a great number of Jews, for the first time "after the millennia of powerlessness," could participate in state-building. Shternberg argued that the ethnography of this total, activist, positive transformation could be only sociological—not in the old sense of evolutionary sociology, but in a revolutionary sense of exploring the entire totality of new Jewish life.[127] The absence of references to race in this project of Soviet Jewish ethnography presented a striking contrast to Shternberg's early writings on "militant" Jewish science. Penned when the Soviet policy of indigenization was coming to an end, and nation-building had stabilized in the form of territorial republics, Shternberg's article reflected the atmosphere on the eve of Stalin's "Great Break." On the one hand, as Sloin writes specifically with regard to the Jewish situation, "the resurgence of antisemitism and racialized thought in the late 1920s triggered a total reversal of Soviet nationality policy, as the Soviet state and the Bolshevik party ultimately decided that the politics of integration mattered more than the politics of nationality."[128] On the other hand, any biological obstacles to the success of collectivization and industrialization could not be tolerated. Mark B. Adams dates to this time the emergence of a new pejorative term in the official Soviet ideological discourse—to biologize (*biologizirovat'*) a political or social problem.[129]

Soon Nazi racial science would completely discredit biosocial theories, providing a pretext for the final assault against disciplines such as physical anthropology or eugenics in the Soviet Union. These considerations explain why Shternberg's last programmatic article spoke in the language of revolutionary sociology rather than an evolutionary biosocial understanding of Jewish groupness. But this does not have to obscure the fact that his earlier texts, written between the revolutionary year of 1917 and the year of his death in 1927, revealed a very different pattern: Shternberg expected the Bolshevik Revolution to unleash the accelerated Boasian evolution of the Jewish biological nature, and the Jews—to adapt to a new and better environment as an objectively existing nation and on their own terms, consonant with their sociobiological nature.

In 1924, after the wars and revolutions were formally over, Shternberg revived *Jewish Antiquity* and assumed leadership of the EIEO (after Dubnov's emigration in 1922).[130] The first Soviet issue of the journal under his editorship featured two remarkable articles, both dealing with the problem of Jewish biological transformation in the revolution through the Boasian model of emigration to modernity. The first article was penned by the anthropologist Boris Vishnevskii and presented the results of an anthropological examination of 289 Jewish children in the inner Russian city of Kazan, the recently established capital of the Tatar Autonomous Republic (formed 1920), located far away from the Pale of Jewish settlement. Anthropological data for the article were collected in the summer of 1922. In trying to make sense of them, Vishnevskii directly followed the blueprint of Boas's study: he divided his sample Jewish group into natives of Kazan and recently arrived refugees from the Pale. One of his main conclusions addressed change in the head index of these refugees and their children. Vishnevskii's data for Jewish children born in Kazan demonstrated a familiar trend: the head index of first-generation Kazan-born Jewish children approached the average for native Russian children of Kazan. Children of Tatars—the second major ethnic group in the city—were tellingly absent from this comparison illustrating the convergence of Jewish and hegemonic Russian racial types and hence the Jewish ability to assimilate and adjust to what was perceived as a more modern and advanced environment. Just as in Boas's report, the head index changed in one generation that remained largely endogamous.

The second article featured in the same issue of *Jewish Antiquity* was by Shternberg, "Questions of Jewish National Psychology," in which he claimed that each people had its own national character made of the traits that "are continuously transmitted from one generation to another and can reveal themselves in different times, under different external circumstances, and in various external forms."[131] Changes in the biological substratum of the national character occurred on an extremely slow path, much slower than anthropometric changes as proved by Boas and much slower than visible cultural changes. The national character thus substituted for "race" as understood before Boas. In accord with classical Darwinism, Shternberg retained sexual selection as a key mechanism of race formation.

> The durability of a biological-psychological complex, much like the stability of some racial features of a physical nature, can be explained by the fact that this complex has been formed in a specific period of development of somatic peculiarities of a given people. . . . A necessary condition for the formation of such a durable racial complex is *isolation*. As biology shows, isolation, that is, a situation in which the process of biological heredity happens without interference from external crossing, is the main condition of race creation.[132]

Simply put, isolation guaranteed racial endogamy and the emergence of the unique national character that continued to define race even when anthropometric indicators were changing. Here Shternberg found himself on familiar territory that bridged his ethnography of Siberian natives and the Jewish anthropology. He referred to the Eskimos, who "for centuries have been isolated on the shores of the Arctic Ocean" and developed unique artistic abilities that were not typical of other Asiatic polar peoples. Their artistic gift was a racial trait.[133] However, "cultured" people such as Jews, especially in the twentieth century, were anything but isolated. For them, the act of sexual selection had become a conscious and well-informed choice as well as an element of affective participation in a national collective. Still, between the two poles—primitive and modern—Jews constituted an exceptional case, an "experiment in the laboratory of world history." As all other races, the Jewish race had been formed under the conditions of cultural isolation and biological endogamy, but unlike most other races, the Jews had always remained "the most endogamous among all contemporary cultured peoples. Hence, biological inheritance of the primordial [psychological] type that had originated during the ancient period of maximum isolation is as apparent in the Jewish people as the inheritance of traits of the physical type."[134]

Having established this on a theoretical level, Shternberg offered a list of specific features of the Jewish primordial, that is, racial, psychological type. The list included intellectualism, rationalism, "intensive emotionality and activity" (the "law of parallelism of psychic phenomena" was supposed to explain the coexistence of rationality and emotionality), prophetism, and optimism. Together, all these flattering "racial" qualities defined the Jewish propensity toward ethical monotheism and a broad humanistic agenda.[135] Surprisingly, the list exposed Jewish fundamental unfitness for the Soviet version of modernity that insisted on "productivization," proletarization, and greater mobility of the population. Shternberg's discussion of Jewish "intellectualism" was particularly telling in this respect:

> There have been and remain among us naive people who dream about creating, in their words, a normal nation out of the Jews, with farmers as its basis. A futile hope it is! As long as conditions allow, Jews will never be satisfied with this numbing occupation [*pritupliaiushchee zaniatie*]. They do this either out of necessity or in a rush of national enthusiasm. Children of enthusiastic Palestinian colonists leave their family nests at the first opportunity for universities in Paris and Berlin . . . American Jewish workers of younger age work hard during the day and prepare for university exams at night, while those who are older send their children to universities instead of factories.[136]

If children of "enthusiastic Palestinian colonists" tended to "return" to their primordial intellectual and rational type, why would children of Russian Jews want to become Soviet agriculturalists or industrial workers? In addition, Shternberg could not reasonably count on the persistence of Jewish "endogamy" after the revolution. He was writing this article in 1924, after many Jews had been deported from their hometowns in western borderlands during the Great War or fled the vicious pogroms of the Civil War (as had the Jewish "immigrants" studied by Vishnevskii). Others jumped on the opportunities provided by the New Economic Policy and moved from small shtetls to bigger cities. It was obvious that Jewish mobility and intermarriages would only increase in the Soviet state.

In "Questions of Jewish National Psychology," Shternberg dealt with this challenge by focusing on the theme of racial crossing much more substantially than ever before, relying simultaneously on the authority of Mendelian genetics and the most recent works by Soviet eugenicists. His main objective remained the disavowal of hybridity as a stable biological phenomenon compatible with national belonging. He understood hybridity in strictly biological terms as a contamination of the purity of race, now transferred to the "national psyche." It had finally become apparent that Shternberg's position on this did not differ too much from the arguments of Zollschan and Zangwill, whom he used to criticize. As they did before 1917, he now took seriously the threat of Jewish assimilation under the conditions of full legal and political emancipation. His first step in protecting "race" was moving it deeper inside the physical body. His second step was engaging the biological law of recessive genes (recessive inheritance), or, as he explained, the law of preservation of the "initial biological complex of mixing elements." Thus, the Mendelian formula of the return of "initial pure complexes" became the centerpiece of the program of Jewish national survival. Shternberg explained that the fewer participants in a crossing (as was presumably the case with the unique Jewish race), the more recessive genes it yields.[137] He read Mendelian genetics through the eyes of contemporary eugenicists, citing their works for the first time in his career: "What is unquestionable for the physical type and physiological characteristics should work as well for a psychic type. This has been explicitly proven by eugenicists on the basis of the recently assembled [family] genealogies, which they evaluated with regard to the [inheritance] of mental abilities."[138]

Building on such studies, Shternberg claimed that a talented minority within a nation had both a biological and a moral obligation to retain its endogamy and thus cultivate the primordial type, that is, the racial character, that would then represent a "race" in interracial encounters.[139] Shternberg did not venture into a discussion of concrete forms of regulation of intermarriages or cultivation of a talented minority within a race. Still, his promotion of expert knowledge and scientific population policies suggested a possible answer.

Published alongside Vishnevskii's article, Shternberg's "Questions of Jewish National Psychology" provided a resolution of the predicament of Jewish assimilation through racial disintegration—a dangerous interpretation of the Boasian revolution, which Shternberg so desperately wanted to challenge. Vishnevskii's article confirmed the Boasian discovery of intraracial changeability and adaptability to a modern and better environment under the influence of some short-term factors, whereas Shternberg's own article explained that these changes could alter only physical type. The foundational racial structure—the national character—remained immune to short-term influences and, in any case, nature created mechanisms of racial self-reproduction, such as sexual selection and Mendelian recessive inheritance laws. If well informed about their evolutionary history and properly motivated, Jews would be able to preserve their national wholeness and specificity in the environment that had been radically altered by postrevolutionary dislocations.[140]

The elaboration of the racialized concept of the "national character" in the domain of Jewish science suggested that the racialization of culture could be Shternberg's general response to the trauma of postrevolutionary cultural iconoclasm. A manuscript titled "The Anthropology of Jews," preserved in Shternberg's archive, seems to confirm this speculation. It can be dated to 1925–1926, as it mentions the 1924 articles from *Jewish Antiquity*. The text itself could be a synopsis of a lecture or a cluster of lectures taught by Shternberg at the Leningrad Institute of Jewish History and Literature (hence didactic statements such as "You will have to learn from anthropology some of its special methods"). Or it could be a draft of an article that was never completed (after Shternberg's death, his colleagues considered translating his Jewish-related articles into Yiddish and publishing them as a collection of "selected writings on the Jewish question").[141] The manuscript opened with explanations of why physical anthropology as practiced by physicians and natural scientists was not "a foreign science" to ethnographers, who, Shternberg insisted, had to master special anthropological skills and methods. The purity of the Jewish race was obvious, he argued, and "a striking physiognomic similarity of Jews with their images on ancient monuments and their centuries-long endogamy force one to see Jews as a completely unique people that has remained racially pure." However, the true racial essence of Jews rested not only or primarily in their physical indicators. "Mendelism has already cleared the [tense] atmosphere surrounding the issue of the purity of the Jewish race," and the future would bring even more clarity on this matter as soon the foundations of "race" were sought in the "domain of 'internal secretion.'" To keep up with modern science, ethnography had to focus on family histories and genealogies, on the study of young people, the

progeny of mixed marriages (to explore recessive genes), on marriages between cousins and other types of marriages among relatives, and on similar eugenics-inspired topics.[142]

In the early 1920s Shternberg actively collaborated with former activists of the Society for Protection of the Health of the Jewish Population,
mostly physicians by training, practitioners of Jewish eugenics and medical self-help. Their influence can be seen not only in the manuscript "The
Anthropology of Jews" but also in the comprehensive "Instruction,"
which Shternberg developed in 1924 for ethnography students at the Institute of Jewish History and Literature (and later the Geography Institute, where he also held a professorship). The "Instruction" attempted to
reconcile different ethnographic approaches and visions from Shternberg's
past and present: his ethnography of "primitive" social organization, kinship, religion, traditions, and languages; a new sociological ethnography
of the total revolutionary transformation of everyday life; and the physical anthropology of the Jewish race and national character. The influence of Jewish race science was especially noticeable in the sections of the
"Instruction" on sexuality, family, and health, and in the discussion of
outstanding and typical Jewishness and of Jewish collective psychology.
"Unable or unwilling to refrain from some wishful thinking when it came
to his own beloved people," Kan wrote, "Shternberg closed the Instruction with a call to collect information on the 'spirit of the Jewish people's
optimism.'"[143] But what Shternberg meant here, most probably, had little
to do with any idealism or wishful thinking. He regarded optimism as one
of the biological traits of the Jewish national character—a racial substrata
of Jewishness that could not disappear as long as Jews continued to exist.

In one way or another, Shternberg's manuscripts and published materials dating from the 1920s reflect the increasing biologization of his general ethnographic discourse that was formerly segregated from the self-
racializing "militant" Jewish science within the domain of normal
academia. Shternberg's signature academic style has always been cultural
in focus, ethnographic in method, sociological in analysis, and universalist
in application. In 1916, he still advocated for a clear distinction between
biological anthropology and ethnography (and, by extension, between imperial ethnography and Jewish science). During the debate initiated in
1916 by the newly appointed liberal minister of education, Pavel Igna-
tiev, the MAE curator voiced his position against joining the two disciplines in the university curriculum. University anthropologists studied
human bodies and races, whereas ethnographers worked with spiritual

and material cultures, Shternberg insisted. He argued against the position that he himself would assume in the 1920s—that is, against the motto that anthropology was not a foreign science to ethnographers and that the latter needed to receive solid training in brain morphology, physiology, psychophysiology, and other natural sciences.

What had changed since the 1916 discussion was the "environment"— the political and academic contexts in which Shternberg developed both his academic ethnography and Jewish race science. He had finally achieved the ideal as exemplified in the figure of his imagined Boas: an outstanding and independent scholar whose personal Jewishness was no handicap in the eyes of the socially concerned and interventionist state. The old split between such a scholar's academic and political personas became obsolete under the new circumstances. Now Shternberg was simultaneously developing a new university ethnography curriculum and an academic Jewish science, and, in the process, he let the racialized Jewish ethnography influence his understanding of the object and method of general ethnography. The reason he did this could be purely subjective: just as was the case with the Jews, he felt the need to protect "culture" from being destroyed by merciless and total revolutionary transformation. Putting objective biological limits on this socially inflicted disaster was his way of responding and adapting. The biosocial understanding of groupness that Shternberg embraced as a Jewish scientist promised to stabilize and safeguard nationality for nonterritorial nations such as Jews, and the ethnographic cultures of "primitive" and "cultured" peoples alike.

On an intrapersonal level, Shternberg's intellectual drift toward a biosocial view of groupness resonated with a trend common to many imperial ethnographers of his generation operating in the rapidly changing imperial context. In the 1910s, scholars such as the curator at the Russian Ethnographic Museum in St. Petersburg Nikolai Mogilianskii (1871–1933) or the younger ethnographer of the Manchu people affiliated with the MAE, Sergei Shirokogoroff (1887–1939), were moving in the same direction. Eventually, they proposed a concept that was fundamentally akin to Shternberg's racialized "national character." This concept was *etnos*— "a single totality of physical (anthropological) qualities, . . . historical destinies, and finally, a communality of language . . . worldview, national psychology, and spiritual culture."[144] Shirokogoroff is considered the official father of the *etnos* concept, and although his and Mogilianskii's relationships with Shternberg were quite complicated, mutual influence can hardly be denied.[145]

They met in the MAE, where Shirokogoroff started working as a student. Later, while conducting ethnographic research in Northern Manchuria, China, and Mongolia, he corresponded with Shternberg as a mentor. In his pioneering study of Shirokogoroff's ethnographic career, David G. Anderson suggests that Shternberg instructed Shirokogoroff to conduct an anthropological survey of Siberian aboriginals, although it remains unclear "if the anthropometric program, which the Shirokogoroffs suddenly pulled out of their saddlebags, was originally part of Shternberg's plan for the fieldwork."[146] The combination of anthropological and ethnographic methods inspired Shirokogoroff's original formulation of the object of his study, which he described in the letters to Shternberg as a "psychomental complex" (of the Tungus and the Manchus) and which evidently resonated with Shternberg's "national psyche." Their correspondence continued after 1922, when Shirokogoroff fled Soviet Russia across the Chinese border. In China he published works that became foundational for the theory of *etnos*, which Shirokogoroff defined by five elements: common physical foundation, one collective identity, common language, shared traditions and historical destiny, and common worldview. He also insisted that an *etnos* could only achieve its equilibrium in a native geographic environment.[147] Such a combination of biological, cultural, and geographic (territorial) elements had to stabilize groupness in the imperial context, on which Shirokogoroff had found himself reflecting during his expeditions to the Zabaikal region. The ethnographer saw only "a 'metisage' of projected physical types and people speaking a 'jargon' of Tungus, Russian, or Yakut."[148] Similar to Shternberg, who in his Jewish science was motivated by nationalism, Shirokogoroff's *etnos* thinking in 1912–1913 and especially later, in the 1920s, reflected his Russian nationalist sympathies and his rejection of imperial universalism and hybridity. The fact that his views had assumed final shape at the Russo-Chinese borderland, "at a time when Japanese and Chinese scholars were indigenizing and redefining European notions of *ethnie*" to suit their own postimperial versions of nationalism and national statehood, reinforced nationalizing tendencies in Shirokogoroff. His ethnographic "categories . . . reflected the debates structuring the new nationalisms on the Eastern Pacific rim," and were meant to bound externally and stabilize internally fluid and multilayered "imperial" identities he encountered both in the Zabaikal region during his first expedition and in the imperial capitals from which he departed.[149]

Shternberg knew about the *etnos* theory firsthand from Shirokogoroff's works and letters, and while not adopting the term itself and criticizing his first book on *etnos* for "provincialism," he used some of Shirokogoroff's findings to back his own ethnographic hypotheses.[150] Kan reasonably suggests that this was one reason for the souring of their relationship. Another reason, of course, was Shirokogoroff's characterization of Jews in one of his early works as a "parasitic etnos."[151] In a letter to a friend written in 1933, Shirokogoroff confessed that Shternberg's signature evolutionism and comparative methodology "of the Frazer type" made him "sick." He concluded the bitter portrait by calling Shternberg "a sentimental Judeophile (believe me, this was a true complex), an idealist seeking improvement of the non-Russians' situation in Siberia by means of embracing them into [the ranks of] 'progressive humanity,' and other things which only interested me in my senior high school years."[152]

Shirokogoroff unmistakably deployed basic anti-Semitic stereotypes by painting the Jewish "psychomental complex" (*etnos*) as irrational, impractical, and foolishly idealistic. In this sense, Shternberg's Jewish "national character" ("psychomental complex") was an inversion of Shirokogoroff's image, as it exhibited optimism, intellectualism, positive idealism, and humanism. Yet, however antipodal these images were, methodologically they both shared a racialized *etnos* thinking. Shternberg, Shirokogoroff, and Mogilianskii, just as late Soviet proponents of *etnos,* such as the director of the Institute of Ethnography of the Soviet Academy of Sciences in 1966–1989, Julian Bromlei, were equally looking for a definition that would combine statics and dynamics in the analysis of a group; universality and exclusivity—in the conceptualization of its distinctiveness; biological determinism and historical variability—in the interpretation of its history; and postcolonial authentic subjectivity and controlled developmentalism—in projecting its future.[153]

Anderson and Dmitry V. Arzyutov revisited the moment of official reintroduction of the *etnos* concept in the Soviet Union in 1964, when it was readopted to "describe the essence of identity, and the object of ethnography, without the taste of fear lingering over Stalin's concept of nation." On the one hand, the concept referenced many of the familiar assumptions about ethnicity, but, on the other, "proponents of *etnos* theory attribute to it a biological foundation, which can serve as a stable core of identity. . . . At best, we translate *etnos* as a type of ethnicity, politely ignoring its marked biosocial elements. At worst, the concept is distinguished as an out-of-date survival from the era of triumphalist theories of cultural evolution."[154] Given Shternberg's prominence in early Soviet academia, he could not but contribute to this result by encouraging *etnos*-like biosocial thinking in Soviet ethnography. Even in 1929, after the official ban on *etnos* as an ethnographic concept that put limits on radical Soviet constructivism in population politics, it continued to live in the professional thinking of pre–World War II Soviet ethnographers.[155] "The early Soviet experience with *etnos* theory was such that it was impossible to live with it openly, but, practically, it was impossible to live without it."[156] Moreover, after its official resurrection in the late Soviet Union, the then director of the Moscow Institute of Ethnography, Bromlei, derived the biological foundations of "his" *etnos* less from anthropometric measurements (as Shirokogoroff would do) and more from marriage choices and endogamy (as Shternberg would insist).[157]

Shternberg also must have influenced the racialization of Jewishness in the official Soviet discourse, although he could not control the specific forms this racialization would assume. His project of modern racialized Jewishness was conceived in the political and intellectual context of late imperial, not Soviet, Russia, and in opposition to—not as a part of—hegemonic science and ideology. Shternberg developed his "militant" Jewish science as an epistemological critique of the assumed political neutrality of knowledge. His critique rested on the embrace of "authentic" knowledge that was being produced through national self-mobilization and self-help. This crucial aspect of Shternberg's racialized Jewishness had completely disappeared in its Soviet reiteration. Shternberg's Jewish science had to empower Jews as agents of their own history; it provided means and forms for conscious participation in Jewish collectivity. By exploring the enigma of Jewish nationhood, his Jewish science was supposed to uncover its objective foundations and offer prescriptions on how to protect Jews from the assimilationist and universalizing pressures of modernity, especially under the Soviet politics of territorial nationalism. The prescriptions offered by Shternberg included endogamy and preservation of kinship ties and authentic forms of social organization (Jewish self-governing communes); cultivation of the Jewish national character—the everlasting racial core of Jewishness; implementation of modern, scientific medical and social population policies, developed exclusively by Jews and for the Jews; cultivation of Judaism as a distinctly national cultural and moral system that reflected the Jewish "national character"; integration into modernity but without neglecting the above requirements; and conscious and affective participation of Jews in their modern national collective, including through the production of modern knowledge about themselves. Eventually, Shternberg argued, only the preservation of the biological substance of the Jewish national-racial psyche by Jews themselves could prevent arbitrary transformation of their groupness by external forces. This original version of racialized Jewishness bore a powerful anticolonial message, which gradually faded away in the early Soviet context and completely disappeared thereafter.

PART II

THE BIOPOLITICS OF RACE

A Necessary Introduction

Jewish Biopolitics as a Victim of Aphasia

Disease leaves its indelible anatomic signs on human organs, writing on them pages that last forever. And like archaeologists who read hieroglyphs or cuneiform inscriptions, we read these pages that record not only an individual's past but the history of the entire kin, tribe, and race to which a particular individual belongs.

[Sergei Petrovich] Botkin addressed his students with the following speech: You see an individual who is extremely emaciated, of puny bodily aspect. His chest is weak, and under one of his ribs some wheezing can be heard. He makes sculptures and spends entire days working with wet clay; his diet is poor. . . . He began throwing up blood from his throat. He is twenty-one years old. If such a patient comes to you, immediately send him away from Petersburg. . . . His illness can develop quickly. But do not be afraid for *this* man, for he is Jewish. . . . They pass down emaciation from generation to generation, but along with it—remarkable endurance. . . . Exactly ten years ago another Jew turned to me; this was the first man's teacher, Antokol'skii, also a very emaciated [individual]. His throat illness was at such an extreme stage that I freaked out and forgot about all these circumstances I have just described [that is, circumstances of Jewish history, conditions of life, and resistance to specific illnesses including tuberculosis]. I predicted his certain death. . . . But he got better and is still with us. So, if such a patient comes to you, remember about his previous life, his origin, and his racial particularity, and do not be afraid. Do not think that he is in danger.

It is self-evident that a sick soul derives content from the material of the past, from an old stock of ideas, concepts, views, and emotions. It is also self-evident that this content is closely connected to language, religion, everyday life, mores, and material and social conditions of life of every people. Someone who is not familiar with Jewish life would not be able to understand a mentally sick Jew, just as we cannot understand a mentally sick Tatar or Chuvash.

These three quotations come from the works of Solomon Vermel, a Jewish physician, literary critic, public activist, and journalist.[1] Born in Shklov in the Pale of Settlement in 1860, he received a university education, lived and worked in Moscow in the Russian Empire, and died in Moscow in the Soviet Union in 1940. Whereas his long life almost coincides with the chronological period covered in this part of the book, the quotations here delineate the main themes being considered: the connection between Jewish race science and the Jewish medical profession and activism; the Jewish physicians' construction of the Jewish body, both individual and collective, as distinctive and in need of national medicine; and, finally, the role of Jewish physicians in putting all these ideas into practice and developing and implementing a far-reaching biopolitical project of national self-help. Like many Jewish physicians, Vermel awaited the end of the old empire with great hope for the change of the "environment." The more democratic postimperial environment, he believed, would favor new and just forms of life. At the same time, along with other nationally minded Jewish physicians, Vermel counted not on a new, postimperial, state to solve the Jewish problem, but on Jewish national self-organization.

In the last years of the old empire, he served at Kazan province's psychiatric asylum, which during World War I became a military hospital. The hospital treated different military contingents, including Jewish soldiers evacuated from the Tworki (near Warsaw) and Grodno psychiatric asylums, and provided psychiatric evaluations of conscripts and ill servicemen from all over the Russian Empire. Vermel believed that Jewish medical statistics that he assembled in Kazan offered "a picture of Jewish psychopathology in Russia in miniature."[2] While exposing the anti-Semitic views of some of his colleagues among the military psychiatrists, he called on rank-and-file Jewish physicians like himself to treat individual Jewish patients as stand-ins for the entire racial group and thus to contribute to the national self-help and self-determination project.[3]

He followed this program selfishly. In addition to working as a physician, Vermel served as the chair of the Society for Mutual Aid of Moscow Jewish Physicians and was a member of the Jewish Historical-Ethnographic Society (Evreiskoe istoriko-etnograficheskoe obshchestvo, EIEO).[4] He also served on the Board of the Society for the Promotion of Culture among the Jews of Russia (Obshchestvo rasprostraneniia prosveshcheniia mezhdu evreiami v Rossii, OPE) and on the OPE school committee, where he oversaw the medical-statistical evaluations of Jewish schoolchildren. In addition, he participated in the Society for the Dissemination of Correct Data about Jews and Jewry (Obshchestvo rasporstrateniia pravil'nykh svedenii o evreiakh i evreistve, founded in Moscow in 1906), where his responsibilities included spreading objective medical information on Jewish health. In 1912, he joined the Society for Protection of the Health of the Jewish Population (Obshchestvo okhraneniia zdorov'ia evreiskogo naseleniia, OZE), and during the war—the Jewish Committee to Aid Victims of War (Evreiskii komitet dlia pomoshchi zhertvam voiny). He published on Jewish topics, from literature to politics to medicine, in various Jewish periodicals. The epigraphs above come from two of his psychiatric studies, *From the Jewish Pathology* (1911) and *Jewish Psychiatric Diseases* (1917).[5] After the revolution he published a number of historical books: *A Historical Overview of the Activities of the OPE Moscow Division, 1864–1914* (1917); *V. G. Korolenko and the Jews* (1924); and *Moscow Expulsion, 1891–1892: Impressions, Recollections* (1924).[6] His early study of Isaak Levitan, the talented painter of Russian landscapes, argued that Levitan "was a 'pure-blooded Jew,' and this fact had inevitably influenced his personality, character, and worldview, and hence his art, too. It is indeed laughable to approach Levitan as a 'Russian nationalist-populist' . . . Levitan was a Jew, and his richly gifted nature was, first and foremost, the result of inborn racial and individual heredity" (1902).[7] In bringing together Jewish art, literature, politics, and physical and mental health—the main areas that interested Vermel personally and professionally—and declaring these a unique complex that required specifically Jewish professional expertise, his work reflected ideas that were quite widespread among Russian Jewish educated elites.

However, in the handwritten autobiography composed for his pension application in 1934 at the beginning of "high Stalinism," the seventy-four-year-old Vermel tried to censor his involvement with Jewish race science and biopolitics.[8] Now he clearly aimed to compose an exemplary autobiography of a progressive Russian physician and a member of the democratic Russian intelligentsia: 1886—graduated from Moscow University's Department of Medicine and interned for a year as a neurologist at Professor A. Ia. Kozhevnikov's medical clinic; 1887–1904—worked as a physician at the Moscow Society for Rubber Goods factory and became a member of the progressive Pirogov Society of Russian Physicians; 1904—drafted to serve in the Russo-Japanese War as a military physician. By the end of the war, he felt "exhausted and half-sick," and in 1905–1912, due to poor health, he was able to do only some unspecified literary work. In 1912—resumed working as a factory physician; 1915—mobilized to service in the Great War and sent to work at the Kazan psychiatric asylum; 1917—discharged due to poor health and returned to the rubber goods factory's hospital. Omitting any information on the revolutionary period, the autobiography proceeded directly to 1922–1924, when Vermel took a position as a psychiatrist in the Office for the Study of the Criminal Personality (Kabinet po izucheniiu lichnosti prestupnika). Finally, in 1924, he became a neuropathologist at the Sanitation Division of the Moscow police department (*militsia*).

One section of the autobiography that was particularly difficult to edit properly dealt with Vermel's "public service." This section was very important for someone claiming a "personal"—that is, a higher than average—pension. Vermel tried to slip this section in between the description of his career as a factory physician and the list of his general medical publications. However, even the incomplete list of his "memberships" and "chairmanships," mentioned above, appeared suspicious. His latest affiliation was with the Council of the Society for Settling Toiling Jews on the Land (Obshchestvo zemleustroistva evreiskikh trudiashchikhsia, OZET)—the only organization, of those listed in the autobiography, still formally active in 1934. It was not too hard to see a common tendency in all Vermel's public engagements: he always tried to contribute to Jewish national welfare as a physician. The list of publications appended to the autobiography was also censored. There were quite a few purely medical or criminological publications, but since the great bulk of Vermel's medical, criminological, or literary and historical works dealt with Jews, the Jewish connection was impossible to conceal.[9]

In Vermel's works published in the 1920s, "nation" (*natsiia*) and "national phenomenon" substituted for "race" and "racial factor" from his earlier publications. Vermel explained that "we now understand the term 'nation' as a complicated cultural-psychological complex, which, besides pure biological elements also includes a people's past, historical experience, religion, everyday culture, mores, views, and in general everything that we call culture and that transforms a primitive man into a particular spiritual-cultural type."[10] This rendering of "nation" had incorporated "race" ("pure biological elements") and, during the years of the Cultural Revolution and the First Five-Year Plan (1928–1932), shielded it from the ideological campaign against biological determinism (as putting limits on Stalin's radical social engineering). As was the case with Lev Shternberg's original definition of race-nation, Vermel's definition of nation ignored class, the key term of Bolshevik biopolitics, which Vermel never learned to appreciate. As we shall see, Vermel's intellectual trajectory in this sense was paradigmatic of the evolution of Jewish science of race and biopolitics from the late imperial period to the early 1930s.

Even more so was Vermel's institutional trajectory. His earliest affiliation was with the Pirogov Society—the progressive professional society of Russian physicians that formulated its goals in a language that was insensitive and indifferent to nationalism or imperialism. Then his main sphere of professional and public activism had shifted toward the Society for Mutual Aid of Moscow Jewish Physicians. Finally, he joined the national movement of Jewish physicians, the OZE, in 1912–1921. Thousands of Russian Jewish physicians went through the same stages and arrived at the Soviet revolutionary reconstruction with their own well-developed biopolitical project. The OZE provided a truly national institutional framework for their activism. During the Civil War—the time of the most profound social disintegration—the OZE network continued to unite Jewish activists around the basic task of physical survival and the broad national eugenic agenda. In these years, Vermel served on the Hygienic Committee of the Moscow OZE. During the cold and hungry winter of 1919, he lectured to the Moscow Jewish public on Jewish psychiatric diseases (a typical title of one of his lectures was "Nervous and Mental Illnesses among Jews, Their Origin and How to Fight Them").[11] In 1920, Vermel worked on the OZE Statistical Committee, processing data on Jewish victims of the Civil War pogroms.[12] Finally, when the official pursuit of the Jewish biosocial project became possible only within the framework of the state-sponsored OZET (1925–1938), Vermel joined this organization.

Neither of these stages was reflected in Vermel's autobiography (except for the OZET membership). While Vermel and his generation cultivated individual and collective amnesia regarding their Jewish biopolitical activism as a survival strategy, the post–World War II generations of Jewish and non-Jewish historians already dealt with aphasia—the structural impossibility of articulating the memory and history of Jewish active and creative engagement with race and the politics of race. This past became displaced and occluded through the loss of access and active dissociation: "Aphasia is a dismembering, a difficulty in speaking, a difficulty in generating a vocabulary that associates appropriate words and concepts to appropriate things."[13] The Holocaust, which had forever changed the relationship between Jews and scientific "race," is to blame for this, but self-censorship as illustrated by Vermel's autobiography and many similar examples in the chapters that follow, coupled with state censorship, greatly reinforced an aphasic effect.

By reconstructing the forgotten story of mass mobilization of Russia's Jews under the slogans of national biopolitics, the chapters in Part II undo at least some of the damage inflicted by the structural condition of aphasia. I argue that thinking of Jewish biopolitics as a form of subaltern politics creates a possibility for this story to be told. It includes the formation of a national Jewish biopolitical movement in the network of Jewish medical societies; the grassroots collection and interpretation of mass Jewish statistics; the creation of the infrastructure of relief and physical and psychological rehabilitation of the victims of wars, pogroms, and famine; the debates about national Jewish medicine; and the development of Soviet Jewish eugenics. In all these spheres, Jewish activists performed collectively like a nation-state without having any Jewish state backing their efforts. I will show that, in the end, their biopolitical project ran into ideological conflict with the Soviet state, which was pursuing its own postimperial and anticolonial agenda of nation-building. An exploration of this conflict also reveals spaces of cooperation and mutual influence on both sides. As a result, by reconstructing, literally in bits and pieces, the story of Jewish biopolitics, we can arrive at a better understanding of imperial and Soviet politics and biopolitics, and the role of Russian Jews such as Dr. Vermel in developing modern national population politics in twentieth-century northern Eurasia and Eastern Europe.

CHAPTER FIVE

Russian Jewish Physicians and the Politics of Jewish Biological Normalization

RUSSIAN JEWISH PHYSICIANS participated in the international professional culture that had been centered in many respects on the concept of race since the end of the nineteenth century. Many of them had graduated from European universities, mostly German and Austrian institutions, where they were directly exposed to racially based medical and anthropological theories. Half of the 156 Russian students attending the Paris School of Medicine at the turn of the century were Jewish. In 1911, Jews accounted for over 90 percent of all Russian medical students in Prussia, and 56 percent of those at the University of Vienna. In 1912–1913, 1,393 out of 1,492 medical students in German universities (93 percent) from the Russian Empire were Russian Jews.[1]

Most Russian physicians were state employees (primarily in the army and navy) or worked for zemstvos.[2] A minority (roughly one-fourth) were in private practice, especially in cities of the central and western regions of the empire. Since the late nineteenth century, Jewish physicians had been progressively banned from making successful careers in the military and they were harassed by zemstvo boards, which were often less than enthusiastic about hiring them. "Being assimilated and ready for full integration into Russian social and political life, these people were still excluded as Jews, and not only by the state," writes Lisa Rae Epstein. Zemstvo medicine was also "affected by the impetus to exclude Jews."[3] At the same time, Epstein quotes many examples of Jewish physicians' successful integration into the all-Russian medical profession and professional movement, and these examples present a coherent trend. Many Jewish physicians were still able to find employment with zemstvos, but even more of them began switching to private practice, which often implied treating primarily Jewish patients. This structural situation influenced their professional ethos, sense of identity, and civic duties, whereas training in medicine and natural sciences made them the best experts on race and medicalized politics. If, according to Veronika Lipphardt, throughout the 1910s German Jewish scientists and physicians were gradually withdrawing from discussions about race because race discourse was used by colleagues as an instrument for segregating Jews, then the Russian case appears to be the opposite.[4]

Russian imperial physicians as a group were dissatisfied with their professional dependence on the state and their low social status. Since the Great Reforms of the 1860s, they had been self-organizing to advance distinct professional demands and, later, political ones, replicating the mechanism of building the intelligentsia as a supranational community united by a common ethos.[5] In 1886, the Society of Russian Physicians in Memory of N. I. Pirogov (known as the Pirogov Society) was established and quickly became the organizational and ideological center of zemstvo medicine and the progressive medical movement in the country. Russian Jewish physicians participated in this professional movement and shared its civic concerns.

The oldest Russian regional medical association that included Jews—the Vilna Medical Society—was founded in 1805 but reached its peak of activity after the Great Reforms.[6] The society's story is emblematic of Jewish participation in the Russian imperial medical profession in general. Historically, the society was dominated by Polish and German physicians: in the late nineteenth century, Jews accounted for 20–30 percent of its members.[7] From time to time they voiced specific national concerns, but until the first Russian Revolution of 1905–1907 never insisted on Jewish exclusivity and never attempted to secede.[8] Despite the strictly professional nature of the Vilna Medical Society, the social identity of its members—nationality, confession, language, and political loyalties—always remained an issue. When they elected their governing bodies, each name on the voting list was accompanied by the nominee's confession: Catholic (which usually meant a Pole), Lutheran (most likely a German), an Israelite (a Jew), Russian Orthodox (a Great Russian or a Ukrainian). This corresponded to the "confessional empire" model promoted by the state, but it was also a meaningful language of nationality in this particular region. Until the 1890s, most of the society's documentation, including the voting lists, was handwritten. Its membership did not exceed 180 active members (if member-correspondents and honorary members were included, the numbers were greater).[9] Approximately half of all active members regularly attended the society's meetings and thus knew one another personally. The fact that they still found it important during the elections to indicate their confession suggests that under some circumstances national alliances played a defining role. Jews in the Vilna Medical Society were never elected to major posts except for the position of treasurer.[10] With very few exceptions, Poles and Germans filled all the other leading positions.[11] Jews, however, were most active as regular presenters at the society's meetings.[12] Specifically, they most actively researched issues pertaining to popular sanitary and medical demography.[13]

The history of the Vilna Medical Society shows how, until the era of rising mass politics at the turn of the twentieth century, the imperial order helped downplay potential national tensions. For example, despite the predominance of Polish doctors in the ruling bodies of the society, they were officially prohibited from record keeping and presenting in Polish. The official language of the society was Russian, and its Polish, German, Jewish, Belorussian, and Ukrainian members used Russian as a professional lingua franca. However, in November 1905, at the peak of the first Russian Revolution, when this restriction was lifted, Polish physicians immediately switched to Polish, while Jewish doctors continued presenting in Russian, their primary language of scholarship and professional socialization.[14] The majority of the society's members must have been bilingual if not trilingual, so the change in the linguistic protocol in November 1905 was clearly politically motivated.

We can follow the delicate political dynamic for managing diversity in the Vilna Medical Society in the story of one of its distinguished members, the Jewish doctor Samuil Iakovlevich Fin, who died in 1899, having bequeathed some of his property the society. Fin established a scholarship for local youth applying to one of the Russian universities to receive medical training. According to Fin's will, and in the spirit of professional solidarity characteristic of the Vilna Medical Society, one year the scholarship was to be awarded to a Jewish applicant and another year to a Christian. A special instruction stipulated that if, contrary to expectation, the Russian universities shut their doors to Jews, the scholarship was to become entirely "Jewish" and used to support Jewish students applying to foreign universities.[15] In 1905, when the imperial Ministry of Education confirmed Fin's donation and the stipend, it perceived this instruction as "being out of place and purpose," as the Russian state had no intention "of banning Jews from the universities."[16] Fin, who died twelve years after the introduction of numerical quotas for Jews in Russian universities, had reason to expect the deterioration of the situation but, as his will makes clear, preferred (just as the official bureaucratic state) the continuation of the status quo. Ten years after Fin's death his former colleagues, Jewish members of the Vilna Medical Society, saw no incentives to preserve or restore the status quo. They seceded and established the Vilna Jewish Physicians' Circle. The erosion of the supranational professional solidarity of the medical profession and the crystallization of a distinct Jewish medical activism paralleled the nationalization of politics in the empire.

This process could assume different forms and temporalities in different parts of the empire. In the Pale, where Jews joined local professional medical societies in large numbers or even prevailed numerically, calls for professional solidarity often coded a newly acquired sense of Jewish professional solidarity. In 1899, in the native city of Samuel Weissenberg, Elisavetgrad, where Jews composed close to 40 percent of the population, the local chapter of the Society for Mutual Aid of St. Petersburg Physicians publicly voiced its concern about the attitudes of the zemstvos and some Christian physicians toward their Jewish colleagues. The Elisavetgrad doctors emphasized religious tolerance and professional solidarity as essential for their society. They did not claim any special status for the Jewish physicians but recognized the ongoing polarization along national lines.[17] On September 20, 1904, physicians of another city in the Pale, Białystok, gathered at the local train station to give a send-off to a member of the Białystok medical society, Dr. Epstein, who, similar to Solomon Vermel and many other Jewish physicians, was mobilized during the Russo-Japanese War. They asked the station's superintendent to unlock the hall for special ceremonies, to which he reportedly replied: "I am not going to unlock the main rooms for a yid" (Dlia zhida ia paradnykh komnat ne otkroiu). This line was picked up by newspaper reporters who confirmed its typicality: "Similar incidents occur so often that they cannot be considered as individual cases."[18] In Białystok, doctors Pines, Divenson, and Weinreich convoked a special meeting of the local medical society and resolved that the superintendent had offended not only Dr. Epstein personally but their society as a whole.[19] But such unanimous determination of the local medical society can be explained by the numerical dominance of Jewish physicians in it. According to the 1897 population census, out of 66,032 residents of Białystok, 40,972 (62 percent) claimed Yiddish as their native language—in other words, Jews were the majority in this city that was also home to Poles (17.2 percent), Great Russians (10.3 percent), Germans (5.6 percent), and Belorussians (3.7 percent).[20] Obviously, Jewish physicians felt more confident there.

In other localities in the Pale, where the proportion of Jewish physicians was smaller, the picture could be different. For example, in the Mogilev Society of Physicians, in which Jews made up only one-fifth of its members, the discussion of "Jewish issues," especially after the revolution of 1905, provoked harsh disagreements. At a meeting on January 12, 1907, the surgeon A. M. Dolgov criticized a colleague for praising the quality of medical assistance provided to Mogilev Jews. Dolgov was infuriated by this assessment and quoted two instances of nonprofessional behavior of Jewish physicians.[21] The issue at stake was not even anti-Semitism: Dolgov and his supporters rejected the very idea of singling out "Jewish medicine," whereas Jewish members of the society perceived the Jewish community of Mogilev as having special needs and special relationships with local Jewish physicians.

Whether endorsing Jewish physicians or sidelining them, in the wake of the 1905 revolution, medical societies reflected the new reality, which exposed and politicized the divisions, identifications, and sentiments that used to be downplayed, blurred, or contextually conditioned. The role of the revolutionary anti-Jewish pogroms was quite significant in singling out the Jewish body as the symbol for a nation, but the nationalization of the Jewish medical discourse was not merely a response to the pogroms. The metaphor of the ill but curable body capable of withstanding the challenges of the epoch of mass politics, resisting anti-Semitism and direct violence, and reemerging for or returning to "natural" life in the democratized and modernized Russian "empire of nations" attracted many Jewish intellectuals, professionals, and political activists. After the 1905 revolution, Jewish physicians began to actively secede from local professional medical societies and to organize their own Jewish medical societies with a distinct medical-sanitary agenda for normalizing the Jewish nation by creating a healthy biological basis for it. Their secession was characterized by a combination of "reactive" and "proactive" strategies. The "reactive" part was conditioned by pogroms, rising anti-Semitism, and the politicization and nationalization of professional discourses and practices, whereas the "proactive" program was inspired by the racialized image of national salvation and revival shared by many Jewish doctors.

From Medical Societies to Jewish Medical Societies

In 1910, the Jewish members of the Vilna Medical Society established their own Jewish Physicians' Circle.[22] Their priority was Jewish national sanitation as a way of forging a healthy, biologically "normal," modern Jewish nation. The statute of the Jewish Physicians' Circle did not formally discriminate on the basis of ethno-confessional status, and its stated goals were formulated in universalist terms, ranging from mutual aid to the promotion of research among local physicians. In reality, however, this was an exclusively Jewish network; its core activists (former members of the Vilna Medical Society such as Ilya Girshovich Shil'dkret, Iakov Grigor'evich Dillon, Il'ia Mordukhovich Blokh, and Abram (Adolf) L'vovich Kogan) were concerned specifically with studying, treating, and improving Jewish bodies.[23] No wonder, then, that many members of the Vilna Jewish Physicians' Circle later became activists of the all-imperial Society for Protection of the Health of the Jewish Population (Obshchestvo okhraneniia zdorov'ia evreiskogo naseleniia, OZE) and remained active participants in this movement throughout the early Soviet period. A notable example would be the charismatic Tsemakh Iosifovich Shabad (1864–1935), "who played a critical role in Jewish medicine and the Jewish cultural world of Eastern Europe in the twentieth century" and inspired the image of the no less famous Doctor Aybolit of the children's book.[24]

Around the time of the Vilna Jewish physicians' secession, in 1909–1910, one of the brightest representatives of the liberal tradition of zemstvo medicine and a proponent of the all-Russian unified organization of physicians, Dr. Moisei Gran (1867–1940), changed his priorities and channeled his energy into a specific Jewish cause—the creation of a national organization of Jewish physicians.[25] This was a highly symbolic transformation, as for several decades Gran had epitomized the very idea of the supranational professional mobilization of liberal physicians in the Russian Empire. But in 1911–1912, he joined the founding circle of OZE activists and played a seminal role in defining the concept of Jewish national medicine and eugenics.[26]

In 1913, Kiev's Jewish doctors made two attempts to establish a Jewish medical society. In February, a group of Jewish physicians wanted to launch the "Society of Physicians and Natural Scientists for the Study and Improvement of Medical-Sanitation Conditions of Life of the Jewish Population" (Obshchestvo vrachei i estestvoispytatelei dlia izucheniia i uluchsheniia mediko-sanitarnykh uslovii zhizni evreiskogo naseleniia). Their application was turned down by the local Kiev authorities, who were alarmed by the project's explicitly national orientation. As the official reply explained:

> Given the fact that doctors Benderskii, Golochiner, and Gurevich want to establish a society that aims at uniting *inorodcheskie* [alien, non-Russian—a legal category in Russian imperial law] elements on the basis of their exclusive national interests, thus leading to the deepening of national isolation and discord, and also given multiple explanations of the Governing Senate (edicts of June 18, 1908, no. 9120, on the Ukrainian society "Prosvita," and of September 4, 1909, on the Polish society "Osviata," etc.) stating that all societies that lead to the deepening of national isolation and discord should be regarded as threats to public tranquility and security, the Provincial Administration [Gubernskoe pravlenie] . . . has resolved: to refuse registration.[27]

In June 1913, Jewish doctors attempted to register the "Society of Academic Conferences of the Physicians of the Kiev Jewish Hospital." The physicians Alexander (Zisel) El'evich Garkavi, Mikhail Borisovich (Mikhel Berkovich) Jukel'son, and Vladimir Markovich Varshavskii only wanted "to research materials from the Kiev Jewish hospitals" and publish the results of their studies.[28] However, even this modest project could not pass without major revisions. It is telling that the official denial of the registration, quoted above, singled out two non-Jewish—Ukrainian and Polish—cultural-educational initiatives as precedents. The difference in the format of national mobilization (cultural versus biomedical) notwithstanding, in these formally different projects, the Kiev authorities identified a common threat to the imperial order. The Kiev Provincial Administration for Public Associations (Kievskoe gubernskoe po delam ob obshchestvakh prisutstvie) requested that the Kiev Jewish Hospital physicians limit their studies to one specific hospital and its patients, and to keep the society's paperwork and publications only in Russian. Apparently, this should have both precluded the society from engaging in Jewish politics in the Pale and secured state control over its activities.[29]

Upon accepting these requirements, the Society of Academic Conferences of the Physicians of the Kiev Jewish Hospital was registered on February 20, 1914.[30] But a few months earlier, in October 1913, the already operating all-Russian OZE simply opened its branch in Kiev, without a struggle with the local authorities.[31] The liberal Jewish periodical *Novyi voskhod* (New sunrise) triumphantly reported that "a new society has been born" in Petersburg and "according to its tasks and statute it can extend its activities throughout all of Russia by using the right to open its chapters everywhere."[32] In this case, the imperial "strategic relativism" helped circumvent local politics: the language of Jewish biomedical self-normalization was simultaneously perceived as a show of modern anti-imperial nationalism in Kiev and as a relatively benign scientific initiative in St. Petersburg, where the OZE was registered in 1912. For decades, Petersburg remained a major locus of Jewish cultural and educational initiatives, mainly aimed at promoting a Russian Jewish identity. This kind of Jewish liberal politics seemed preferable to the particularistic imperial regime that tried to accommodate the challenge of nationalism by tolerating less politicized initiatives. Imperial strategic relativism ultimately worked in favor of the Jewish biopolitical project: the newly opened Kiev Committee of the all-Russian OZE brought together all the main activists of the two local Jewish projects (doctors Avadii Davidovich Birshtein, Vladimir Markovich Kudish, Semyon Solomonovich Rybakov, Semyon Isaakovich Fleishman, Ilya Iosifovich Frumkin, Vladimir Markovich Varshavskii, and Mark Moiseevich Gershun).[33] In the same way, the OZE movement integrated activists of local medical societies throughout the empire and stimulated the opening of new local societies. Unlike the public sphere of party politics, the all-imperial network of Jewish physicians absorbed people of different political persuasions as long as they shared a common concern over the normalization of the "Jewish body" as a necessary precondition for any kind of modern politics.

The *Jewish Medical Voice* from Odessa

Unlike the Vilna and Kiev societies, the professional Society of Odessa Physicians (1848–1924) had been dominated by Jews since the mid-nineteenth century.[34] Its ranks featured luminaries of Jewish politics such as the author of *Self-Emancipation,* Lev Pinsker, and one of the founders of the journal *Zion*, the secretary of the Odessa division of the Society for the Promotion of Culture among the Jews of Russia (Obshchestvo rasprostraneniia prosveshcheniia mezhdu evreiami v Rossii, OPE), Emmanuil Soloveichik. Like other Russian medical societies, the Society of Odessa Physicians supported the scholarly pursuits of members who were especially prominent as contributors to vaccination theory and practice, research on epidemics and measures to deal with them, as well as studies in medical demography and social sanitation.[35] In February 1886, the society sponsored a three-month trip to Paris of one of its members, the future founder of Russian modern epidemiology, Nikolay Fyodorovich Gamaleya, to learn from Louis Pasteur the method of producing rabies vaccine. After his return, Gamaleya vaccinated the first Odessan (and the first Russian!) bitten by a rabid dog.[36] The same year, he founded Russia's first bacteriological laboratory, the so-called Pasteur station, and Pasteur was elected a member of the Society of Odessa Physicians.[37]

By the turn of the century, the Society of Odessa Physicians was enjoying great popularity in the city: its meetings were open to the public and attracted many curious outsiders. This public nature of a professional society was formally recognized by its March 1888 statute, upon the proposal of its eleven leading members.[38] Since 1891, the Society of Odessa Physicians had published the *Iuzhno-russkaia meditsinskaia gazeta* (Southern Russian medical newspaper), which also contributed to its public visibility and influence. The society served as a major channel for disseminating race science in Odessa.[39] On October 5, 1881, when the famous German physician, race scientist, and politician Rudolf Virchow, who was highly regarded by the Russian anthropological community, arrived in Odessa on his way from Tiflis (where he had attended the Sixth All-Russian Archaeological Congress), the first Odessan to greet him onboard ship was a representative of the Society of Odessa Physicians.[40] The society gave a grand dinner in Virchow's honor and awarded him an honorary membership. In his acceptance speech, the German anthropologist proclaimed, "There is no German, or French, or Russian science: we all have only one science that holds us together as a family."[41] To the attendees, this was not sheer rhetoric: Virchow's words resonated with the beliefs and professional values shared by most members of the Society of Odessa Physicians. Jewish physicians, who had dominated its ranks numerically and shaped its discourse intellectually for many decades, successfully downplayed the political implications of race thinking within its public sphere.

The events of 1905 dramatically offset this delicate intellectual and political equilibrium. When the revolution erupted in January, Iakov Iul'evich Bardakh, member of a well-known Jewish Orthodox Odessa family, was head of the society's board.[42] The other board members were also predominantly Jewish: the physicians I. G. Mandel'shtam, Ia. V. Zil'berberg, S. G. Shteinfinkel', L. B. Bukhshtab, and L. M. Rozenfeld.[43] By 1910, 70 percent of all medical personnel in Odessa were Jews, but even the overwhelming numerical preponderance of Jewish physicians cannot alone explain the exclusively Jewish composition of the board.[44] It seems more plausible to suggest that Jewishness was a less significant factor for Odessa physicians than shared professional and liberal-democratic political values. Simply put, the society members were indifferent to the ethno-confessional identity of their elected leaders.

In the spring of 1905, when the professional mobilization of physicians in the empire had reached its peak and the Pirogov Society was demanding the radical democratization of the regime, the Odessa Society turned into a political club. Following the Extraordinary Congress of the Pirogov Society in Moscow (April 3–6, 1905), which had resolved that "no cultural and peaceful work was possible under the existing political regime," the Odessa Society's delegates scheduled an open information meeting in Odessa on April 9.[45] So many people came that the organizers decided to hold it the next day in a bigger hall. The Odessa police, not without reason, classified the event as a political rally. They did not let the public disperse after the meeting, and they made arrests. The society's board protested in local and central newspapers: "By making public these new and shameful acts of police atrocity, the Society of Odessa Physicians expresses its profound outrage and protests against a regime that enables such lawless mistreatment of defenseless citizens."[46] The Jewishness or non-Jewishness of the Odessa physician-activists meant very little under these circumstances.

However, in the fall of 1905, the Jewishness of Odessa physicians had acquired new connotations. Now it suggested a higher-than-average correlation with a leftist political orientation, membership in a Jewish political party, and general disloyalty to the state in response to the rising wave of pogroms in the wake of the emperor's October Manifesto. The Eleventh Pirogov Congress was scheduled to take place in Odessa in 1909. Kirill Mikhailovich Sapezhko, head of the Odessa Committee for Preparation of the Pirogov Congress of Physicians, assembled the committee, consisting of fellow members of the Society of Odessa Physicians. All of them shared the views of the left-leaning segment of the political spectrum (conservative members refused to participate) and many of them were Jewish. Knowing what to expect, the Odessa city governor (*gradonachal'nik*) anxiously inquired: "Are you considering political speeches and reports for the congress?" Sapezhko's response was the best illustration of the confluence of political and professional activism in the medical profession: "Popular hygiene and medicine are so closely related to the issues of popular and political life that it seems quite natural for the congress to debate political topics."[47] Meanwhile, the Odessa secret service unit (*okhrannoe otdelenie*) informed the city governor that members of the congress's Organization Committee were involved in a range of antigovernment activities, many of them being members of Jewish parties. For example, a doctor at the Jewish Hospital, Konstantin Nikolaevich Purits, a Bundist, participated in fundraising to purchase weapons for Jewish self-defense groups. The head of the society, Iakov Bardakh, was described by the police as a "leftist liberal"—he was active in the Imperial Novorossiisk University Academic Union, which supported student and revolutionary activism during the revolution of 1905. In 1905, a physician of the City Disinfection Board, Alexander Markovich Vel'shtein, participated in the Southern Regional Committee of the Russian Social Democratic Labor Party and in the Soviet of Workers' Deputies, and also supported Jewish self-defense. Dr. Grigorii Il'ich Gimel'farb belonged to the Zionist-Socialist Party and in 1906 presided over all its committee meetings.[48]

Conservative physicians, who all happened to be non-Jewish, constituted a minority in the Odessa medical profession and kept a low profile until 1905. They socialized in their own Society of Russian Physicians registered in Odessa in 1892, which existed in the shadow of the much more popular and numerous Society of Odessa Physicians.[49] By 1905, the Society of Russian Physicians was on the verge of self-liquidation; its few members were not conducting any research or supporting their society financially.[50] On March 17, 1905, the society's board scheduled what it thought was to be the final general meeting of the "Russian Physicians":

> In the course of the year 1904 and until March 1905, the Society of Russian Physicians held no regular meetings. Thus, in this period, the Society did not show any signs of life. . . . Certainly, the reasons for this inactivity should be looked for in the Society itself, in its members' indifference toward supporting a regular and uninterrupted flow of its scholarly studies.

Likewise, in the preceding year, low attendance at the Society's meetings made them uninteresting and boring and forced even the most active members to reconsider their plans to make scholarly presentations to the Society. In the course of the past year, not a single proposal for a scholarly presentation to the Society was received.[51]

However, the revolutionary polarization and national mobilization of Odessa Jewish physicians saved the Society of Russian Physicians from an imminent inglorious fate.

The turning point for both societies was the Jewish pogrom on October 18–20, 1905—an unexpected event that shocked the city. The post-revolutionary activity of the Society of Odessa Physicians centered largely on the trauma of the pogrom, even though its Jewish leaders avoided this term, using instead a characteristic medicalized euphemism—"a horrible traumatic epidemic" (uzhasnaia travmaticheskaia epidemiia).[52] The Odessa pogrom was most violent in the poor districts, where few middle-class educated Jews, including members of the Society of Odessa Physicians, lived.[53] It is unclear whether anyone from the society had experienced direct physical violence, but the psychological trauma inflicted by the symbolic violation of the collective Jewish body was universal. A medicalized account of this traumatic experience was left by a Jewish mathematician and adjunct of Imperial Novorossiisk University, Khaim Gokhman, whose social status in Odessa was close to that of the well-integrated and educated physicians.[54] In a 1914 autobiography, Gokhman described the psychological breakdown he suffered in 1905: "The reason for my sickness was the Jewish pogrom. I suffered emotionally and materially. There were Jews who were killed or wounded, and who were ruined financially or robbed. These all affected my nervous system very strongly. In addition, for a few months I was afraid for my life and the life of my family, until we moved to Dresden (I was expecting a second pogrom)."[55]

The pogrom had profoundly stunned the entire city, so the ideological reorientation of Odessa physicians comes as no surprise. They did not limit their activities only to victims' relief and medical assistance but took proactive policy measures. Using the new possibilities that emerged with the lifting of some of the old restrictions on social and professional organizations and freedom of speech, members of the Society of Odessa Physicians went on to develop a far-reaching program of Jewish biopolitics.

Those physicians (mostly non-Jewish), who felt alienated by this ideological shift, joined the conservative Society of Russian Physicians. In

1906, the latter rose "like a phoenix from the ashes" in the words of its new chair, Imperial Novorossiisk University professor Sergei Vasil'evich Levashov, who had joined the society only in 1905.[56] Not only were there no Jews on the revived society's board, but its members had reputations as reactionary Russian nationalists. Levashov, a native of Kaluga province in central Russia, was a prominent leader of the monarchist and anti-Semitic Union of the Russian People. Vice-chair Dmitrii Pavlovich Kishenskii—also a professor at Imperial Novorossiisk University, born into an ancient gentry family of Tver province near Moscow—was characterized by the Odessa Police Administration as a "reliable conservative." The remaining board members had also moved to Odessa only recently from ethnoculturally homogeneous regions of the empire and therefore were strangers to a specific Odessan intercultural local society: secretaries, doctors Vladimir Ionovich Rudnev (son of a priest in the Don Cossack region) and Grigorii Lorentsovich Ioanno; treasurer, Alexander-Avgust Martinovich Rosenberg (of Derpt [now Tallinn] University); and librarian, the military doctor Leontii Leontievich Mikhnevich.[57] At the same time (in 1908), the Board of the Society of Odessa Physicians exclusively featured physicians of Jewish background, prominent scholars, lifelong members of the multicultural Odessa local society, and holders of leftist political views.[58]

Even this superficial comparison demonstrates the effect of the surge of mass politics and its nationalization on the Odessan medical community. Political and social differences became marked as Jewish and non-Jewish, thus creating two semi-isolated professional spheres. The Society of Russian Physicians kept growing. In 1906, its ranks had reached the unprecedentedly large number of 70 registered members, while the membership of the Society of Odessa Physicians dropped to an unprecedentedly low number of 178 active members.[59] The drop was the result of the political arrests and pogrom that compelled many to leave Odessa; however, in 1908 the trend reversed, and the Society of Odessa Physicians claimed 200 active members.[60] The Society of Russian Physicians was less numerous, yet it had acquired more respectability in 1913 when it registered under a new name and with a new affiliation as the Medical Society of Imperial Novorossiisk University.[61] Now the ratio of members in both societies closely corresponded to the ratio of Jewish and gentile physicians in Odessa (about 70 percent and 30 percent, respectively).

The progressing nationalization of the medical profession in Odessa led to the founding of the journal *Jewish Medical Voice* (*Evreiskii meditsinskii golos;* 1907–1917) by doctors Srul'-Khaim Adesman, Isaak Solomonovich Geshelin, Samuil Efimovich Mar'iashes, and Iankel Meer Raimist.[62] Jewish physicians from all parts of the empire read and contributed to this journal, so its reach far exceeded that of the local Society of Odessa Physicians. Assessing the journal's contribution to the Jewish national cause in early 1909, after its first full year of existence, Dr. Moisei Sherman noted that the two hundred pages constituting the annual volume were filled primarily with statistical data on "national Jewish health." He especially welcomed the position taken by the editors as not merely "observers and registrars of Jewish sanitation life" but activists eager to "energetically intervene and influence the resolution of the burning questions of our life."[63] The journal played a key role in articulating the ideas of Jewish medical mobilization, eugenics, and biopolitics as a version of scientific politics capable of overriding the ideological differences between the many Jewish parties and movements and engaging Jews of different ranks, regions, and persuasions. The pages of the *Jewish Medical Voice* brought together leading Russian Jewish race scientists such as Samuel Weissenberg, Arkadii El'kind, and Lev Sheinis; medical activists and Jewish publicists such as Solomon Vermel; and ordinary Jewish physicians from the Pale and Russia's inner provinces. They studied Jewish race, "pathology," statistics, school hygiene, and patterns of morbidity. Sheinis called on the Jewish doctors to develop a "comparative racial pathology" in order to help lessen some biological patterns and reinforce others by influencing "social factors."[64] Other contributors to the *Jewish Medical Voice* prioritized eugenic measures, such as regulating marriages and prescribing strict medical norms for children, in combination with the development of Jewish national medicine. It was in the *Jewish Medical Voice* that Dr. Mikhail (Moses) Shvartsman published a programmatic article, "Jewish Public Medicine" (1910), which "received a sympathetic response from Jewish medical circles in Petrograd, Moscow, Kharkov, Vilna," and Kiev—all big university centers with active Jewish medical societies.[65] Two years later, at the founding meeting of the all-Russian movement of Jewish physicians in St. Petersburg (October 28, 1912), Shvartsman's article and the *Jewish Medical Voice* were cited as the main inspirations for the project of Jewish national medicine the movement embraced. The new organization became known as the Society for Protection of the Health of the Jewish Population (Obshchestvo okhraneniia zdorov'ia evreiskogo naseleniia, OZE). The keynote speech delivered by its elected chair reflected on the experience of Jewish medical societies and concluded that isolated local professional societies of physicians were "incapable of implementing the very significant goals and tasks that the newly founded society was hoping to implement with the support and help of the broad strata of Jewish society."[66]

Birth of the Movement

The program of biological self-improvement appeared to be a more realistic alternative to political revolution, which had failed once and, as many believed on the eve of World War I, had little chance of succeeding in the near future. The implementation of the medicalized revolution did not directly depend on the resilience of the political regime, the support of all-Russian or Jewish political parties, or the mobilization of the anti-Semitic or philosemitic Russian intelligentsia. All the project required for success were the consolidated efforts of the Jewish public, and it promised in return to resolve the central problem of Jewish modernity by improving the inadequate state of the collective Jewish body and psyche.

As an authentic project of self-help and self-modernization, the OZE and the public movement it generated affirmed Jewish biopolitical subjectivity. They rejected both the position of passive victim of oppression and the role of subordinate partner in the common revolutionary struggle against the imperial regime.[67] Together with OZE supporters from the ranks of teachers, lawyers, and other educated and semi-educated members of Russian Jewish society, Jewish physicians expanded the sphere of practical politics. Their medicalized and racialized version of organic nationalism contrasted with the differentiation of conventional Jewish party politics. At the heart of the OZE ideology was not a political doctrine but a belief that the coordinated treatment of individual physical Jewish bodies could yield control over the collective social body of the nation. Within the framework of the OZE, Jewish experts on race and medicine worked toward this grand political task by means of apolitical politics, whereas their patients and objects of study acquired political significance as junior partners in this collective endeavor. Thus, grassroots biopolitics became the politics of a stateless nation whose sizable professional elite perceived the current situation as colonial. Determined to recover the nation's authentic Self and make it modern, this elite turned to the means of statistical self-cognition, the development of national medicine, and the eugenic engineering of the enhanced national body.

Dr. Moisei Gran, one of the founders of the Pirogov Society, who became a member of the OZE founding group and dedicated his energy to Jewish medicalized politics, pioneered the concept of Jewish national medicine. Gran studied medicine at Kazan University, where his mentor was the famous neurologist Vladimir Bekhterev. A bright student, Gran nevertheless postponed graduation to participate in the public relief campaign during the famine and cholera epidemic of 1891–1892. He then became a zemstvo physician in Samara province and later in St. Petersburg. During these years, Gran remained indifferent to Jewish politics but actively participated in the progressive intelligentsia political and professional medical activism. At the Pirogov Society's seventh congress in Kazan (1899), for the first time in the history of Russian medicine, Gran articulated the idea of establishing university chairs in public medicine. He championed the agenda of public medicine thereafter, although after the revolution of 1905 his understanding of "public" had changed. Now Gran applied the experience he had acquired in zemstvo medicine and in the Pirogov Society to serve the Jewish "public," together with other Jewish physicians of similar background, advancing a daring project of Jewish biopolitical nationalism. In his opening speeches at the OZE founding meeting on October 28, 1912, and later at the first OZE Conference in 1916, Gran admitted the OZE's debt to "zemstvo and city medicine."[68] The project of "Jewish public medicine" borrowed from them the principles of self-funding, public participation, and local and broadly accessible medical facilities. But the OZE wanted to adjust "the general principles of public medicine to the national and everyday life [*bytovoi*] requirements of the Jewish population."[69] Gran and his colleagues were openly and consistently nationalizing "the road that had been trodden and tested by Russian zemstvo medicine half a century ago."[70] They also afforded national eugenics a place and scale it had never enjoyed in the ideology of the Pirogov Society or zemstvo medicine. At the OZE founding meeting, Gran instructed his colleagues to "explore the problem of Jewish vitality, a biosocial problem of Jewish heredity, history of medicine, sanitation, and hygiene. . . . We should study in detail the Jewish type and physical constitution, the [Jewish] main sanitation indicators, sickness and mortality rates, and causes of psychic instability alongside the special predisposition of Jews toward nervous diseases."[71]

No wonder almost all the physicians whose names have appeared in these pages sooner or later joined the OZE. Weissenberg joined at its inception, remaining a member until the OZE's formal liquidation by the

Bolsheviks in 1921. Besides being a local activist, he served the cause as a national celebrity facilitating fundraising for OZE initiatives. In this capacity, Weissenberg offered lectures in St. Petersburg, Moscow, and the Pale. For example, his talk "Jews as a Race and a People" at a big OZE fundraising event in St. Petersburg on January 12, 1914, covered the following: "Racial mixing and the purity of Jewish race; Jewish ethnographic type; Jewish anthropological type; Jewish types: Yemenite, Mesopotamian, Persian, Black, Indian, and so on; Jews of Eastern Europe; typological hybridity of Russian Jews; the initial Jewish type—Semites; causes of Jewish typological changeability; and conclusion: race and the people."[72]

In early 1915, the Elisavetgrad newspaper *Golos Iuga* (Voice of the South) published announcements of the same lecture, "Jews as a People and a Race," now "in support of kindergartens [*detskikh ochagov*] organized by the local chapter of the Society for Protection of the Health of the Jewish Population."[73] The latest mention of Weissenberg's OZE activity that I was able to find is dated June 6, 1920.[74]

Members of Kiev, Vilna, Odessa, Moscow, and other Jewish medical societies, in the Pale and in the capitals, joined the OZE, which makes it hard to outline a demographic or political profile of this association. Samuil Kaufman (1840–1918), the first OZE chair elected at the founding meeting, was seventy-three when he assumed his post, the only member of the Central Committee who shared right-liberal political views.[75] Uniquely for a Russian Jew, Kaufman had achieved a spectacular military career, starting as a ten-year-old *kantonist* (child military cadet) in 1850 and retiring from the navy in 1902 as a rear admiral. At the OZE founding meeting, Admiral Kaufman confessed: "Every time that sons of the Jewish people appeared before the [military draft medical] committee, I suffered painfully. What a people these were! Narrow-chested, bloodless, miserable. . . . This should not continue."[76] Kaufman's colleagues on the committee described their collective ideological profile as national-populist or even "public populist":

> In terms of their public and political views, the group of physicians and public activists who became OZE founders, represented a "populist" trend with a nationalist orientation; among this group were all shades of Jewish populist thought, from Zionists to national socialists. Physicians as a group were all public populists [*obshchestvennye narodniki*] physicians, who in Russia united under the banner of the Pirogov Society and Pirogov Congresses.[77]

A modern historian put it differently, stating that the OZE "was shaped by elements of liberal politics, through figures like Gran, and by the nationalism of Zionism and territorialist politics, through Shvartsman and others."[78]

The movement planned its first all-Russian congress for 1914 but had to reschedule due to the outbreak of World War I. When the delegates finally got together in Petrograd in November 1916, they were officially prohibited from designating their meeting a "congress" and instead had to agree on the more modest "conference."[79] Regardless of the name, the real achievements of the OZE, presented in the Central Committee's report, testified to the growing power of the new movement and the importance of its role in ameliorating the refugee crisis created by the Russian military authorities and the government. According to the report, by August 1, 1916, the OZE founded and operated 84 clinics; 19 hospitals (507 beds); 19 feeding stations; 9 canteens for children; 115 kindergartens for 9,552 children; 11 milk kitchens for nursing mothers, known as "A Drop of Milk"; 20 playgrounds, used daily by 6,180 children; 14 summer camps (colonies) that accepted 1,240 children before the summer of 1916; 2 sanatoriums (both in Crimea: one in Alushta, for 100 people, and one in Evpatoriia, for 50 people).[80] The OZE stuffed 27 "flying brigades" to accommodate refugees on their way to places of exile and to help them settle in new homes (assisted 30,000 refugees), and it served hot breakfasts to 9,160 schoolchildren. Overall, the report reflected activities in 102 cities, in 32 provinces.[81] In addition, in Petrograd, the OZE ran gymnastic clubs, a shelter for wounded soldiers, and a program that coordinated visits by Jewish physicians to wounded Jewish soldiers treated in general hospitals. Furthermore, the OZE founded a central Jewish medical library in Petrograd and launched its own publishing program. In 1914–1916, the OZE published: 9 popular brochures ("for the people") on contagious diseases (110,000 copies); a popular book by Dr. Lioku-movich, *School and Health,* and 10 posters with hygienic instructions—all in Yiddish. The scholarly series *OZE Proceedings* (*Trudy OZE,* 3 volumes) was published in Russian, which remained the main language of scientific research, whereas Yiddish was becoming the main language of dissemination of sanitation knowledge.[82]

The overall OZE budget was 245,000 rubles in 1915 and 943,883 rubles in 1916.[83] When evaluating these numbers, it is important to keep in mind that during the Great War, specifically in early 1915, the OZE had joined other leading Jewish relief organizations under the administrative and financial umbrella of the Jewish Committee to Aid Victims of War (Evreiskii komitet dlia pomoshchi zhertvam voiny, EKOPO), which received considerable funding from the state. In this alliance, the OZE was responsible for medical and sanitation services and information.[84]

Having so much to report, the OZE leaders nevertheless saw their first conference as an opportunity for strategic planning in anticipation of the great changes to come after the war. No one perceived the OZE emergency wartime activities as an end in itself. In Gran's plain language, the two years of war contributed to "strengthening the vitality [of the Jewish population] . . . for the conditions of peacetime." He suggested turning healthcare and childcare institutions established as relief centers during the war into "regular institutions of local Jewish communities and public organizations."[85] Gran imagined the postwar OZE as a surrogate Jewish ministry of public health and expected "all currently active Jewish communal, public, and charitable organizations involved in Jewish medical care" to voluntarily incorporate themselves into the OZE and accept its leadership. The system had to work "from the center to the periphery," with the OZE Central Committee developing policies that would then be transmitted down to "the regional and district periodic Conferences [soveshchaniia] of Physicians and Public Activists of Jewish Healthcare."[86] If the OZE was a "ministry," the Jewish nation was a "state" that had to budget its centralized healthcare system. Gran and others called for establishing mutual aid and mutual insurance societies, practicing communal self-taxation, and requesting funding from local zemstvo and city self-governments in addition to soliciting private donations. With sufficient funding, the OZE leaders wanted to build a national medical system that would encompass hospital medicine, outpatient treatment and home visits, pharmaceutical services, psychiatric help, support for the "deaf-mute," epidemic prevention, and sanitation control over synagogues, mikvahs, and other ritual spaces. In addition, the OZE wanted to assume the medical-sanitation supervision of schools, heders, preschool and summertime facilities, pediatric medicine, childcare, and vaccination. Finally, the popularization of medical knowledge and support of scientific research were seen as integral parts of this imagined Jewish national health system. Everyday management of OZE medical, educational, and other facilities was to be entrusted to local OZE cells or, if these were absent in a particular locality, to the specially created local sanitation committees staffed with physicians and pedagogues as well as representatives of Jewish communities and organized groups of the local public.[87]

Among the most pressing priorities debated at the first conference were psychiatric services, especially the establishment of the OZE Jewish psychiatric hospital.[88] Dr. Mikhail Shvartsman, a Zionist, announced the Central Committee's vision of the Jewish national psychiatric service in the paper "The Neuropsychiatric Health of Jews and the Tasks of Its Protection." He reminded readers that the rate of psychiatric and nervous diseases among Jews was higher than in the surrounding population, and that Jews obviously exhibited "some physical signs of degeneration." Shvartsman operated with West European and North American statistics, comparing them to Russian and Habsburg data and suggesting a "Western" psychiatric norm as an orientation for the Jewish project of medical self-help. He blamed the unfavorable economic and social environment for reinforcing "individual and racial degeneration" in Jews. The effect of the world war on the Jewish population did not promise quick changes for the better, and Shvartsman warned about the danger of further weakening of the Jewish "neuropsychiatric constitution."[89]

The establishment of the central Jewish mental hospital could prevent a worst-case scenario. According to Shvartsman, this facility was to combine a hospital, research center, and asylum for the isolation of dangerous elements. Shvartsman stressed that due to the "special liability and difficulty" of the task, only the OZE leadership could handle it. Next, the plan projected the establishment of two more research hospitals: one for the northwestern region and another in southern Russia. Obviously, Shvartsman not only imagined an intact postwar imperial polity but also expected the majority of Jews to remain in the Pale or return to it.[90] The locations of the three "central neurological-psychiatric institutions" reflected the old imperial Jewish mental map. The northwestern and southern regional psychiatric centers had to coordinate a network of psychiatric emergency rooms set up in each Jewish hospital in the Pale, offering exemplary treatment and isolating the untreatable from the healthy body of the nation. All three psychiatric hospitals had to advance a broad scientific agenda and develop and test "methods of prevention of neuropsychiatric degeneration" that were specifically suitable for the Jewish race. This was an extremely ambitious and very expensive plan. Shvartsman proposed to immediately establish a special psychiatric foundation to initiate fundraising on the OZE's behalf among Russian and American Jews.[91]

Such plans projected boundless optimism regarding the power of Jewish self-organization, the internal cohesion of the Jewish society, and its universal support of Jewish biopolitics. Indeed, OZE activists, who had already accomplished so much under the least favorable conditions of war and rising anti-Semitism, anticipated the postwar period with confidence. The February Revolution of 1917 that emancipated Jews and granted them full political citizenship made the OZE's grand vision even more plausible. The OZE even succeeded in meeting a few of its financial goals such as attracting state subsidies and international financial support and relegating some of the financial responsibilities to local communities. It opened new offices and recruited more local physicians.[92] No wonder that decades later some former OZE activists recalled the period leading up to the Bolshevik takeover as a ride on a powerful train, racing full speed through the most difficult periods of World War I only to end in a deadly crash in October 1917:

> The 1917 Revolution in Russia caught the OZE at the peak of its activity, when it served tens of thousands of children; when its 700 qualified employees[93]—physicians, social workers and nurses, teachers, and others—were energetically implementing, in a few dozen towns and shtetls of the country, the OZE's comprehensive program of physical regeneration of the needy Jewish population. At this time, the annual OZE budget was, in round numbers, 1,000,000 U.S. dollars. Part (35 percent) of this money came from foreign Jews through Joint, [another part] was from state subsidies, but the bulk was donated by the domestic Jewish population. This is strong evidence of the OZE's popularity in Russia.[94]

Steven Zipperstein has argued that relief networks such as the OZE, the EKOPO, and the Society for Handicraft and Agricultural Work among the Jews of Russia (Obshchestvo remeslennogo i zemledel'cheskogo truda sredi evreev v Rossii, ORT), introduced Jewish communal leaders and ordinary relief recipients to the world of modern ideologies and thus prepared them for the postwar mass politics.[95] Echoing his observation, Polly Zavadivker, a historian of the EKOPO, has stressed the role of nationalism in this strain of mass politics: "The EKOPO's relief workers and leaders . . . defined the war as a 'national catastrophe' (*natsional'naia katastrofa*) for Jews, and they framed their relief work not as a humanitarian campaign but a national mission to provide material and spiritual support by and for fellow Jews."[96] Both observations are relevant, but the difference between the model of apolitical politics as practiced by the OZE and other Jewish relief organizations and Soviet-type mass politics or contemporary nationalist politics in postimperial East-Central Europe or Turkey should not be discarded. The OZE did not mobilize masses around a certain ideology and did not have the state to back its national politics. It emphasized self-organization without and beyond any binding political vision or national state structure. As the fate of the OZE under the Bolshevik regime shows, while preparing masses of Jews for the world of modern ideologies and expert knowledge, the apolitical politics of the OZE best suited the context of the Russian Empire, which itself lacked any single hegemonic ideology or modern interventionist state.

Reclaiming Eugenics as the OZE Philosophy

By 1918, the usual channels of OZE financial support had been interrupted and the former empire was disintegrating together with its Jewish community. The Bolshevik regime did not tolerate any apolitical politics or noncommunist self-organization. The collapse of the old life, the Civil War, and genocidal pogroms in Ukraine took a heavy toll on the collective Jewish physical body and psyche. Already in early 1918 it was decided to send two OZE founders, Gran and Shvartsman, to the United States to fundraise, but the plan did not materialize due to the impossibility of foreign travel. A special committee elected on March 21, 1918, that included Gran, Shvartsman, S. G. Frumkin, and Veniamin Binshtok started work on the itinerary and information materials for the American Jewish public.[97] They decided to compose an overview of OZE history for an audience that did not understand the Russian context in which the Jewish medical movement had been born. The historical account they produced included information that was absent from synchronous narratives. For example, the authors revealed that the initiating group, in contemplating the creation of the OZE back in 1912, had to deploy personal connections and play "various tricks and [ride on the] coattails that were in broad use in tsarist Russia" to get the OZE statute registered.[98] More important, they confessed that to make the statute passable, its authors adapted the text "to the conditions of Russian reaction and antisemitism," so it did not honestly reflect the true intentions of the OZE founders.[99] Presumably, the founding fathers would have preferred "completely different formulations and statements," which were coined at semi-conspiratorial meetings in the process of long and heated debates. One such "formulation" was unrestricted freedom regarding Jewish communal self-organization. Another set of censored "statements" developed a far-reaching program of Jewish eugenics as the key concern of Jewish physicians. Instead, the final statute advanced the rhetoric of serving the immediate needs of Jewish healthcare that was acceptable to the authorities. However, the authors of the document added, "long debates about the OZE ideology and tasks" were not in vain. They de facto informed the practical and scientific work of Russian Jewish physicians.[100]

This emphasis on eugenics might have been intentionally tailored for the intended American audience, which does not negate the actual focus on eugenics in the work of Jewish medical societies that had preceded the formation of the OZE and the role of the eugenic vision in defining the priorities of Jewish population statistics and medical practices since 1912. After all, the OZE project was developed by Jewish physicians who had long socialized in the culture of race and eugenics.

The 1918 presentation to the Americans did not overuse the word "race" but clearly exposed racial thinking when it traced Jewish otherness from the early Middle Ages to the present, ascribing it indiscriminately to all Jews. This ultimate otherness justified the need for Jewish national medicine and eugenics. Moreover, the physical and psychological Jewish difference was not expected to disappear in the future. Instead, it had to be explored, urgently mitigated where necessary (signs of degeneration, physical weakness, and medical pathology), and reinforced and embraced where it presented advantages ("sharp and natural" physical vitality and durability, intellectualism). The OZE, the 1918 presentation explained, was conceived as a movement for the "comprehensive scholarly-popular [nauchno-obshchestvennoe, i.e., conducted by intellectuals outside of formal academia] investigation of Jews in general, and Russian Jews in particular, as a sociobiological phenomenon, and of the question of the psychophysical regeneration of Jews." This was "the most pressing and urgent national-popular slogan for the whole of Jewry."[101]

If we accept this account of the OZE founding group as truthful, then the rapid transformation of the OZE into a broad horizontal network open to new members appears in a new light. Apparently, this was rather an unintended consequence of staging an ambitious social engineering initiative in the conditions of World War I and the enormous refugee crisis that it produced, followed by the February Revolution, which had radically changed the political environment of Jewish activism. These developments had contributed to the OZE's democratization and elevated the "public" component in the initial design to counterbalance the original emphasis on expert authority. It is not accidental that many OZE accounts stress the role of Jewish youth who joined the OZE during the Great War, and credit them with making the OZE professional network a truly national movement. The OZE activists admitted that "the flow of events and the February Revolution of 1917 had irreversibly fixed this democratic outlook," so that the OZE methods became "accessible and closer to the masses."[102]

The entire OZE story, from its inception in 1912 to its formal banning in 1921, is a case of compressed, telescoping development ("on steroids"). Its liberal (elite) and democratic (popular) phases took place almost simultaneously. Then, as we will see in Chapters 7 and 8, the broad network was scaled back to its barebones elite-expert frame during the social disruption of the Civil War and the regime of proletarian dictatorship, when it became impossible to sustain a vast network of progressive activism both logistically and ideologically. The movement assumed a format of semi-isolated cells of motivated experts and physicians, determined to continue the project no matter what. They refused to see relief work, however important at the moment, as an end in itself, and they also refused to see their actions as determined by immediate necessity on the ground and being isolated from the common grand cause. In 1921, after the OZE had been banned by the Bolsheviks as a nationalist noncommunist Jewish organization not approved by the Jewish Section of the Bolshevik Party, the OZE founding fathers in Petrograd reflected on their recent experience:

> Even when the center was denied any opportunity to continue its activities, local [OZE] chapters expressed their strong desire to preserve it. Even in a state of inactivity, the center remained a symbol of the organization's unity. This unity—not just a formal togetherness but a real profound unity—was one of the most enduring principles that remained important at the time of greatest disintegration and ruptures.[103]

The unity evoked in this quotation implied, among other things, a vision of Russian Jews as one body that was impossible to split into parts following the logics of imperial disintegration.

CHAPTER SIX

Medical Statistics and National Eugenics

T HE EXODUS OF Jewish physicians from general medical societies to
Jewish medical societies to the national medical movement was accompanied by growing public awareness of the importance of medical
statistics for the functioning of the Jewish "national organism." Debates
about Jewish "degeneration" among physicians in the professional and
popular press and Zionist criticism of Diaspora Jews made Jewish "physical defects" more apparent as a problem and underlined the "absence of
objective numerical data for countries in which the predominant mass of
Jews concentrate, Russia in particular."[1] Statistics became a code word
for adepts of Jewish biopolitics who complained that "statisticians turned
to this question only recently, and the data that they have accumulated
are not sufficient for serious conclusions."[2] In fact, in the early twentieth
century, works of Jewish economic and demographic statistics produced
by Jewish scholars already existed, and a complaint about the "recent"
nature of this statistical endeavor sounded strange unless its author meant
a different kind of statistics.

Indeed, as far back as 1884, Warsaw Jewish businessman Jan Bloch
established a statistical bureau to gather materials on Jewish economic
and everyday hardships in the Pale. The materials collected for Bloch filled
five Russian-language volumes published in 1891, *A Comparison of the
Physical Subsistence and Moral Condition of the Population in the Jewish
Pale of Settlement and Elsewhere* (very few copies survived due to a fire

that happened shortly after the book was published).[3] One of Bloch's re-searchers, economist Andrei Subotin (not a Jew), published his own two volumes, which combined statistics on the economic conditions of Jews of the Pale with descriptive observations of the day-to-day hardships of Jewish life. His statistical work simultaneously evoked the literary genre known in Russia as "physiological essays" (*fiziologicheskie ocherki*) and reports of the government envoys to the Pale with the assignment to col-lect information and observations on Jewish economic activities (which the imperial authorities could have considered a priori as harmful).[4] The next generation of Jewish demographers and economic statisticians drew on these earlier studies as well as on the data of the first imperial census of 1897 and the census organized by the Jewish Colonization Society in 1898. The very fact that the data, and often the goals and objectives, of Jewish and state demographers were complementary underscores the dif-ference between the trend they exemplified and the growing demand for authentic Jewish science.

In the years after the first revolution of 1905–1907, when Jewish phy-sicians were complaining about the absence of national Jewish statistics, two Jewish demographers, Liebmann Hersch and Jacob Lestschinsky, were advancing their own versions of these statistics. Their studies tackled specifically Jewish social and economic challenges such as the problem of rising emigration, on the one hand, and the inadequate—in Marxist terms—pattern of Jewish proletarization, on the other. Hersch, a math-ematician by training and a Bundist by political affiliation, studied and then taught at the University of Geneva (as an instructor in the Depart-ment of Statistics and Demography). His works on Jewish emigration had appeared since 1908 in Polish and Yiddish, in addition to other European languages. He explicitly rejected racial (the race of wanderers) and ter-ritorial (a minority inevitably marginalized by national capitalist econo-mies) explanations and insisted that the reasons for Jewish emigration from the Russian Empire were exclusively political and legal—that is, po-litical persecution and discriminatory laws.[5] The elimination of these ex-ternal conditions would empower the Jewish proletariat and hence Jews as a modern nation—such was Hersch's message formulated in the lan-guage of population statistics.

The works of Lestschinsky, who later became known as the "dean of Jewish sociologists," exposed a peculiar biological determinism charac-teristic of positivistic Zionist science in general.[6] In his study "The Statistics of One Town," published in Hebrew in 1903, Lestschinsky referred to the "biological law" that allowed him to extrapolate from "individual cells" (that is, from one town, which happened to be his native town, Horodishche in Kiev province) to the entire organism.[7] This rhetoric notwithstanding, the main focus of the study was not the Jewish national organism but Russian Jewry's abnormal economic structure.

Lestschinsky identified as a Marxist and a Zionist, and later as a territorialist, which in combination produced socialist Zionism (he was a member of the Zionist Socialist Workers' Party). He published primarily in Yiddish and Russian, and used statistics to prove that to engage Jews in mass productive labor a national territory was needed, not necessarily in Palestine. For Lestschinsky, as for Hersch, Jewish statistics acquired meaning only in the larger external context of general undemocratic politics and/or economics, which made a racial epistemological framework secondary. This type of demographic statistics found outlets in initiatives such as the Bureau for Jewish Statistics and Economics at the Jewish Library in Vilna, which Lestschinsky headed in 1913–1914, or in the pioneer Yiddish academic periodical *Pages for Demography, Statistics, and Economics* (*Bleter far demografye, statistic un ekonomik*), which he co-edited in 1923–1925 in Berlin.[8] Hersch's and Lestschinsky's works were not entirely immune to racializing the Jewish collectivity: it was apparent, for example, in their framing of Jewish behavioral and psychological peculiarities and in the propensity to construct Jews as a separate group regardless of a specific time and context. That said, the two leading Jewish demographers and the sociological tradition they represented remained characteristically disinterested in Jewish biological statistics and the Russian Jewish medicalized discourse of Jewishness, self-diagnostics, and self-transformation.

Unlike Hersch, Lestschinsky, and many non-Jewish demographers and statisticians in the empire who worked with social categories such as class, profession, or confessional group, Jewish physicians pioneered nationality-based surveys of Jews as one social-biological organism. They started with Jewish students as better organized and educated and the most urban segment of the Jewish society. From a technical viewpoint, student statistical surveys were easy to implement because of students' universal literacy. Conducting student surveys did not require any specially trained personnel—the respondents themselves were perfectly capable of filling out questionnaires, and willing to do so. As Dr. Moisei Sherman from

Nikolaevsk wrote in the *Jewish Medical Voice* (*Evreiskii meditsinskii golos*) in 1909,

> The data collected among the Jewish students, regardless of their insignificant numbers, are especially valuable. For the first time, we can relatively easily obtain material that throws light upon the life of our young intelligentsia. It is the first time that we can compare these data collected from Jewish students with the data received from students of other nationalities (Great Russians, Poles, Georgians) and derive some conclusions about our national psyche, or cultural stability, and on many other issues.[9]

The suggestion to compare data on Jewish students with similar statistics on other nationalities was, of course, wishful thinking. Such data did not exist. Although student statistical surveys had been conducted in the Russian Empire as early as 1872, they rarely contained an entry for nationality, and even when they did, it was never self-sufficient.[10] At best, it helped explicate the connection between one's nationality and political preferences.[11] In 1909, so-called sexual surveys (*polovye perepisi*) were conducted among students of Kharkov, Yur'ev, Tomsk, and Moscow Universities. In Moscow, the questionnaire included a question about nationality, which revealed that out of the total number of 2,150 responses, 117 (5.6 percent) were submitted by Jewish students.[12] This was rather an exception, as student surveys paid very little attention to either nationality or health issues, which interested Jewish physicians the most.

All-Jewish statistical surveys were initially tested in the university centers of the Pale before being administered in other university cities. The first such survey was conducted among the Jewish students at Kiev University and Kiev Polytechnic Institute in 1909. Another survey targeted Jewish students in all Kiev high schools (legally, Jews could constitute up to 10 percent of Kiev students). In 1911–1912, Jewish students in Odessa participated in a similar survey. In 1913, it was repeated outside the Pale, in Moscow, where *numerus clausus* allowed Jews to make up no more than 3 percent of the students.[13] The same year, a statistical survey of Jewish students was conducted in Kazan at the university and the Veterinary Institute. There, the official cap on admitting Jewish students was set at 5 percent, but in reality, up to 9 percent of the students admitted by Kazan University were Jewish.[14] All these surveys combined traditional questions about social and economic status and political preferences with questions concerning health issues. The 1913 Moscow survey asked respondents to share, in addition to information on their physical state and illnesses, facts about their heredity.[15]

David Isaevich Sheinis (1877–1945?) was the mastermind behind the Kiev Jewish student survey and later helped to prepare for publication the results of the Moscow Jewish student survey.[16] He trained at a pharmaceutical school, then graduated from the Kiev University law department, and described himself as a Marxist and an autonomist. Neither of these could explain his preoccupation with social and medical statistics, the basics of which he mastered by himself.[17] The other Kiev Jewish survey, initiated and implemented independently of Sheinis by a group of Jewish students attending Alexander Rusov's statistics seminar at Kiev Institute of Commerce (1910), was not methodologically different from the one designed by Sheinis.[18] Both surveys constructed nationality as an essentialized form of groupness that determined social and cultural identity and even the economic behavior of an individual.[19] The opening section of Sheinis's 1909 questionnaire, "General Information," included questions on health in addition to the usual entries concerning age and social and marital status. The combination of basic social and medical information put into a certain perspective the responses in the next sections dealing with students' budgets, academic life, and national self-identification. Sheinis elaborated on this connection between the "material and hygienic conditions of life," health, and the "cultural-national image of Jewry" in the introductory chapter to the published materials of the 1913 Moscow statistical survey.[20]

The Moscow questionnaire went further in the use of medical statistics as an explanatory device. The health section now included questions concerning heredity, which many students did not know how to answer because they did not possess the information. Still, such questionnaires popularized the concept of heredity among the Jewish students by instructing them to think about their own physical and mental problems as part of collective Jewish biology in the *longue durée*. Whereas not all students were familiar with this approach, they were well prepared to analyze personal medical conditions. Overall, students' assessments of their health, both physical and mental, were rather pessimistic and, one could say, unusually detailed for young and busy people. In addition to diagnoses considered at the time typically "Jewish," such as myopia, neurasthenia, and general physical weakness, the Kiev students mentioned hemorrhoids, constipation, cough, tuberculosis, sore throat, and headaches. The Moscow students added to this list insomnia, vertigo, physical and mental exhaustion, rheumatism, and a number of other maladies.[21] The repetitive nature of many answers and their excessive details produced the picture of a generation and a nation in need of special medical attention. The accumulated responses painted a grim picture of a physical and mental body exhibiting clear signs of "degeneration" (a word broadly used by the respondents and Sheinis alike). The questions and answers reflected a conscious and consistent tendency to medicalize and racialize social problems:

"I support assimilation and find it to be deeply needed and useful for the recovery [*ozdorovlenie*] of the Jewish nationality" [in his comments, Sheinis indicated that the responding student self-identifies as a Social Democrat];

"I consider the Jewish nation sickly and its entire culture unhealthy; support assimilation by means of mixed marriages" [Sheinis: medical student, son of a "rich merchant"];

"I support assimilation as a natural evolutionary process" [Sheinis: a Social Democrat];

"I find that racial differentiation is an abnormal and harmful phenomenon" [Sheinis: son of a capitalist; married to a non-Jewish woman, does not know Yiddish];

"I support any Jewish movement as a force that sustains and defines features of national originality. It [Jewish national originality] is valuable as such, but for me, of course, it is also a matter of blood kinship";

"Under the current conditions in Russia, Jews do not assimilate so much as they degenerate physically and morally."[22]

Questions pertaining to one's social and economic life conditions, cultural patterns, especially the use of Jewish languages, and ideological preferences still dominated Jewish student surveys. However, these individual details were aggregated by a statistical approach that treated "spirit" and "consciousness" as collective entities, as attributes of a common national body. Since Jewish students were assumed to belong to this body by virtue of their race, questionnaires measured degrees and forms of their deviation from the scientifically posited norm. These were both physical deviations—produced by hard and unhealthy life conditions, heredity, or even some racial weakness—and deviations of "consciousness." Sheinis clearly perceived some responses as "right" and others as "wrong," framing differences as deviations. He called some statements "naive": "There should not be Jews or Russians. Everyone should be [considered] a human being in the first place," attributing this sort of "naivete" primarily to female Jewish students (*kursistki*), who "in general exhibit less maturity in [understanding] the national question than male students."[23] This and other statistically "established" gender disparities between young Jewish men and women on issues pertaining to health and social maturity were presented as yet another national problem to deal with. Surveys thus allowed their organizers to "trace threads *objectively* connecting student youth to their nationality" and to cultivate or medically influence them.[24]

Paradoxically, what Sheinis and others called objective belonging, that is, the assumed sociobiological symbiosis of physical health and identity, turned out to be weakly developed among the Jewish students, even among those with the most traditional Jewish upbringing. From the perspective of national eugenics, the surveys confirmed that the current generation of young Jews had already been lost for national biopolitics. This, in turn, meant the need to concentrate statistical efforts and eugenic engineering on children. But unlike students, children were not easily accessible, did not form a coherent group sharing similar urban experiences, and could not answer complex questions by themselves. Moreover, Jewish medical activists could not rely on the Russian imperial state to organize mass school checkups (especially since only a small number of Jewish children attended state schools) and promote centralized statistical surveys. A mechanism was needed to initiate a mass statistical study of Jewish children as a precondition for subsequent medical treatment and physical and mental improvement as the necessary steps toward the normalization of the future Jewish nation.

The Jewish Child as a Problem of Medical Statistics

Before the statistical turn, the Jewish child had been most visible in debates about the reform of Jewish education, the heder (Jewish religious elementary school) versus Jewish secular schools.[25] Since the 1840s, Jewish modernizers (maskilim) criticized the heder as an obstacle to the successful integration of Jews of Russia, claiming that "nothing was learned there;

the *melamed* beat the children, destroying them physically and spiritually; it was one of the vilest of religious institutions."[26] At the beginning of the twentieth century, this attitude began to change "among some of the most important specialists in the field of Jewish education in Russia. Some of these experts discovered in the *heder* previously unnoticed dimensions that could be salvaged in future schools while others saw parallels between the values of the *heder* and the new nationalist-leaning Jewish institutions. Still others were impressed by the *heder* longevity."[27] This partial rehabilitation of the heder reflected the rise of nationalism accompanied by the demand for cheap and accessible education in the emerging mass society. Zionists, for their part, hoped that the modernized heder, still popular among Jewish parents despite the proliferation of other forms of schooling, would facilitate the revival of Hebrew. These considerations added new urgency to the old discussions of Jewish education, making the problem of upbringing and schooling of Jewish children a matter of national concern.

By the turn of the century, the number of newly established *hadarim metukanim*—improved, modernized heders—still remained insignificant.[28] According to an 1895 survey conducted by the Society for the Promotion of Culture among the Jews of Russia (Obshchestvo rasprostraneniia prosveshcheniia mezhdu evreiami v Rossii, OPE) together with the Russian Imperial Free Economic Society, a typical heder teacher was either affiliated with the synagogue or had turned to this occupation after failing in another profession.[29] It was hard to imagine these people as leaders of national education and promoters of the Jewish project of medical revival. The very method of the 1895 OPE study was indicative of the old type of Jewish philanthropy and was pronouncedly nonexpert in nature: instead of surveying children according to some expert-designed program and studying the actual conditions in which their learning occurred, the survey organizers sent questions to several hundred state rabbis and communal leaders, soliciting their observations and opinions.[30] By contrast, in 1911, when the St. Petersburg OPE established a heder commission tasked with exploring these types of schools, it dispatched four modern educational experts to different regions in the Pale (South Russia, Volhynia, Lithuania, and Poland) to personally collect statistics and examine the situation on the ground.[31] The change in method reflected a more fundamental move toward a modern, expert-based population politics and a perception of the heder as playing a formative role in nationalizing the Jewish children.

Forging a harmoniously developed Jew both physically and spiritually was now declared to be "the key practical goal [of the school], however idealistic this goal may seem from a philosophical point of view."[32] New specialized periodicals such as the monthly magazine the *Jewish School* (*Evreiskaia shkola*; 1904–1905) promoted the view that "the Jewish mass is degenerating physically. Therefore, it is especially important to pay attention to the hygienic situation in schools and to children's physical development."[33] By the 1910s, Jewish physicians had acquired such authority among the nationalizing Jewish public that they could impose their specific understanding of the Jewish child as a national problem.[34] The ensuing medicalization of the discourse of Jewish childhood received additional inspiration from new and progressively popular human sciences such as experimental psychology, modern pedagogy, and physical anthropology, as well as different branches of medicine, especially neurology and psychiatry. In the early twentieth century, all these disciplines identified in children exemplary subjects of development:[35]

> [Children] were turned into particular biopsychosocial *embodiments* of development itself, the broader social meanings of which went beyond sheer ontogenesis or childhood as a special period of human life. The metaphor of "development" was extended as a normative and teleological structure of "life" in general—biological, social and historical, including the life of "the human race" or the life of a given "nation." It was in this context that children, taken as a generation of future citizens, became the target population not only for eugenic nurture, but also for mass normative evaluation, categorizing and streaming in "developmental" terms. . . . While "development" remained fundamentally teleological and normative, it was understood as "natural," objectively given and, crucially, measurable. As a consequence, *norms* of development came to be represented as *probabilistic laws,* to be arrived at by positivist experimental and statistical methods.[36]

The national project of Jewish physicians was based exactly on this philosophy of scientifically guided development. The nation's children had to overcome their problematic heredity in the name of the future nation, to develop strong and healthy bodies and nervous systems, in addition to well-trained minds. For this, Jewish physicians needed precise and measurable information about Jewish children's capacity for resistance, adjustment, and participation in all spheres of modern life, from sport to industrial work and from military service to politics—because this was what "normal" nations did.

Before 1905 had inaugurated the era of mass politics in Russia, only a few systematic studies of this type were conducted by Jewish race scientists. Afterward, the situation began to improve gradually, but physicians undertaking medical and anthropometric evaluations of Jewish school-children in specific localities recognized that their data were "extremely insufficient" and complained about the "complete absence in our litera-ture of the data on health and body development of school-age kids of Jewish nationality."[37] For almost two decades, until about 1909, the standard reference and main academic study of this sort remained *Physical Development of Children in Moscow's Institutions of Secondary Educa-tion: Materials for the Evaluation of Students' Sanitation State* (1892). This was the published dissertation of Naum Vasil'evich (Nahum Wulfo-vich) Zak (1860–1935), a Moscow physician and privatdozent of Imperial Moscow University, a specialist on school hygiene.[38] Zak insisted on the importance of studying "physical development by nationality," but in the early 1890s, his language of nationality reproduced the imperial confes-sional worldview and hence identified nationality in the modern sense only in the case of Jews. He compared medical statistics for three groups of Moscow schoolchildren: "children of Russian Orthodox [parents], who are in the absolute majority of cases of Russian nationality"; "children of foreigners (Catholics, Lutherans, Englishmen [!], and so on"; and Jews. Confessional taxonomy was converted into nationalities without cumber-some additional clarifications only in the case of Judaism, which auto-matically implied Jewishness. Jews were also the only group whose medical and anthropometric statistics Zak collected personally—for the rest, he borrowed data from secondary sources. He perceived Jewish students as representing a separate race and interpreted their unfavor-able height and chest circumference measurements as racial indicators.[39] Regardless of scholars' attitudes toward Zak's study, they had to rely on it for many years until greater numbers of physicians began measuring, testing, and examining Jewish children en masse, thus providing the sci-entific foundations for making some normative judgments and setting developmental goals.

The expectation that Jewish educated professionals would do this reflected the general pattern of development of the "science of the child," which everywhere began as the grassroots mobilization of professionals, scientists, and bourgeois parents.[40] Next, modern states, especially nationalizing ones, interested in deploying a knowledge-as-power mechanism to acquire social and political control over their populations, would normally join the trend by supporting coordinated interventions in the spheres of education, health, and the psychiatric supervision of children.[41] In the Jewish case, the initial stage was similar, only the role of experts was much greater than that of the "bourgeois" parents; the second stage, however, was absolutely different. Being in many respects no less "modern" than its European counterparts, the Russian imperial state did not conform to the normative Foucauldian model of state–expert relationships based on the idealized case of the French Third Republic.[42] The imperial bureaucracy was split along regional and ministerial lines and often disagreed with the dynastic regime regarding whom to consider legitimate objects of nascent social policies and with respect to the patterns of governmentality in general. The broad human diversity of the population further complicated the workings of the empire's bureaucratic mind.

Already in the late 1870s, the Ministry of Education had attempted to carry out some random sanitary surveys of schools, and in the 1890s it considered forming a commission to supervise the work of physicians monitoring school hygiene. This plan was implemented only in 1904; the actual work did not start before 1906, when the ministry's commission directed school doctors to perform general physical checkups of all schoolchildren in secondary schools twice a year, focusing especially on measuring height, weight, and chest circumference.[43] And it was only in the 1910s that the Ministry of Education started showing a more sustained interest in the development of hygienic and sanitation norms for schools, commissioning special psychiatric and experimental pedagogical studies. In 1916, the newly appointed minister of education, Count P. N. Ignat'ev, announced a plan to set up the School Hygiene Laboratory, staffed mostly by doctors and psychologists. Among other assignments it was entrusted with conducting, on behalf of the ministry, a mass psychological study of schoolchildren and working out developmental norms by age, gender, class, geographical region, and ethnicity—an approach that reflected the imperial condition of the deployment of modern population politics.[44] The plan revealed the minister's understanding of the fundamental conflict between the universal nature of scientifically informed models and the specificity of religious, ethnic, and regional circumstances in the diverse empire.

The Jewish project of collecting medical data on Jewish children was easier in the sense of dealing with a less heterogeneous populace. It aspired to establish national developmental norms specifically for the "Jewish child" and was to be carried out independently of the state, which clearly lagged behind, and conducted exclusively by Jewish experts and volunteers. Jewish medical societies, especially the Society for Protection of the Health of the Jewish Population (Obshchestvo okhraneniia zdorov'ia evreiskogo naseleniia, OZE) as an institutionalized all-imperial Jewish health movement, proclaimed the collection of statistics on Jewish children a national priority. The initial commission established by the OZE in the first month after its official registration (in late 1912) was the Commission for Sanitation Statistics. Veniamin Binshtok, one of the most experienced sanitation physicians in St. Petersburg and a member of the OZE Central Committee, took the lead and invited a non-Jewish medical statistician with a solid scholarly reputation, Sergei Novosel'skii, as vice-chair. Via members of the OZE professional medical network, Binshtok and Novosel'skii immediately distributed medical-statistical questionnaires among 125 Jewish hospitals in 501 localities in the Pale, and swiftly prepared the assembled materials for publication in volume 2 of the *OZE Proceedings* (*Trudy OZE*) in 1913 (under the title "Medical-Sanitation Organization among the Jewish Population").[45] In parallel, the OZE launched a number of systematic statistical projects, including on Jewish children. In 1916, the commission was renamed a bureau—a permanent government-like institution. The same year, the OZE leadership reported that everywhere the "sanitation survey of schools and the physical state of schoolchildren is under way."[46] The bureau developed statistical cards and forms, monitored their distribution to local OZE agents and institutions, analyzed and published the results of surveys, collected published materials on Jewish statistics for the statistical division of the OZE central library, and communicated with non-Jewish statistical organizations, such as the Pirogov Society statistical commission and zemstvo statistical bureaus.[47]

Among the most original achievements of the OZE Statistical Bureau was the standard medical-statistical card for Jewish childcare facilities that became known as the "kindergarten passport." In developing this statistical document, the OZE relied on the approaches of experimental pedagogy and on the practical experience of members of Jewish medical socie-

Table 6.1   Card for the study of schoolchildren by L. L. Rokhlin (1903)

#	Heder	Sex
Time of the checkup	Duration of education [in heder]	
First and family names	Age	
Parents' occupations and titles (*zvaniia*)		
General health condition		
Height in cm.	Chest circumference	
Weight in grams	Ratio of chest circumference to ½ of height indicator	
Academic performance	Smallpox vaccination	
Condition and color of: Hair	Eyes	Teeth
Notes		

Source: Rokhlin, "O fizicheskom razvitii uchashchikhsia v khederakh mestechka Krasnopol'ia Mogilevskoi gubernii," 124.

ties, such as Dr. Lazar L'vovich Rokhlin, who used his own statistical card (see Table 6.1) for the study of heder students in the shtetl of Krasnopol'e, Mogilev district (sixteen heders, 183 students).

In addition to such cards developed by individual Jewish activists, the creators of the "kindergarten passport" were inspired by experimental psychiatry. They followed the work of Vladimir Bekhterev's Pedological Institute—a small lab-nursery with an ambitious research agenda. The leading Russian neurologist and founder of so-called objective psychology established this institute in 1907 as a branch of his avant-garde Psychoneurological Institute in St. Petersburg—a private enterprise financed by private donors and student tuition. Every discipline taught at the Psychoneurological Institute was construed as biosocial by nature. For example, sociologists were required to study anatomy, physical anthropology, criminal anthropology, the theory of degeneration, pedology, and social sanitation. Biological sciences, in turn, encompassed much sociology and law, and pedagogy was replaced with pedology, a discipline blending nature and nurture and promoting a eugenicist perspective.[48]

As a private institution, the Psychoneurological Institute accepted students regardless of gender or religion, attracting especially those who were either banned from imperial universities for political reasons or whose admission was restricted by numerus clausus (Jews). The student register of the Psychoneurological Institute in 1909–1917 has survived only partially. It contains 5,120 names, of which up to 40 percent could be identified as belonging to ethnically non-Russian students: Armenians, Jews, Tatars, Germans, and so on.[49] There is much direct and indirect evidence supporting the impression that pedological know-how from the institute's lab and courses penetrated Jewish medical and statistical circles and influenced the OZE statistical and eugenic approaches to the "Jewish child" problem.[50] After the Bolshevik Revolution, when the institute lost its educational departments in 1919 and became a research institution only, the OZE leader, Moisei Gran, still perceived it as the best platform to host the three-month OZE emergency courses on "childhood protection in school and in the family."[51]

The medical part of the OZE's "kindergarten passport" (Table 6.2) resembled a compressed version of the "diaries" kept by the personnel of the institute's Pedological Lab. These diaries reported reactions to external stimuli and contained anthropometric measurements; information on heredity; records of bodily temperature, pulse rate, respiration, and sleep patterns; and changes in children's' appetite and mood. In addition, the diaries registered their illnesses.[52] Similarly, the "kindergarten passport" included family information, anthropometric measurements, and the results of psychological and medical evaluations.[53] If one considers that in the course of just over a year the OZE opened 125 kindergartens (detskie ochagi) in 100 different localities and attended by 12,000 children, each of whom was supposed to receive such a "passport," the mass scale of this statistical operation becomes apparent.[54] At the First OZE Conference in 1916, its Statistical Bureau had already attempted to review the statistics accumulated in the kindergartens, requesting that the "passports" and "sanitation forms filled out by physicians" be delivered to the committee.[55] Dr. Mikhail (Moses) Shvartsman, one of the key presenters at the conference, mentioned the ongoing statistical survey as the main element of the future system of prophylactics of neurological and psychiatric disorders among the Jewish children.[56] Another presenter, Dr. Iakov Eiger, also hoped to use these data as the basis for a comprehensive program of physical and mental improvement of the Jewish child.[57]

The "Defective" Jewish Child

Another identifiable source of influence on the medicalization of the discourse on the Jewish child was the Moscow school of psychiatry represented by Alexander Bernshtein and Georgii Rossolimo, who worked in

Table 6.2 Kindergarten passport, Society for Protection of the Health of the Jewish Population

---

#

City, shtetl	Province
Name of the institution	
Year	Month
Family name of the child	First name of the child
Age	
Sex: Male	Female
When admitted; when discharged	Reason
Are mother and father alive?	Causes of death
Occupation of father; mother	Of other family members
Number of children in the family	
Had [the child] attended heder before and for how long	
Heredity: tuberculosis, syphilis, alcoholism, psychiatric illness	

Periodic checkups, every three months	Height
	Weight
	Chest circumference during normal resting respirations

Body type: strong, satisfactory, weak
Nutrition: good, satisfactory, bad
Smallpox: vaccinated, not vaccinated, survived smallpox virus
Revaccination: received, did not receive
Skin on the face, body, and scalp: clear, eczema, furunculosis, scabies, shingles
Glands: normal, moderately enlarged, greatly enlarged
Traces of rickets: yes or no
Anemia: yes or no
Scrofula: yes or no
Curvature of the spine: yes or no; to the right, to the left, forward, back
Hearing: normal, impaired
Ear diseases: absent, suppuration, perforation of the membrane
Nose and throat diseases: absent, chronic runny nose, enlarged adenoids, polyps
Teeth: cavities, missing, abnormal growth
Eyes: normal, conjunctivitis, trachoma, corneal ulcer, strabismus
Visual acuity: normal, farsightedness, shortsightedness
Internal organs: heart, lungs, spleen, liver
Speech: normal, stuttering, tongue-tie, nasal, lisp
Nervous system: headaches, dizziness
Hernia: yes or no
Illnesses experienced while in kindergarten (*vo vremia prebyvaniia v ochage*)
Notes
Physician's signature

Source: DAKO, f. P-4018, op. 1, d. 16, 22, ll. 1–22; here l. 1.

the first specialized clinics assessing children with behavioral or developmental problems. In the 1910s, they began developing mental tests and statistical cards for diagnosing "deviance" and establishing the criteria of a "defective" child. Following their example, the OZE physicians and pedagogues defined "defective children" as "lazy, dysfunctional, morally corrupt and strange" and traced the causes of "defectiveness" to both heredity and environment. Medical professionals were instructed to study and treat such children separately and to partner with parents and educators in special schools and sanatoriums and on medical-pedagogical advisory committees.[58] "Defective" children were regarded as identifiable and treatable, unlike those who were degenerate and sick and had to be isolated from the healthy national body.

In the wake of World War I and the revolutions of 1917, in the context of the Civil War and collapse of the old social order and norms, OZE activists began criticizing the prewar scientific notion of "defective" child as framed by that obsolete norm. The initial medical bias embedded in this concept now acquired the connotation of an explicit social stigma after becoming applicable to too many Jewish children who had experienced wartime displacement, loss of family members, physical and psychological violence, and other deep traumas. The OZE physicians insisted that hundreds of Jewish children as well as adults were deeply traumatized and did not fall under the notion of "normal" and treatable in the old sense. They felt that the limited resources of the Jewish nation had to be directed at rehabilitation of the traumatized and severely physically weakened children, who now constituted the majority of the Russian Jewish population of children. Ideally, in the name of the future, all such children were to be saved "from the environment of their unstable parents" by being placed in special facilities for "nervous and retarded children."[59]

The OZE activists assessed the situation as "desperate" and "tragic" because the plan to separate children from adults in the nonexistent special facilities was unrealizable. Others complained that "the center did not prepare us for work with defective children." Listening to such complaints by local practitioners, representatives of the OZE Central Committee decided to embrace a twofold strategy for dealing with the "huge percentage of Jewish children who deviate from normal development": to initiate and coordinate yet again the "organization and collection of statistical data" and to train school physicians and educators for the new tasks. In addition, a special resolution, "On Defective Children," passed by the First All-Ukrainian OZE Congress in Kiev (November 1–5, 1918), decreed that a journal be launched to explore the problem scientifically.[60] A sense of deep despair permeates the deliberations regarding the criteria of "defectiveness" as documented in the minutes of the congress. At the same time, they convey an unconditional belief that medical science and medical statistics are capable of helping the nation in the face of the unfolding catastrophe.

Another effect of the Great War was normalization of the Jewish national eugenics project focused on children. After 1918, only the lack of government support distinguished the Jewish statistical and eugenic project from similar efforts initiated in many European states, including the new post-Versailles nation-states of Central and Eastern Europe. In the United Kingdom, the leading British anthropologist and advocate of positive eugenics, Karl Pearson, and his colleagues undertook research on the effects of the war on English children. One outcome was Pearson's own study, *On the Relationship of Health to the Psychical and Physical Characters of School Children,* published in 1923.[61] In Germany, Dr. Rudolf Martin, who from 1917 held the chair in physical anthropology at Munich University, launched a major anthropological study of Munich schoolchildren.[62] European scholars were assessing the quality of "biological material" required for the national revival after the war and arguing the need for national eugenic institutions and programs. Jewish medical activists were asking such questions since at least 1909–1910, further revising their research program during the war. Compared to similar statistical and eugenic projects in Western and Eastern Europe or the United States, since 1912, little had changed in structural terms for the Russian Jewish activists of medicalized politics. No postimperial nation-state took on their mission, and the end of the military confrontation did not ease the systematic collection of statistical data or provide Jewish communal financial resources to support it. Moreover, the war as a biosocial disaster was not over for Russian Jews in November 1918, both hampering the activities of the OZE and adding urgency to them. Throughout the Civil War, despite violence, destruction, hunger, and pogroms, OZE documents continued to call attention to "kindergarten as a stable institution that allows for the detailed and in-depth study of the child's physical and psychological nature."[63]

Whereas political and cultural disagreements regarding scenarios of the Jewish national future persisted well into the postwar period, the need for urgent relief and medical intervention to restore the Jewish national body and especially the bodies of the children—the future of the nation—was unquestionable. The OZE reflected this tension between potentially divisive cultural, and potentially unifying biopolitical, nationalisms. Its activists unanimously supported measures aimed at creating infrastructure for the treatment of children in kindergartens, sanatoriums, and schools with special medical supervision. They also agreed on the need to train physicians and pedologists knowledgeable in the specifics of the Jewish physical race and racial psyche. But they continued to disagree on cultural matters and called the language question a "sore point" in their work. One local OZE activist reported at the First All-Ukrainian OZE Congress in 1918:

> Especially in cities such as Kharkov or Kiev there are children whose mother tongue is Russian and whose environment is also Russian. To mix all the layers [of Jewish children] and care for them in a Jewish language [that is, Yiddish] is prohibitive from the pedagogic standpoint. Having children in one institution speak different languages is undesirable, for they influence each other negatively. We thus have to separate Jewish-speaking children from the Russian speakers.

In response, another OZE activist asserted that Jewish kindergartens had to accept children of different classes "but only the Jewish speakers." Yet another contended that there should be separate kindergartens operating in Russian and in Yiddish languages, but the program of education should be "national in content."[64] The language question turned out to be the sole issue on which the congress failed to pass any general resolution. By contrast, decisions regarding medical statistics or the founding of a new scientific periodical were passed unanimously.

Even considering the disproportionately high number of physicians among the Jewish professionals and adding to them a new cohort of wartime nurses, who also joined the ranks of the OZE, it was unrealistic to expect that every Jewish school and kindergarten could be staffed with a medic capable of compiling "passports" on each child and responding to each new OZE Statistical Bureau initiative.[65] This structural deficiency was further exacerbated during the years of the empire's disintegration and the Civil War, when communication, supplies, and communal structures were disrupted, even more Jews became refugees, and pogroms destroyed the lives and livelihoods of hundreds of thousands of Jews in the former Pale, while the Bolshevik authorities did not tolerate "liberal" public activism in the territories under their control. But even under these circumstances, dedication and the mass participation of Jewish physicians and nurses in statistical and medical work on children was unprecedented. One local report from the Zhitomir OZE kindergarten stated that in 1917–1918, "for most of the time, things with medical help were bad. We did not have a kindergarten physician. Only recently have we managed to find one. By now, he has been able to examine all the children. He took their weight, height, and head circumference measurements. Now, almost all the cards sent to us from Petrograd are filled out."[66] Even after the OZE had been officially outlawed in 1921 and was operating semi-legally, it continued to collect statistics on children. For example, in 1923 it organized a survey of children-returnees to Brest-Litovsk and homeless Jewish children of Ekaterinoslav.[67] These surveys produced pessimistic results, confirming that Jews had carried into the new postrevolutionary life traces of their old racial negative distinctiveness, "defectiveness," and "degeneration." Moreover, Jewish physicians diagnosed a "worsening of race" of the Jews who had survived revolutions and wars.[68]

There was a paradox in that the growing number of statistical data did not produce the expected qualitative shift toward a better understanding of how to reach the desired goals of Jewish physical improvement and national-racial normalization. In a report composed by the OZE leadership in 1921, the summary of the statistical information collected over the period from 1914 to 1920 occupied 136 typewritten pages, a truly solid pool of data, organized chronologically and thematically. In terms of scale and richness, these Jewish statistics superseded data available for other postimperial nations considered "normal"—potentially territorial, generally healthier, with less troubled histories. Nonetheless, in the first paragraph of the report the authors complained about "the absence of fundamental materials to characterize the psychophysics of Russia's Jewish population, whereas one cannot doubt the conclusion that the psychophysical and sanitary state of Russian Jews is catastrophic."[69]

The OZE leadership cited the same justifications for collecting further Jewish statistics as they had in 1912: Jewish anthropological statistics and information about Jewish heredity and pathology could cast light on many complex problems of modern science in general; Jews represented a unique "sociobiological experiment" in the history of humanity; finally, biological statistics provided a basis for national self-cognition and self-normalization. The authors of the report insisted that understanding Jewish social biology was key to organizing the modern life of "the Jewish people in all countries and states." The two communities that, according to the authors, deserved the most extensive statistical exploration were Russian and American Jewish communities.[70] This characteristic comparative perspective unmistakably evoked the methodological framework Lev Shternberg once offered for Russian Jewish anthropology as an exploration of the Jewish "emigration" to modernity (as an adaptation of Franz Boas's study of Jewish immigrants in New York). When Shternberg introduced Boas's study to the Russian Jewish audience in 1912, the champions of Jewish biopolitics in St. Petersburg were preparing for the OZE founding meeting. Nine years later, Shternberg and the OZE leaders in Petrograd worked together and advanced, with minor modifications, the same conceptual solutions to the predicament of Jewish modernity in Soviet Russia. The solution was predicated on medical, anthropological, and psychological statistical self-exploration and national eugenics. This meant they continued to perceive the Jewish physical body as the only space of authentic Jewish modernity under the Bolshevik political regime and in a world where the national-territorial principle had triumphed over the old empires.

The accumulation of Jewish statistics provided a way to substantiate abstract ideas about modern national Jewry, express them in developmental terms, and foster eugenicist thinking among educated Jews. At the same time, in the epoch of radical socioeconomic breaks and bloody wars, statistics could never be complete. It never provided the sought-after sense of achieved stabilization of the Jewish body—first, because this body never existed as a homogeneous whole, and second, because even as an imagined entity it was constantly pulled apart, moved, and decimated in the course of wars, revolutions, forced population resettlements, pogroms, and economic and political disintegration. At the same time, in the new age of total ideologies, the language of medical statistics remained the only version of nonideological, "objective," scientific nationalism and modernity. This ability to generate consensus amid cultural and political wars explains the perpetual nature of the Jewish statistical project. By the mid-1920s, under the Bolshevik policy of "indigenization" and territorial institution-alization of nations, biosocial statistics, understood as immune to ideolog-ical pressures, became even more attractive as an independent language of Jewish nationhood.

CHAPTER SEVEN

Civil War and the Biopolitics
of Survival

RECENT SCHOLARSHIP has problematized the conventional wisdom about revolution as the gravedigger of empire. Taking a cue from the historian of Iberian empires Jeremy Adelman, who coined the concept of the "imperial revolution," Ilya Gerasimov has examined the views on empire during the revolution of 1917 and its immediate aftermath.[1] To his surprise, none of the parties in the revolution that toppled the Russian Empire showed much concern about empire; moreover, empire seemed to become invisible in the discourses of revolutionary-era political movements and regimes:

> The Provisional Government occasionally referred to the country as "empire" in its proceedings, but the September decree announcing the republic identified the existing polity simply as "Motherland" (thrice within a paragraph-long text). It stated laconically and tautologically: "the state order that governs the Russian state is the republican order." At the same time, the Provisional Government did not denounce tight imperial control even over autonomous territories such as the Grand Duchy of Finland (appointing a new general governor of Finland several days after the declaration of the republic), not to mention Ukrainian lands or Transcaucasia. . . .
>
> Curiously enough, the deposed emperor himself did not think much about the fate of empire. In his paragraph-long statement of abdication, Nicholas II did not say a word about the empire. He referred to "motherland" four times and as many to "the state," framing his decision in terms of renouncing "the throne of the Russian state." . . .

... The Bolsheviks spoke a lot about "imperialism" (international or domestic), but recalled "Russian Empire" only twice during their first year in power: on August 29, 1918, in the decree denouncing all international treaties of the "government of the former Russian Empire" with the governments of the Central Powers, and on September 19, 1918, regulating the procedure for changing one's citizenship, which affected "former subjects of the Russian Empire, residing within the Russian Republic" who wanted to become Ukrainian citizens.

... Leaders of all the "counterrevolutionary" movements similarly ignored even mentioning empire in their programmatic statements—whether Lavr Kornilov, Anton Denikin, or Pyotr Vrangel. True, the slogan of "Russia, one and indivisible" was prominent during the revolution and the civil war, and various regimes wishing to establish their capital in Petrograd or Moscow (both Reds and Whites) fought against separatism of national movements and regions. Yet there was nothing specifically "imperial" in insistence on the integrity of Russia, or the perception of the country as a nation-state. In the 1917 Revolution, the multifaceted conflict was framed in terms of sovereignty, political regime, economic reform, and social order, while "empire" remained all but irrelevant to the main political actors.[2]

The analyses of Adelman on Iberian empires, Pieter Judson on the Habsburg Empire, and Gerasimov on the Russian Empire reveal a common pattern: political revolution in empires happens as a response to a radically upset status quo in the old imperial system, when a powerful minority in the metropole (Russian nationalists and the upper classes in the Russian case) threatens to usurp the entire empire.[3] The revolutions begin as attempts to restore the initial promise of the empire to provide a common social sphere of belonging to the commonwealth. Exacerbated by wartime hardships, the escalating political crisis and the predominance of nation-centered social imagery that perceived the social sphere as construed of internally homogeneous groups (nations and classes) contribute to the failure of the imperial revolution. This explains the persistence of old imperial spatial boundaries, including in pro- and antirevolutionary projects, despite the disappearance of empire from the articulated discourse. It was only after the Bolshevik October coup, which brought the country to the brink of open civil war and economic catastrophe, that national regions and movements hastened to distance themselves from the crumbling imperial state by proclaiming independence.[4]

The special nature of the Jewish case became particularly apparent at this moment, as the lack of "national" territory left Jews only the alternative of continuing to operate within the paradigm of the "imperial revolution" and working to renegotiate their collective status as a nation in the emerging new supranational (spatially still "imperial") polity. In terms of post-Wilsonian international law, Jews constituted a legal minority in the new nation-states of Eastern and Central Europe. This status did not go well with the way in which ideologues of Jewish anti-imperial nationalism saw their postimperial future. Even in 1921, during the most troubled stage in the relationships between the Society for Protection of the Health of the Jewish Population (Obshchestvo okhraneniia zdorov'ia evreiskogo naseleniia, OZE) and the Bolshevik regime, OZE leaders essentially credited Bolsheviks with having reassembled the imperial political space: "The civil war that had reached its point of highest tension in 1919, ended with the full victory of Soviet power. Under its rule, Great Russia, Ukraine, a part of Belorussia, and Siberia were again unified. It is true that peace with Poland was yet to be achieved, but military actions on this front were already dying out."[5]

The OZE's vision of the Jewish physical and social body retained the old imperial scope and scale. In the spring of 1918, after the signing of the Brest-Litovsk bilateral treaties with Germany by the Ukrainian Central Rada and the Bolshevik government, when Ukraine was cut off from the rest of the former empire, the OZE Central Committee complained:

> Territory on which Russian Jewry used to concentrate turned out to be split between a few states whose borders and internal [political] regimes still remain undetermined. To the extent that the current transitional situation allows us to predict the future, our perspectives look grim. Instead of living in the multinational state that Russia was before the Brest peace partition—that is, in a state in which, under the democratic regime, hegemony of one nation over another is unthinkable—Jews of the former Pale will now find themselves in the situation of Polish Jews. They will become a minority in a titular nation—Lithuanians in Lithuania, Belorussians in Belorussia, and Ukrainians in Ukraine. And these titular nations will jealously guard the "national" character of their new states.[6]

The November 1918 OZE Congress convened in a region de facto disconnected from the rest of the former empire (the Hetmanate regime of Pavlo Skoropadskyi in Kiev did not recognize the Bolshevik authority). Ukraine remained home to a great number of former imperial Jews, who suffered enormously during World War I and who became targets of the new, postrevolutionary wave of anti-Semitic violence. Ukrainian Jews constituted a central concern of the OZE leadership in the former imperial capital, Petrograd. Although the congress was branded the "First All-Ukrainian Congress," and the spreading Civil War made travel to Kiev extremely dangerous, the perception of the Jewish people as integrated and of the OZE movement as pan-imperial was not altered a bit. A few representatives of the OZE Central Committee did manage to attend the congress: Abram Bramson and Maria Shteingauz (a pedologist and activist of the OZE children's summer colonies program) from Petrograd, and another pedologist and future children's writer, Elizaveta Shabad from Moscow.[7] The bulk of the delegates were from Ukraine, representing twenty localities in total: twenty-two participants were from Kiev, two from Kharkov, six from Ekaterinoslav, one from Odessa, and so on. They demonstratively defied the ongoing Ukrainization of "Southern Russia." The current government of Skoropadskyi and the reality of independent Ukraine were never mentioned at the congress. Instead, OZE activists referred to unspecified "government institutions." All presenters at the congress described the community for whom they were working as "Russian Jews," and understood the language problem in Jewish society as a choice between Russian and Yiddish (and sometimes Hebrew). Ukrainian national demands and Ukrainian language were never mentioned. Ukraine featured only in rhetorical formulas such as the "physical rejuvenation of Jews of Ukraine and Southern Russia."[8]

The unanimous and hardly always conscious refusal to explicitly deal with the new reality exposed the paradox of empire's simultaneous persistence as a dominant context-setting category and its omission as a political reality of the recent past. The First Ukrainian OZE Congress introduced major administrative changes, but they were not framed ideologically as adaptation to the ongoing political and territorial nationalization of the former imperial society and the ensuing inevitable fragmentation of the Jewish "national body." It was only the break of effective communications with Petrograd that prompted the creation of the Ukrainian OZE Central Committee in Kiev, closer to the population that needed urgent relief. The committee was mandated to administer OZE facilities on the ground and organize financial self-help by working with local Jewish communities and "government institutions." At the same time, the Petrograd Central Committee retained responsibility for general ideological coordination and scientific guidance. This ad hoc redistribution of responsibilities did not rely on any consistent vision of postimperial sociopolitical realignment.

Since the February Revolution of 1917, OZE's political ideal was Jewish communal autonomy in "democratic Russia" (which meant the Russian Empire). Moisei Gran and Mikhail Shvartsman officially confirmed this position on behalf of the OZE leadership in July 1917, at a meeting in preparation for the All-Russian Jewish Assembly.[9] At this point, the grandiose OZE plan for a Jewish national health system was still a priority. Democratically reconstituted and empowered Jewish communities were expected to financially support institutions of national healthcare and medical and eugenic research, whereas Jewish physicians had to join the national medical system voluntarily for the sake of the physical recovery of the nation. "We are living through a period of world-wide upheavals," Gran explained. Jews had to respond to these revolutionary transformations with immediate national medical mobilization. He continued:

> As this revolutionary and evolutionary process unfolds, the fates of peoples and nations are being forged. All peoples and nations have endured innumerable sacrifices to this great process of regeneration, and the Jewish nation—due to its historical fate—took upon itself an incommensurate, lion's share of these sacrifices. The greater the sacrifice, the more rights and reasons the Jewish nation has to believe and hope that its political and social life will now assume a new course. To initiate it, the Jewish nation needs, in the first place, to make the basic foundations of its national life significantly healthier. In this process of rejuvenation, physical recovery and national popular revival are possible only by means of Jewish national public medicine. For the mass-scale implementation of this plan, a dedicated staff of physicians is required.[10]

The OZE Central Committee called on the movement's members to get involved in local communal politics and thus to facilitate the reorganization of Jewish communities on democratic principles and encourage them to start contributing organizationally and financially to the project of national medicine.[11] Political differences among Jewish parties and movements, as well as disagreements between the OZE and communal leadership on what exactly constituted the most pressing medical needs of the Jews, cooled down the initial enthusiasm. But it was the Bolshevik coup that put a decisive end to these plans.

The Retreat to Apolitical Politics

When the All-Russian Congress of Jewish Communities had finally assembled in Moscow (June 30–July 4, 1918), the OZE had changed its position on transferring its facilities to the newly formed civic Jewish communities. In the view of the movement's leadership, local Jewish communities did not possess real autonomy under the Bolshevik regime. From the OZE's standpoint, the chance of a political solution to the Jewish nation's predicament had been lost, and the OZE quickly retreated to its old strategy of apolitical politics focused on healing, stabilizing, and preparing the Jewish physical body for the next political opportunity. The statistical project continued too, and as Veniamin Binshtok explained in March 1918, "All these data are practically needed for sanitary purposes; but one cannot do without them to understand Jewry scientifically, in terms of their pathology, specifics of heredity, and so on."[12]

The OZE refused to take a clear political stance toward the Bolshevik regime. It needed to cooperate with the regime to help the Jews, and, in any case, the Civil War context altered the very understanding of political neutrality. What used to qualify as apolitical social engineering now was perceived as undermining the regime's monopoly on class-based politics, and the notion per se of Jewishness became politicized. The OZE leadership had less trouble finding a common language with Soviet central agencies interested in modern social policies, especially with the People's Commissariat for Health (Narodnyi komissariat zdravookhraneniia, abbreviated as Narkomzdrav), than with the Jewish Section of the Bolshevik Party (Evreiskaia sektsiia, abbreviated as Evsektsiia), established in the fall of 1918 and entrusted with elaborating and implementing the Sovietization of the Jewish population.

From the OZE leaders' perspective, their confrontation with Evsektsiia was an uncompromising struggle between a truly national organization that aided all Jews and schismatics, who pitched one Jew against another in the name of doubtful ideological considerations.[13] From the perspective of contemporary scholarship that explores the economic, generational, and political differentiation of the Jewish society during the broadening crisis of wars and revolutions, this is not necessarily so.[14] After all, the Evsektsiia vision of the Jewish nation, as defined in class categories and expected to recover through the demiurgic power of "productive work," was equally authentic and nationally Jewish. Jews-as-a-proletarian-nation was a legitimate version of modern nationhood, although much more exclusive than the OZE class-blind approach.

Less dependent on the scientific expert discourse, and more explicitly ideological, the Evsektsiia vision of Jewishness shared with OZE medicalized nationalism a concern about the collective Jewish body and psyche, which were to be transformed for modernity. According to OZE records, in the course of 1918–1919 Evsektsiia confiscated four-fifths of OZE facilities (kindergartens, medical stations and colonies, and so on) to administer them according to class principles. Evsektsiia prevented OZE personnel from traveling on OZE business by not issuing official travel passes (*komandirovka*), which were required to board trains without risking arrest.[15] In general, Jewish Bolsheviks were no different from gentile Bolsheviks in denying any independent "classless" self-organization, even when aimed solely at providing relief to the population. However, because it was also an internal Jewish conflict, it resembled a bitter family feud between close relatives—two types of modern Jewish elites—over the right to mold Jewish masses into a modern nation.

While the Evsektsiia's position was incomparably stronger politically, in early 1919 the OZE received some political support from Narkomzdrav, which sanctioned its work with refugees and war prisoners. In the summer of 1919, the People's Commissariat for Education (Narkompros) of the Northern Region (Petrograd) signed a similar document legalizing the OZE's medical and psychological assistance to Jewish children—refugees, victims of pogroms and hunger, and orphans from Ukraine. In the early fall of 1919, the OZE rescue teams composed of Jewish physicians were able to travel to Ukraine officially (unofficially, they had been working there under the auspices of the Ukrainian OZE and the Jewish Committee to Aid Victims of War [Evreiskii komitet dlia pomoshchi zhertvam voiny, EKOPO]). As news about the devastating pogroms in Ukraine reached the OZE headquarters in Petrograd, many more Jewish physicians, especially medical students, volunteered for relief work. In the fall of 1919, in Petrograd alone one hundred medical students signed up as OZE volunteers. Narkomzdrav issued official passes authorizing their travel on OZE business.[16] Together with the activists of EKOPO and the Ukrainian OZE, who were already active on the ground, they provided medical help and delivered medications and, when possible, food to the people whose health and homes had been ruined. They also collected the orphans and sent them to Petrograd. While doing this important and urgent work of helping individual victims of anti-Semitic violence, the same Jewish physicians and other volunteers labored for the future of the collective Jewish body. In a study of the narratives of pogrom violence, Harriet Murav describes their efforts as a response to necropolitics of the Civil War with positive biopolitics endowed with a "providential dimension, as a means of overcoming the chaos and abandonment." "This focus did not permit attention to the individual," Murav adds.[17] Indeed, statistical discourse and a biopolitical approach were the most adequate means to carry on this mission amid the unfolding catastrophe.

The OZE and EKOPO activists collected the most precise and systematic demographic and medical statistics on pogrom victims. In addition to personally taking measurements, counting, and questioning, they used any data on population that happened to be available and concerned Jews, such as materials of the all-Russian agricultural census (conducted in April 1920), pertaining to the rural Jewish population in the Ukrainian countryside. Jewish statisticians had obtained these materials well before the census had been processed and officially published by the Soviet government. Most probably, Jewish activists working on the ground received the raw census data directly from Soviet provincial executive committees (*gubispolkom*). These data were compared with similar data from the 1917 agricultural census and with statistical data assembled by the Jewish activists themselves, which allowed conclusions to be made about demographic changes in the Jewish population. Unlike prerevolutionary Jewish social statistics (associated with Jacob Lestschinsky and Liebmann Hersch), the Civil War statistics no longer aimed at a scientific comprehension of and response to the problems created by political persecutions and economic deformities. Now, the goal was to document the nation's genocide (the term yet to be coined by Raphael Lemkin in 1944), comprehend the reality of necropolitics, and scientifically analyze its effect on national biology and the probability of Jews as a modern nation in the future.

In one archival collection that I studied, the reverse, blank, sides of the handwritten regional reports of the all-Russian agricultural census were covered with questions that Jewish relief workers intended to ask pogrom victims. They inquired about specific circumstances surrounding pogroms, places, and names and invited them to identify the perpetrator. The latter task required selecting from a list, as in a multiple-choice test, containing the names of guerrilla commanders (atamans) and main regular armies in the region: the army of the Ukrainian People's Republic, Nestor Makhno's Revolutionary Insurrectionary Army, the Red Army, and so on.[18] The collocation, literally on the same page, of general population statistics and records of genocidal violence is highly symbolic of the context in which the adepts of national biopolitics conducted their relief campaign. They had to reconcile the *longue durée* temporality of eugenic social engineering with the short-term temporality of genocide destroying the national body. They also had to account for each individual victim and survivor, trying to help the latter while assessing the toll taken by their physical and mental traumas on the collective national body. Both parts of the mission were equally important, and hence, for the sake of the nation's future, the deeply traumatized recipients of aid and medical assistance were simultaneously debriefed about the horrors of recent pogroms.

Under the circumstances, the only way for the OZE Central Committee to continue its mission was to cultivate relationships with the Soviet government. Besides the obvious need of obtaining immunity from political persecution, it was also a matter of financial sustainability. The policy of war communism ruined individual sponsors and the economy in general, leaving the OZE without domestic sources of funding. The OZE leadership placed their hopes in foreign financial help, which could not be procured without the government's permission and cooperation. When the capital of the Russian Soviet Federative Socialist Republic (Rossiiskaia sovetskaia federativnaia sotsialisticheskaia Respublika, RSFSR) moved from Petrograd to Moscow in spring 1918, the OZE, following the government, began transferring its administrative offices to Moscow as well.[19] In 1920, its Central Bureau was finally established in Moscow, leaving in the purview of the old Petrograd Committee finances and scientific work.[20] By the end of the Civil War, when the economy was in complete ruin, and famine was acquiring horrible proportions in the Volga region and Ukraine, the Soviet authorities finally agreed to negotiate a permanent financial arrangement with the Jewish public organizations that were still alive and active: EKOPO, the OZE, the Society for Handicraft and Agricultural Work among the Jews of Russia (Obshchestvo remeslennogo i zemledel'cheskogo truda sredi evreev v Rossii, ORT), and the Kultur-Lige (Cultural League).

An Unhappy Compromise: The OZE under Evobshchestkom

Such was the situation in late May 1920, when representatives of the American Jewish Joint Distribution Committee (JDC), the judge Harry Fisher and the trade union leader Max Pine, came to Moscow to negotiate help for Jews of Soviet Russia.[21] During these negotiations, the OZE leaders from Moscow, Petrograd, and Ukraine formed a joint front with other old Jewish organizations. The Soviet government and Evsektsiia demanded that financial help be channeled either through a special state agency created for the purpose or directly through Evsektsiia. Representatives of the Jewish organizations insisted that only they had the organizational capacity, vision, and moral authority to distribute the JDC funds to all Jews in need, regardless of their class status, and they categorically refused to work under Evsektsiia. The JDC representatives were leaning toward the old Jewish organizations but tried to remain pragmatic. They were somewhat shocked to realize that the OZE vision went far beyond immediate relief. "I must say without the slightest exaggeration," concluded one of the JDC representatives, "that we Americans have a lot to learn from you. . . . In America, the Jewish public does not possess an ideological vision such as the one we witnessed in your midst. Our activities are mostly focused on philanthropy. . . . But your organizations aim to solve entire Jewish problems. Your Russian (Jewish) public (activists) are incomparably deeper than our American Jewish public."[22]

An OZE report described the negotiations as "the fiercest struggle," which ended with an unhappy compromise. The Soviet government, the JDC, and Russian and Ukrainian Jewish activists agreed to form the Jewish Public Committee for Relief of Pogrom Victims (Evreiskii obshchestvennyi komitet pomoshchi postradavshim ot pogromov, abbreviated as Evobshchestkom in Russian or Yidgezkom in Yiddish). Evobshchestkom included Evsektsiia-affiliated representatives of Jewish communists and delegates from four Jewish public associations (OZE, EKOPO, ORT, and Kultur-Lige). The first group outnumbered and hence outpowered the second (accounting for two-thirds of the overall membership). The Evobshchestkom's Presidium of three included two Jewish communists and only one representative of the four Jewish organizations.[23] This structure predictably had already collapsed in January 1921, when the Evobshchestkom's majority made it completely impossible for Jewish organizations to operate while complying with the principles imposed by its Presidium. The OZE and EKOPO had to withdraw, after which they were officially banned in the RSFSR. Thus, the period of the OZE's legal functioning as a part of the Soviet state, under the auspices of Evob-shchestkom, lasted for only half a year, from the early summer of 1920 to January 17, 1921. Over this brief window of limited opportunities, the OZE managed to accomplish many of its goals, especially in the spheres of Jewish national statistics and eugenics.

Particularly spectacular was the instantaneous reemergence of the OZE statistical machine, coming to large-scale operation from its previous semi-hibernate mode in no time. Four OZE departments that resumed their activities immediately after the legalization were the Medical-Sanitation Department, the Department for Children's Relief (*otdel detskoi po-moshchi*), the Department for Sanitation Education, and the Department of Supplies and Bookkeeping. They set up special subdivisions, such as the commission for the protection of motherhood and childhood and the commission for work with "defective children." "The old workers and activists of the OZE" resumed their work with redoubled energy; the OZE Central Bureau "received dozens of offers [of services] from physicians and teachers daily."[24] Together, the OZE and EKOPO established the Sta-tistical Department headed by Dr. Ilya Pevzner, and the OZE took upon itself the processing and interpretation of medical and sanitation statistics for the department. This work was entrusted to the old OZE leadership in Petrograd. As an OZE report detailed:

> Assignments were planned, programs of surveys developed, and the final data processing accomplished in Petrograd. However, to make the work coherent, all [statistical] initiatives were run by the Central Statistical De-partment in Moscow. From the very beginning, Jewish organizations paid special attention to statistical data and surveys. They prioritized surveys of the localities where pogroms took place, surveyed families of the vic-tims, and composed lists of those who had been murdered, wounded, raped, and orphaned. The corresponding survey forms and cards were de-veloped collectively by members of the Statistical Department and the OZE—V. Binshtok, M. Gran, S. Novosel'skii, and I. Pevzner.[25]

Pevzner was a St. Petersburg/Petrograd psychiatrist and OZE member. The others, including the non-Jewish founding member of the OZE Sta-tistical Committee, Sergei Novosel'skii, who would soon assume the first Soviet university chair in sanitation statistics in Leningrad, had been col-lecting Jewish medical and anthropological statistics from the first days of the OZE.[26] Now they played leading roles in what was at the time, perhaps, the most comprehensive synchronous study of the human consequences of genocidal violence. Nahum Gergel, the EKOPO chairman since 1918 and a participant in the Jewish statistical project under the Evobshchestkom's auspices, published the aggregated data on pogrom fatalities after he emigrated to Germany in 1921. According to Gergel, between 1917 and 1921, fifty thousand Jews had been killed in pogroms.[27] To deliver this number to the world was already quite a significant result, but for the OZE activists, the physical and psychological rehabilitation of the survivors was not less important than counting the victims. Pogrom survivors, especially women and children, carried scars of their traumatic experiences into the future on their bodies and in their psyches. If left to their own devices, this generation had the potential to compromise the ability of the postimperial Russian Jews to develop into a strong modern nation, as envisioned by the architects of Jewish biopolitics. Unlike the communist part of Evobshchestkom, they did not believe that productivization and proletarization were the answer to the fundamental predicament of the Jewish collective body, which they continued to frame in racial terms—as a consistent and steady deterioration of the Jewish race that had intensified tremendously after 1914. Relief efforts, statistical self-exploration, and eugenics remained tightly intertwined in the minds and activity of the OZE members and their many Jewish contemporaries. A regular contributor to the Zionist newspaper *Sibir-Palestina* (Siberia-Palestine), published by Russian Jewish refugees and émigrés in Harbin (1920–1943), explained this connection to his audience in 1923:

> To secure a place among civilized humanity, a nation needs more than to be up to the mark of modern culture. The stock of life forces of a great nation should exceed the basic needs of the moment. Even the most remote future is nourished by present-day life juices. The roots of the nation's future existence [are in the present] and are nourished only in the present. The entire historical past confirms this law.[28]

The tropes of Jewish longevity, of the Jewish body evolving from common roots through centuries (being split territorially but connected spiritually and physically), and of the present that will directly determine the nation's future, resonated with different political visions. They "nourished," to use the metaphor from the *Sibir-Palestina* article, the mass reception of demographic, statistical, and eugenic knowledge as capable

215

of predicating the future and fixing problems in the present. The willingness of pogrom victims to reply to the extensive questionnaires and the readiness of ordinary Jews to offer their bodies for anthropological measurements were the results of not only their dependence on relief workers but also a broad public consensus about the power of Jewish statistical knowledge to save the nation.

The Afterlife of the OZE

In early 1921, Jewish communists succeeded in forcing out representatives of the old elites from Evobshchestkom. Simultaneously, they were attempting to eliminate the JDC's intervention by developing an independent fundraising infrastructure in the United States. The OZE was banned, and it dissolved as an organization but apparently did not disappear as a community of like-minded activists. Tracing the afterlife of the OZE illuminates the power of the social imagination to enable a coordinated action under even the strictest ideological regimes, and the function of Jewish national biopolitics as politics of resistance.

In 1922, former OZE leaders founded the Society for the Study of Social Biology and Psychophysics of the Jews, which in 1924 was transformed into a committee of the same name under the auspices of the Jewish Historical-Ethnographic Society (Evreiskoe istoriko-etnograficheskoe obshchestvo, EIEO), now headed by Lev Shternberg. Thus, the old Petersburg/Petrograd OZE committee de facto remained a coordinating center of the grand Jewish medical-statistical project. As we shall see, its activities helped to keep alive a nonterritorial, race-based, eugenically inspired vision of Jewish modernity.

Although the OZE representatives had left Evobshchestkom, the committee continued the OZE-inspired medical and scientific initiatives. For example, in 1922, the Medical-Sanitation Division of Evobshchestkom contacted its district committees with a request to "immediately" initiate a comprehensive medical-statistical survey of Jewish children—refugees and pogrom victims—thus "easing our medical work and enriching our statistics": "Evobshchestkom feels confident that the physicians and personnel of Jewish institutions for children . . . understand the importance of this survey and, despite the difficult conditions under which they work, will fulfill their duty for Jewish children and science."[29] This plea was accompanied by a number of questionnaires. The "Experimental-Psychological

Survey of Children" consisted of two long parts (forty entries in the first part and twenty in the second) and reflected the approaches tested by Vladimir Bekhterev's pedologists and prerevolutionary experimental psychologists mentioned earlier in these pages. The questions in the first part were formulated so as to explicitly divide the life of a traumatized Jewish child into the past, present, and future, and to contrast dreams with real experiences. This chronological differentiation of life experience implied the possibility of programming the future by curing (or censoring) negative content from the past:

6. Whom did you most fear in childhood and why?
7. What did you most fear in childhood and why?
8. Whom do you most fear in the present and why?
9. What do you most fear at present and why?
21. Talk about the most terrible events in your life.
22. Talk about the happiest events in your dreams.
23. Talk about the saddest events in your dreams.
30. What or whom do you want to become when you grow older?
31. Is it better—to be Jewish or non-Jewish?
32. Do you want to be—Jewish or non-Jewish—and why?
39. Where would you like to be now and why?
40. Where do you want to be when you become an adult, and why?[30]

This questionnaire was part of the large-scale "Medical-Characterological Survey of Children," which included the collection of general biographical information on each child and his or her parents and relatives, and documentation of the child's pogrom experience ("how many, when, where, and under what circumstances" the child had survived pogroms).[31] Then, there was a second form, purely medical. It was intended as a real mass document: unlike the first questionnaire, it was bilingual—that is, printed in Yiddish and Russian—so not only physicians but every provincial nurse could use it, and each child could answer the questions about heredity, past diseases, and suffering. The instructions required physicians to take all measurements from the children's naked bodies, paying attention to abnormalities in the forms of their skulls and registering color indicators of skin, eyes, and hair.[32] The entire project exposed the signature OZE approach and style.

Other archival documents confirm this impression. Since the end of 1921, the OZE Petrograd committee members, who now theoretically could not affiliate themselves with the OZE, had lectured young physicians

visiting Petrograd for additional medical training. According to one report composed most probably in 1922 by Dr. Abram Bramson, an OZE
cofounder and lifelong activist, these lecturers "familiarize the audience
with former OZE activities, OZE ideology, and with the future work necessary among the Jewish masses." Bramson expected these physicians to
compose "the new core from which a renewed OZE will emerge."[33] When
in 1922 the Medical-Sanitation Division of Evobshchestkom issued its call
to Jewish physicians cited above, Bramson was personally mobilizing old
and new OZE cadres in its support in his capacity of a lead Narkomzdrav specialist on tuberculosis.[34] A letter to Bramson from one of his cor-
respondents, the physician M. Vidershtain from Simferopol in Crimea,
who worked at two local Jewish facilities (an orphanage for sixty children
and a kindergarten for fifty children) suggests that Bramson and Evobshchestkom were distributing identical medical statistical forms, and thus
the OZE-Evobshchestkom collaboration continued.[35]

In February 1923, two years after the OZE had been officially banned,
its former Petrograd activists Moisei Gran, V. I. Rabinovich, Veniamin
Binshtok, and Iakov Eiger prepared a report for the Petrograd Region
Executive Committee's Department of Nationalities (Jewish Section),
explicitly mentioning the OZE in the title: "Report on the Activities of
the OZE Committee for the Study of the Questions of Jewish Healthcare." Apparently, the group of old OZE leaders did not hesitate to officially identify themselves as the OZE Committee even in a document
addressed to the OZE's archenemy, the Jewish Section (Evsektsiia).
Moreover, we learn from the report that it was this OZE committee that
had initiated the 1922 Evobshchestkom medical-sanitation survey, and
since then had worked with the accumulated data. In addition, they continued studying child victims of pogroms housed in Jewish orphanages in
Petrograd. The report mentioned a paper by Dr. Rabinovich (the head of
the committee) titled "From the Psychic Life of Pogrom Victims," presented at a February meeting of their research group. Other directions of
the committee's research included studying tuberculosis in Jewish orphanages (supervised by Bramson), processing the OZE statistical archives
(under Binshtok), and collecting data on the "medical-statistical" situation of the Jewish population of Soviet Russia (under Gran).[36] Another
OZE document preserved in Lev Shternberg's archive mentions the same
group of four as the authors of the "plan for medical and psychological
and physical study of children-refugees and pogrom victims residing in
Petrograd orphanages" and the questionnaire used in the study (most

likely the one cited above).[37] This research group or "committee" would be active in the Soviet Union into the early 1930s, although the OZE affiliation completely disappeared from the titles of their publications and reports.

Outside of Petrograd, the situation was more complex. Just when the Russian OZE was dissolved in 1921, famine struck the Volga region and Ukraine, thus making the need for relief efforts critical. According to official Soviet statistics, by 1922, 20 million people were starving in the RSFSR, and 3.7 million more in Ukraine. Nearly 11 million of those starving were children.[38] Those OZE activists, who by that time had emigrated to Europe, formed the OZE Bureau in Berlin aiming to substitute for the outlawed Russian Central Committee as a coordinating center and to attract and channel foreign funding. At the OZE conference in Berlin in August 1923, the organization had reconstituted itself into a worldwide network of autonomous OZE societies—OZE Weltverband (World Union). This organizational diversification notwithstanding, the target of their relief and research activities was defined as the old "Russian"—that is, imperial—Jewry.[39] The imperial mental map of "western borderlands" persisted and overcame the new reality of sovereign countries, and the OZE committees from Lithuania, Poland (the Society for Protection of the Health of the Jewish Population [Towarzystwo ochrony zdrowia ludności żydowskiej, TOZ]),[40] and Latvia began coordinating their work with OZE representatives from Soviet Russia, Ukraine, Bessarabia, and Belorussia. Only Caucasian Jews remained beyond purview of the Berlin Bureau as the center of the worldwide network, which was generally consistent with the vision of the rejuvenated Jewish national body as belonging to Western modernity.

Gran became the main liaison between Berlin and Moscow. The archival documentation he generated after the formal liquidation of the OZE suggests that the Moscow OZE Bureau (the former political center) continued its work informally, while the Petrograd group (the academic core) quite openly identified with the OZE. Gran tried to negotiate a formal reinstatement of the OZE as an autonomous organization but was again told that any Jewish public activism was possible only under Evobshchestkom, and that the Russian OZE could not be approved anyway.[41] However, the Soviet authorities were willing to collaborate with the Berlin OZE Bureau. The Moscow and Petrograd activists split over the decision: the majority were ready to compromise in view of the humanitarian catastrophe decimating the Jewish population. A minority rejected the very idea of resuming legal activities under Evobshchestkom.[42] While the discussions continued in Berlin, Gran lobbied with the Commissar (minister) of Public Health, Nikolai Semashko, on behalf of the Berlin OZE, to allow Berlin-sponsored rescue teams into Ukraine. Like many other leading OZE members, Gran found employment with Narkomzdrav, where the discourse of sanitation, medical aid, and eugenics united experts with different political views. Since 1920, he had held two appointments with Narkomzdrav in Moscow: as the chair of the Committee for the Study of Sanitary Consequences of the War and the chair of the Committee for the Relief of the Starving in the Volga Region.[43] Both positions directly related to his activities as a Jewish activist and corresponded to the interests of both the OZE and the Soviet government. Exploiting his official status, Gran facilitated the organization in Ukraine of the "circles of assistance" (*kruzhki sodeistviia*) to medical teams dispatched by the Berlin OZE. These "circles" were based on a model already tried in Khimki near Moscow, where the Berlin OZE Committee supported a sanatorium for Jewish children who were victims of pogroms, and where the Moscow OZE Bureau organized former local OZE members as a "circle of assistance."[44]

In 1922, the JDC started its own relief operation in Ukraine under the auspices of the American Relief Administration (ARA). A Moscow Jew who had emigrated to the United States in the early 1890s, Boris Bogen, came to the region to act as a liaison officer between the JDC, the ARA, and the Soviet authorities. Bogen made visits to Moscow and Petrograd, where he also met with Jewish activists. He found the prerevolutionary leaders of the Moscow Jewish society in a state of depression and desperation, whereas in Petrograd the Jewish intelligentsia were busy working on various cultural and scholarly projects. According to Bogen, they seemed to have little concern for the sufferings of Ukrainian Jews. As he later recalled:

I went to Leningrad [until 1924 the city's name remained Petrograd] and there the leaders among the Jews were more realistic and were willing to compromise and walk with those irritating Communist Jews. The stifling influence of the Evsektsiia . . . which almost choked Moscow Jewry to death was not so oppressive in Leningrad, where independent groups felt free to call and discuss with me the Jewish problems.

Though most of the lights of Jewry in Leningrad were not Communists they were working for the Communist Government, which respected them for their abilities and their characters. . . . So I found Dr. Abraham Bramson, who before the revolution had been the national president of the OZE, running the tuberculosis section of the Petrograd Board of Health. . . .

When I came to Professor Tan Bogoras his mind had little use for the troubles in hand and his eyes were filled with dreadful calamities to come. . . . Now he was the director of the ethnographic museum in Leningrad in which place he brooded upon the future of the human race, being sure that the white section thereof had no future whatever. "The future," he said, "belongs to the yellow peoples." . . . "The Caucasian race has finished its run and the time is at hand for the yellow peoples. . . ." And he had only shrugs of the shoulder for the temporary alleviations in which I was interested.[45]

Bogen left Petrograd under the impression that "none of the Jewish scholars of Leningrad cared greatly about our efforts against hunger in the Ukraine. The hunger of the Jewish bodies did not concern them."[46] From his account it follows that besides Tan-Bogoraz, whose dedication to ethnographic study and cultural uplifting of the indigenous peoples of the North could have provoked Bogen's misunderstanding, he met with Bramson and Shternberg ("And when I spoke of the hunger in Ukraine, they spoke rather of their library with its rich collection of Judaica that was almost destroyed"[47]), Gran, and other Jewish intellectuals who had long been active in relief operations. Bogen's meeting with Gran took place just before the described Petrograd visit, in Kovno (Kaunas) in April 1922 at a train station: Bogen was departing for Moscow and Gran for Berlin. They chatted informally, Bogen described his mission in Ukraine and Gran his responsibilities with the OZE. After the meeting, Gran hastened to write a letter to Bramson: "They [the JDC] did not plan for medical-sanitation activities. I think, we should try to append ourselves to their program of feeding children, develop medical organization, and help Jewish children."[48] How could this and other meetings with members of the Jewish Statistical Bureau working under the EIEO, presided over by Shternberg, convince Bogen that "the hunger of the Jewish bodies did not concern them"?

One possible explanation is that at the time, the Petrograd Jewish activists themselves did not want to collaborate closely with Bogen's JDC. On the basis of the available evidence, we can speculate that at this point they expected some recognition of the Berlin OZE from the Soviet authorities and hence they rejected an arrangement that was reminiscent of the unhappy Evobshchestkom compromise. Bogen's JDC in Soviet Russia acted as an ARA department, whereas the ARA accepted the demands of the Bolshevik authorities, including the request to hide the Jewish "identity" of JDC aid and provide it equally to Jews and their gentile neighbors, many of whom were recent pogromists. When interviewed in Riga in 1924, the former EKOPO-OZE leader in Ukraine and now a Berlin OZE representative, Nahum Gergel, explicitly formulated these arguments. Gran happened to be in Riga at the same time (both were there for the opening of the OZE exposition on the physical and sanitation state of the Jews of the former Russian Empire) and was also interviewed. Their interviews appeared side by side in a newspaper of the Jewish Popular Democratic Party, but Gran, for obvious reasons, avoided criticizing the Soviet authorities and focused exclusively on the need for "physical rejuvenation of the Jewish masses" as "the foundational stone of the modern social rejuvenation of Eastern [European] Jewry."[49] Gergel took the burden of criticism upon himself. Soviet Jewish relief organizations and their partners, he claimed, conducted "party politics" by appointing their own people to distribute the aid intended for the Jews among gentiles, many of whom were implicated in anti-Jewish violence. And this was only one aspect of the "antinational" policy of Jewish communist organizations.

> It is not a secret that the Yidgezkom canteens and feeding stations, purposefully and consciously, served non-kosher food to refugees and the starving; that schools for the children of the starving work systematically to "internationalize" (the true words of *Emes*)[50] their students. These children . . . had to attend [mock] trials of religion, *heder*, Zionism and the Jewish language [in this case, Hebrew]. [It is not secret] that they operated refugee trains only on Saturdays, that only on Saturdays they took orphaned pogrom victims to bathhouses, and so on.

Gergel asserted that neither he nor other Jewish OZE activists were guardians of "kosher and Shabbat." "We are left untouched by the religious pathos of these religious holy symbols," he said. But the "elementary democratic legal consciousness" protests against the violent suppression of the "religious conscience of dozens of thousands of people" by Jewish charity organizations.[51]

Such were the explicit and implicit considerations informing the OZE activists' views on the legalization of their work under the Soviet regime. Famine coming on the heels of pogroms did not afford the option to withdraw, whereas working under the Berlin OZE recognized by Evobshchestkom or under the JDC/ARA similarly controlled by Evobshchestkom was a choice between bad and worse. Finally, the Petrograd, Moscow, and Berlin groups decided not to sign an official agreement with Evobshchestkom and instead to act in Soviet Russia through the mediation of the so-called Nansen's Mission and the JDC while continuing to seek autonomous status for the OZE.[52] On December 11, 1923, the Berlin Bureau wrote to Gran:

> While actively developing our current work under the banner of Nansen and opening facilities under our own conditions, we ask you to continue negotiations both with Narkomzdrav about a [special] agreement and with Evotdel (the Jewish Department of the Narkomat [Ministry] for Nationalities) about registering the Statute of the Russian OZE, so that the Berlin OZE will be able to conclude an agreement with it.

The same letter reported that OZE emissaries had departed for Latvia, Poland, Romania, and Bessarabia—the primary destinations of Jewish refugees fleeing hunger—to collect data: "We need precise examinations [of refugees], materials, and statistical data to be able to understand the scope of the problem [of providing] medical-sanitation services for immigrants."[53] Typical of other instances of urgent OZE relief, this work facilitated eugenic concerns. In Poland, the Section for Social Hygiene and Eugenics (Sekcja higieny spolecznej i eugeniki) was formed within the local OZE organization (since 1922 known as the TOZ).[54] In Latvia, eugenic rhetoric helped publicize the OZE public health initiatives, and Lithuania was no exception either.[55]

A brief report on OZE activities composed in 1924, most likely in Berlin, states that in 1922–1924, the OZE (meaning OZE Weltverband) had established in Soviet Russia 15 ambulatory and outpatient facilities; 23 first-aid stations affiliated with charitable canteens; 11 hospitals, including one eye hospital and one maternity hospital; 7 sanatoriums; 15 kindergartens; and 12 vaccination and disinfection teams.[56] In fact, all of these were made possible by the JDC's presence on Soviet territory, and most of the listed facilities were functioning in Ukraine and Belorussia. The historian of the OZE in East-Central Europe in the interwar period, Rakefet Zalashik, explains: "From 1923 onward, OZE acted through the Soviet-based offices of the JDC-OZE committee, supervised by the Soviet health authority (Narkomzdrav). For this, JDC and OZE signed a collaboration agreement valid for two years until 1925, according to which JDC was to allocate funds for OZE's activities in Russia."[57]

In 1925, the OZE network sent a three-member delegation headed by Gran to the American JDC headquarters to negotiate a new agreement. The delegation asked for a subsidy of no less than $300,000 and up to $700,000 (just over $10 million in 2020), which was "in accordance with the agreement with the Soviet government."[58] But the OZE's shaky legal status in the USSR worked against it, and the opinion prevailed that resources would instead be allocated to the Agro-Joint program that handled the resettlement of Jews on farms in Ukraine and Crimea and whose official status in the country was more certain. The OZE delegation then embarked on an independent fundraising tour in the United States, which turned out to be quite successful.

This partial financial success did not change the general situation with the OZE's legal standing in the USSR, and in the second part of the 1920s its activities centered on scholarship. Historians still have little knowledge about what was happening during these years outside the major urban centers, where hundreds of physicians and teachers with OZE experience operated undercover or semi-legally, and provided medical and social help to Jews who, due to their class origin, economic status, or religious views, were legally disenfranchised and socially discriminated against by the Soviet regime as *lishentsy* (dispossessed). They were denied not only voting rights but also the free medical and social services that were available to others. While assisting them and participating in the Soviet project of indigenization (*korenizatsiia*), former OZE activists continued collecting medical and anthropometric statistics, which they sent to the former OZE scientific center in Leningrad, along with articles and reports on Jewish health and the effect of productivization on Jewish bodies. Elissa Bemporad points to the ambiguity of Jewish politics under the JDC:

> First, by engaging in rescue and relief activities and (at least on paper) conforming to the nonsectarian principle agreed upon with Soviet authorities, JDC actively supported the creation of a new Soviet Jewish way of life, the essence of which was communism. In the specific Soviet context, this inevitably implied the promoting of Bolshevization of Jewish life and the institutions actively cooperating with the Soviets. . . . Second, JDC's official and unofficial activities in the city [of Minsk] were closely and inextricably intertwined. The more JDC engaged in official forms of relief and welfare and funded Soviet cultural institutions, the more it could secretly—unbeknownst to Soviet authorities—sponsor underground endeavors.[59]

Thus, it would not be an exaggeration to say that from 1922, the project of Jewish biopolitical national rejuvenation developed in a gray zone. This was not a real underground but rather gradients of more or less unauthorized activities, which the Soviet obsession with control rendered more or less illegal. Some shades of this gray zone were lighter, especially in the capitals, while others were darker and bordered on political crime. To historians, this also means that the primary sources on this topic are much more obscure, scarce, and inaccessible. In order to assess the actual spread and influence of the biosocial view of Jewishness and the racialized practices of body politics in the 1920s, we now turn to one of the relatively well-documented cases from central Russia.

CHAPTER EIGHT

The Gray Zone of Jewish Biopolitics and Its Place in Soviet Modernity

W E NOW RETURN to where we earlier left Lev Shternberg, in Petrograd in 1924, where he revived the journal *Jewish Antiquity* (*Evreiskaia starina*), the mouthpiece of the Jewish Historical-Ethnographic Society, established in 1909 but interrupted by the beginning of World War I. The very first Soviet issue of the journal featured two articles on race. One was by Shternberg himself, in which he advanced an extravagant theory of the Jewish racial psyche. The other, "On the Anthropology of Russia's Jews" by Boris Vishnevskii, will structure the investigation in this chapter. It presented the results of the anthropological examination of 289 Jewish children in the inner Russian city of Kazan, then the recently established capital of the Tatar Autonomous Soviet Socialist Republic (Tatarskaia avtonomnaia sovetskaia sotsialisticheskaia Respublika, TASSR, formed in 1920). At the time of this examination, in the TASSR starvation was total, that is, 96 percent of its population was starving.[1] The American Relief Administration (ARA) played a decisive role in hunger relief there, and a comprehensive anthropological survey used by Vishnevskii was conducted over the summer months of 1922 by the ARA.

Although the description of the survey implied that anthropometric statistics were collected for Kazan children of different nationalities, the article focused exclusively on Jews. Its author, Vishnevskii, then a professor of anthropology at a number of Kazan institutions of higher education and head of the Kazan University Medical-Anthropological Society, was soon to become one of the leading Soviet race scientists.[2] According to Vishnevskii, in the summer of 1922 there were only 412 Jewish children in Kazan, which never had a significant Jewish community. The survey involved 70 percent of the children, ages six to eighteen. The selection included those born in Kazan as well as refugees from former western borderlands of the Russian Empire who arrived in Kazan during the years of World War I and the Civil War. Both groups were represented more or less equally in the survey: out of 135 Jewish boys, 65 were born in the former Pale of Jewish settlement, and 70 outside it, including Kazan; out of 154 girls, 75 came from the Pale and 79 had never lived there.[3] Vishnevskii analyzed their anthropometric measurements and indexes, comparing children measured in Kazan with Jews of Southern Russia and with German and Austrian Jews, and reaching conclusions pertaining to the general state of the Jewish race. His two main classificatory categories, however, reflected the realities of Soviet Russia in the year 1922—the Jews of Kazan were divided into refugees and locals. Likewise, the political context influenced the shift of the focus from establishing stable features of the "type" toward understanding the mechanisms of racial transformation and the role of the environment and medicalized population politics in this process.

Vishnevskii's main conclusion referred to Franz Boas's famous study of descendants of immigrants (Jewish and Italian) to New York, popularized in Russia by Shternberg. Following Boas and Shternberg and embracing the model of "emigrants to modernity," Vishnevskii specifically addressed change in the head index, which Boas had observed among the first generation of American-born immigrants. Vishnevskii's data for the Jewish children born in Kazan demonstrated the same trend as in Boas's report: the head index of first-generation Kazan-born Jewish children approached the average for native Russian children of Kazan. Children of Tatars—the second major ethnic group in Kazan—were tellingly absent from this comparison, revealing its dependence on the view of the Jewish race as exclusive, unrelated to other races of the empire. The convergence of Jewish (the children of refugees, first-generation Kazanians whose parents came from the Pale) and hegemonic Russian racial types was a significant proof of having overcome alleged Jewish racial backwardness and "degeneration" with the change of environment.

But why did the ARA care about Jewish racial normalization, especially in a non-Jewish region and amid an unthinkable humanitarian disaster? The very collection of racialized anthropometric statistics by the

ARA comes as a surprise to anyone familiar with the modus operandi of this organization. In all European countries of its charitable operations, and especially in Soviet Russia, the ARA explicitly requested the right to provide food to starving children "without regard to race, politics, or religion."[4] Recall that this was the main complaint of the Society for Protection of the Health of the Jewish Population (Obshchestvo okhraneniia zdorov'ia evreiskogo naseleniia, OZE) activists who had to act under the American Jewish Joint Distribution Committee (JDC)/ARA in Ukraine.

The ARA arrived in Russia in September 1921, when the Civil War had already ended and the Bolsheviks were consolidating state authority. Seven years of wars and revolutions—as well as the economic regime of war communism that prohibited free market transactions and relied on forced requisitions of foodstuffs with subsequent centralized redistribution to the population—had led to a dramatic reduction of crop areas and primitivization of agricultural techniques. Exacerbating the situation was the severe drought that struck the entire Volga region, the basins of the Kama and Ural Rivers, the Don region, parts of Kazakhstan and Western Siberia, as well as Southern Ukraine in 1921. The fact that grain requisitions targeted the most accessible regions, along rivers and railways, explains why the most fertile and rich agricultural territories suffered the most from hunger. The famine spread to thirty provinces with a total population of thirty million people.[5] Allowing the ARA, an organization based in a capitalist country that did not recognize the Soviet regime, to help those in need testified to the ultimate failure of the Bolshevik government and the horrific magnitude of the humanitarian catastrophe. At the same time, the Soviets regarded the agreement as a way to foster diplomatic recognition of their regime, while Herbert Hoover, the head of the ARA, welcomed the opportunity to extend the ARA's relief operations to Bolshevik Russia as a long-awaited step toward domesticating and eventually eliminating the threat of Bolshevism. Hoover believed that the communist malady was caused by hunger and that the Russian Revolution itself was essentially a "food riot."[6]

The ARA transmitted to its overseas projects the approaches and social models of American Progressivism, which were absolutely consonant with the patterns of Jewish self-mobilization: without openly taking any political stance (Hoover promised that the ARA's "representatives and assistants in Russia will engage in no political activities"[7]), the ARA encouraged local initiatives and resolutions of concrete problems by means of mobilizing local civil society.[8] In all countries of its operations, the ARA employed only skeletal staffs of Americans to supervise large numbers of independent committees of local citizens, thus turning charity into self-help. It was some time before ARA operatives realized that any public committee, any public mobilization independent of the state and the ruling ideology, any form of progressive politics even at a local level were impossible in Soviet Russia, at least legally. In November 1921, Hoover simply directed that native specialists be hired as paid ARA workers, and their independent initiative was expected to be next to nonexistent.[9] Naturally, physicians made up the most promising pool of potential hires: they were well qualified to select the most undernourished children for the ARA's feeding program. On September 6, 1921, one of the ARA officials reported back to the United States: "There are about two hundred doctors in Kazan and registration of children can be easily carried out, and is the best possible way to answer the charge of favoritism."[10]

Vishnevskii's article gives the most striking evidence that physicians in the ARA service, unlike their employers, did not perceive themselves as just technical personnel. He mentions the local head of the ARA medical division, Dr. Ol'ga Mikhailovna Voidinova, who in the summer of 1922 had finally managed to overcome the "considerable rigidity of practical Americans" and was able to include on the form for recording medical observations, in addition to a few measurements necessary for the standard ARA pelidisi test, a list of anthropometric measurements, such as full height, eyes, hair and eyebrow color, hair structure, eye shape, forehead form, facial features, nasal measurements, ear measurements, skull measurements, and so on.[11]

Not a single ARA document, either published or internal, required physicians to take into account the nationality or race of the starving population. Official medical statistics prepared by Russian physicians and their American supervisors for the *American Relief Administration Bulletin* in 1921–1922 also never mentioned nationality or race. At the same time, the original reports composed by the ARA physicians in Kazan divided all children and all their physical measurements, even those collected for the pelidisi test only, by nationality, thus not only de facto overcoming the "rigidity of practical Americans" but also directly violating the main ARA principle that disregarded race and nationality.[12] Moreover, local physicians made nationality the main structural element of their statistical tables, such as "The list of children of Muslim nationality from Orphanage #3 in the city of Sviiazhsk, who were medically examined according to the method of Professor Pirquet on October 22–24, 1921," "The list of children of Russian nationality from Orphanage #1 in the city of Sviiazhsk, who were medically examined according to the method of Professor Pirquet on October 22–24, 1921," and so on.[13] Not only children from Russian and Tatar villages were immediately identified by nationality, but also children from urban schools, refugee camps, and orphanages, who often came to the TASSR from other regions or whose nationality was not necessarily evident from their names, were just as readily ascribed a nationality in the tables of Kazan's ARA physicians.[14]

This consistent nationalization of the initial medical statistics revealed an important aspect of the physicians' independent agenda: they tended to see scientifically verified and manageable collective bodies as the main actors in the seemingly unstructured postimperial sociopolitical environment. Their insistence on using "nationality" as a category that made medical and anthropometric statistics meaningful and useful for the distribution of aid mirrored the Bolshevik approach that relied on "class" for the same purpose. Moreover, this dichotomy reproduced the conflict within the Jewish Public Committee for Relief of Pogrom Victims (Evreiskii obshchestvennyi komitet pomoshchi postradavshim ot pogromov, Evobshchestkom) between the Jewish activists who wanted to save the "nation" understood as a community of origin and the Jewish activists who wanted to save the potential "proletarian" Jewish nation. In Kazan, when the data on starving children were produced not by physicians, but by, say, heads of factories (in the form of requests to the ARA), nationality was never mentioned.[15] Class, on the other hand, could be indicated directly or assumed (in a clause such as "children of workers of . . ."). The ARA's resolute rejection of the class approach created a space of relative freedom for its medical personnel, who, while sharing in the groupist social vision, made a conscious choice in favor of a racialized national stratification of individuals.

Among those in the network that had formed around Vishnevskii, he was definitely the most qualified academic race scientist prepared to work with the category of race, and he was the only one without medical training. Vishnevskii graduated in 1916 from Moscow University's Department of Physics and Mathematics, where he specialized in physical anthropology and geography.[16] In 1918, he earned a graduate degree in anthropology from his alma mater, under the mentorship of the academician Dmitrii Anuchin, a recognized leader of the network of prerevolutionary Russian liberal race scientists.[17] When in 1919, after a year of teaching anthropology at Kostroma University, Vishnevskii applied for the position of anthropology professor at the newly established Kazan North-Eastern Archaeological and Ethnographic Institute, his affiliation with the Moscow school of liberal anthropology played a decisive role. The best way to introduce a candidate in a letter of recommendation was to say that "Vishnevskii received good training from our venerable anthropologist D. N. Anuchin and belongs to the so-called Moscow anthropological school."[18] During his stay in Kazan (1919–1923), he taught physical anthropology at the North-Eastern Archaeological and Ethnographic Institute, Kazan University, and the Kazan Pedagogical Institute. In addition, Vishnevskii held positions as director of the Statistics Office and head of the Tatar Republic's Demography Section of the Statistical Bureau, of the Scholarly Division of the People's Commissariat for Education (Narkompros), and of the Division for the Study of Productive Forces of the State Planning Committee (Gosplan) of the TASSR.[19] The combination of high-ranking academic and nonacademic jobs allowed Vishnevskii to make a normal living during those difficult years, but his choice of nonacademic positions does not seem accidental. It suggests that Vishnevskii naturally connected his scholarly expertise as an anthropologist with the new scientific population politics promised by the revolutionary regime. Vishnevskii also chaired the university Medical-Anthropological Society, which, among other goals, worked toward influencing official policies toward nationalities of the region.[20] As Vishnevskii wrote in the society's journal in November 1921, "Life itself insistently puts on the agenda the task of preparing such researchers-anthropologists who would be fully armed with modern scientific methods and would start a systematic study of the peoples of Russia."[21]

Vishnevskii used the structural situation created by the ARA's presence to advance his social vision.[22] Most probably it was he, the leading Kazan anthropologist, who designed the anthropological program lobbied by Voidinova, and it was he, who used the chance to show the political and social relevance of the assembled data (in the article for *Jewish Antiquity*). Vishnevskii's bold intention to carry out a new, scientific, population politics should have been shared by other members of the medical network affiliated with the Kazan ARA. Only the solidarity of the local staff can explain how the research initiative that directly contradicted the ARA's principles became possible: Vishnevskii did not have the institutional authority to override the organization's instructions.

But the active medical-anthropological social agenda still does not explain the prominent role of Jewish medical statistics in this story: Why select a relatively small group with no significant political presence in the region for special analysis? Why did only Jewish medical statistics from Kazan find their way to a broader audience, and why did the clandestine progressive agenda of local physicians and anthropologists become acknowledged publicly only in the Jewish case?

The ARA's "Jewish Question"

There were two categories of the workforce the ARA preferred not to employ in Soviet Russia: American women and American Jews.[23] These were regarded as the most likely victims of gender or anti-Semitic violence.[24] In October 1921, the ARA signed an agreement with the JDC, which started providing limited humanitarian aid under the aegis of the ARA. Jews constituted only 2 percent of the population of the Volga region, so the initial contribution of the JDC was rather modest—$675,000 (the equivalent of about $9 million in 2016).[25] However, as we know, in September 1922, the JDC and the ARA initiated a semiautonomous JDC unit in Ukraine that did hire Jewish officers.[26] This was not, however, the case in Kazan, where the JDC was not present and the ARA could run the relief operation according to its own principles.

The 182 Kazan ARA employees that I was able to identify present a diverse group. They described themselves as Russians, Great Russians, or Russian Orthodox; Ukrainians and Little Russians; Baltic Germans or "German from the Lithuanian Republic"; Tatars and Poles; Serbs, Estonians, and Chuvash. The group even included an Englishwoman and a Frenchwoman. Out of these 182 paid employees, 10 were Jewish (5.5 percent), and all of them, regardless of their background and position held with the ARA, had medical training: Veniamin Mikhailovich Orlik was a Jew from Siberia, a medical student, who worked as an ARA district inspector; Berta Il'inichna Grin-Gnatovskaia from Kovno province was a dentist; Moisei Abramovich Iglitsin and Abram Isaakovich Shvartsman from Warsaw both graduated from the Medical Department of Kazan University; Evgenii Markovich Konstantinovskii was still a senior student in the same department; Berko Tsalelevich Perel'man was trained as a pharmacologist; Sofia Iakovlevna Plenzitser from Mogilev was a doctor's assistant; Talid Saulovna Brondberg, who came to Kazan from the same region, was a nurse; and so on.[27] Jewish employees represented a tiny minority in the local organization, but their presence was more visible and significant than is suggested by the sheer numbers. Jewish names could be spotted in any of the ten ARA departments, but in one department—the Department of Medical Statistics—all the leading positions were occupied exclusively by Jewish employees, with the important exception of the "manager of the department," Ol'ga Mikhailovna Voidinoff (this spelling was used in the English-language internal ARA documentation). Non-Jews held only junior positions in this department, which included a typist and two clerks doing calculations (*shchetchiki*).[28] It could well be that the candidacy for the managerial position occupied by Voidinova had to be approved by or at least be acceptable to the local Soviet authorities, while the appointment of other employees in the Department of Medical Statistics was not as strictly regulated.

The archival documents clearly show that the initiative to "nationalize" the ARA's medical statistics in Kazan (against its official policy of ignoring all the social and national differences of the relief recipients) originated in the Department of Medical Statistics dominated by Jewish physicians. It is therefore plausible to assume that these physicians shared a specific experience that had shaped their understanding of nationality as expressed in "objective" anthropometric features and indexes. A common methodological stance that allowed them to ascribe a certain sociological meaning to individual medical statistics and a willingness to pursue additional research in this direction present the department's physicians as members of a distinctive network of Jewish professionals-activists. How can this network be located on the social map of early Soviet Kazan?

Kazan was one of the main battlefields of the Civil War in 1918–1919. According to the 1920 census, after the revolution, the population of Kazan had declined by almost one-third (to 60,000), and even with the influx of World War I refugees, it constituted only 146,495. Population growth resumed in 1922 with the introduction of the New Economic Policy and became apparent by 1923 (when the number of Kazanians reached 158,085).[29] Against these numbers, 5,081 Jews in 1920 (3.5 percent of the total) decreasing to 4,038 Jews in 1923 (2.5 percent) seem to constitute a small community, and, indeed, most local Jews and refugees were interconnected.[30] As everywhere in Russia, Kazan Jews were divided politically, but just as elsewhere, the refugee crisis in the wake of World War I provided a strong impulse for communal consolidation.[31] Physicians, who were not only overrepresented among the Russian Jewish population but also highly professionally connected and socially integrated, played the leading role in Jewish relief organizations.[32]

One Kazan physician who championed extensive scientific analysis of the local medical hunger statistics, a professor at the V. I. Lenin Kazan Clinical Institute (established in April 1920), Roman (Ruvim) Albertovich Luria, during World War I cochaired the Kazan division of the Jewish Committee to Aid Victims of War (Evreiskii komitet dlia pomoshchi zhertvam voiny, EKOPO) and the OZE.[33] Similar to Moisei Gran and Abram Bramson, Luria graduated from Kazan University in 1897 and defended his dissertation there in 1902. For a short time, he worked as a zemstvo physician and later taught at the Kazan school for doctors' assistants. One of the two Jewish statisticians from the ARA's Medical-Statistical Division, Liubov' Abramovna Person, daughter of the last prerevolutionary chair of the Economic Board of the Kazan Jewish community, Abram Person, studied at the school for doctors' assistants where Luria was teaching.[34]

A member of the ARA's administrative committee, Dr. Efim Moiseevich Lepskii, worked at the Clinical Institute under Luria and also had previous experience in Jewish relief work.[35] During World War I, Lepskii headed the Kazan EKOPO Medical-Sanitation section, and in early March 1916 he contacted the OZE Petrograd Committee about the intention of local Jewish physicians to establish their own chapter of the OZE. With his wife Revvekka Israilevna Lepskaia, also a physician, and eight other physicians Lepskii established the first OZE committee, headquartered in his family apartment.[36] According to the November 1921 issue of the *American Relief Administration Bulletin,* Lepskii was among the first local physicians the Americans met. Will Shafroth, who came with the ARA scouting party to the Volga region, recalled: "We first visited the Second Children's Hospital. The director, Dr. Lepskii, speaks some English. He seems to be the type who would be valuable as a member of our local committee."[37] It might well be that Lepskii recommended the rest of the Jewish physicians for positions in the ARA's Medical-Statistics Division. Archival documents show that Lepskii—director of the children's hospital, head of the Narkomzdrav's Department for the Protection of Infancy and Motherhood, and a typical socially engaged Jewish physician—turned out to be an active and useful adviser to the ARA. Among other important initiatives, he volunteered to develop "clinical principles"—that is, a scientific and systematic approach to the selection of children for the ARA's food program in the villages, where no physicians were able to perform the pelidisi test or collect accurate anthropometric statistics.[38] There is no evidence that he ever questioned the practice of nationalization of the medical statistics or refused to rely on it himself.

Voidinova's deputy in the department was Moisei Abramovich Iglitsin, born in 1887 in Chausy, Mogilev province, in the Pale of Jewish settlement. He graduated from the Medical Department of Kazan University and was drafted into the Red Army as a military physician.[39] His personal contacts with other future ARA Jewish physicians, if any, are unknown. His role in nationalizing the ARA's medical statistics is much better documented. Many of the archival original handwritten tables of medical statistics for specific nationalities are either composed by Iglitsyn or approved by his signature.[40] From these documents, it appears that Iglitsyn shared the idea of "nation" as an objectively existing biological entity, which could degenerate under unfavorable circumstances, but could also be cured under healthy conditions as identified by trained experts. He shared Lepskii's concern about evaluation methods in the villages. The ARA, Iglitsin argued, could not rely on superficial observations by state officials who visited villages and selected the most undernourished children based on their subjective impressions. He called this method "inadequate." Instead, he proposed locating educated people in each village and organizing them into a network. "Even if these are not physicians (one person per district)," he wrote, it was possible to teach them how to collect medical statistics for the pelidisi test specifically and broader statistics, if needed.[41]

Iglitsyn's colleague in the department, Abram Isaakovich Shvartsman, born in 1886, came from Russian Poland. He began his medical education in Warsaw but graduated from Kazan University. He, too, was mobilized to serve in the Red Army, and then worked in Kazan at the Tatar Ministry of Health (Tatnarkomzdrav).[42] We find Dr. Shvartsman's name in the list of Kazan agents of Evobshchestkom.[43] The name of one of the two Jewish clerks who assisted Iglitsyn and Shvartsman in the department, Fanni L'vovna Elovtsan (the other was Evgeniia Medved'), is mentioned in a report on the activities of the Kazan OZE childcare center (*ochag*).[44]

Thus, all local Jewish physicians in the service of the ARA appeared to be connected through the Jewish relief network, and most of them—due to their professional training—were affiliated with the OZE movement. This experience and the type of social thinking that they had embraced with it prepared the Jewish ARA physicians for collaboration with academic race scientists such as Boris Vishnevskii and motivated them to pursue an independent medical program that had social and political implications absolutely alien to those of their ARA employers.

The Gray Zone Becomes Visible

Moisei Gran was the key link between the intellectual and organizational centers of the larger OZE world and Kazan's Jewish physicians. A graduate of Kazan University, he cultivated connections with local physicians. It was hardly accidental that after Stalin's "Great Break" Gran retired from his administrative positions in Kazan (with a brief interlude in Moscow), where he received a university chair in social hygiene.[45] During the Civil War years, Gran often represented Kazan colleagues in the OZE committee. For example, in the summer of 1918, when the Kazan OZE experienced difficulties with securing a location for a sanatorium for tubercular Jewish war veterans, the joint meeting of the Moscow and Petrograd OZE committees decided to send Gran to "examine the opportunities existing for this [facility] in Kazan or its vicinity."[46] In 1921–1922, the years of the ARA's activities in this troubled area, Gran frequently visited Kazan as the chair of the Narkomzdrav's Committee for the Relief of the Starving in the Volga Region. He personally knew Lepskii and Luria, both of whom participated in the OZE and worked for the Tatarnarkomzdrav (and as consultants for the ARA). After the OZE was banned in 1921, Gran acted as the Berlin OZE agent in Soviet Russia, or, as he was introduced in OZE documents, the "authorized representative of the Central Committee for organizational work in the western provinces" (another telling instance of using the prerevolutionary imperial nomenclature in the postimperial context).[47] He often visited Ukraine as a Narkomzdrav officer and a high-ranking OZE activist, and informed Jewish physicians working in different parts of Soviet Russia, including in Kazan, about what was happening in the big gray zone in which they all operated.

Through mediators like Gran it became apparent that medical workers fighting famine in different regions of Soviet Russia were facing similar organizational and methodological problems. Thus, in 1922, physicians in the Volga region began criticizing the Pirquet (pelidisi) system as being misleading under local conditions.[48] As it turned out, the arguments that they advanced repeated or paralleled the criticism that the OZE- and Evobshchestkom-affiliated Jewish physicians discussed in their own circles as pertaining exclusively to Jewish victims of hunger and pogroms.[49] Thus, Dr. Kh. L. Vilenkina's medical examinations of children in Petrograd who had been rescued from the pogrom-stricken areas of Ukraine, discovered in early 1922 that only 15 percent of these Jewish children qualified to receive American food relief on the grounds of their pelidisi scores. "Our needs, as follows from the results of the survey, are much greater," Vilenkina concluded and demanded that additional food aid for children be requested from Evobshchestkom. The existing norms of food rationing, she believed, were "threatening for our weak, malnourished children."[50] The problem with the pelidisi, as the Jewish and ARA physicians agreed, was that it had been developed to ameliorate the consequences of hunger (in Central and Eastern Europe, where the test had been successfully applied) and it had proved inadequate to deal with total starvation accompanied by total impoverishment and war violence. This was the case in Soviet Russia, and it produced physical and psychological trauma (widespread cannibalism in the Volga region) that required measures far exceeding typical ARA aid.[51]

Through people like Gran, Vishnevskii, Lepskii, and other OZE activists holding positions in Soviet health and academic institutions, the invisible "gray zone" actively participated in the exchange of knowledge regarding the human cost of starvation, economic devastation, and violence, and thus influenced debates in the mainstream medical community. But the cross-fertilization also had its limits. The ultimate goal of the Soviet authorities was, first, to provide support to workers and orphan children in the cities and, second, to save peasants, when possible. They were regarded as the base of the regime and the material from which a future proletarian nation would be engineered. The ARA's ultimate goal was famine relief defined indiscriminately. Whereas Soviet supervisors tried to influence aid distribution in favor of the "proletariat," the ARA's supervising officers modified the interpretation of certain indicators to better serve their practical objective in Russia. Against this background, the tendency of Jewish physicians in the Kazan ARA to nationalize their statistical data and use these simultaneously in practical relief work and scientific research on the Jewish racial type, in accordance with the OZE philosophy of national medicine, exposed a different view of population politics. Jewish physicians claimed that their modifications to the pelidisi test, as well as all their medical, eugenic, and hygienic scientific programs, were specifically developed for starving, weak, and traumatized Jewish bodies. This philosophy equally informed the work of Kazan's Jewish physicians and the members of the Society for the Study of Social Biology and Psychophysics of the Jews, which Gran and the rest of the old OZE leadership established in Petrograd at the same time (1922).

In 1923, Vishnevskii left Kazan and moved to Leningrad, where he assumed the position of director of the Anthropological Division of the Anthropology Museum. This is where he met one of the main protagonists of this book, Lev Shternberg, who had worked as the museum's leading ethnographer since 1901. Shternberg might have known about the Kazan survey via the Jewish network, in particular through Gran, or he might have learned about it directly from Vishnevskii. In any case, Vishnevskii's article "On the Anthropology of Russia's Jews" appeared in *Jewish Antiquity* in 1924 alongside Shternberg's own "Questions of Jewish National Psychology."[52] Both Vishnevskii and Shternberg shared the views expressed by Gran and his colleagues—physicians and statisticians—in their own venues. Race, according to Gran, was a "stable type of physical-biological nature," which nevertheless could change under the influence of "specific socioeconomic and biological conditions."[53]

Jewish Biopolitics and Soviet Modernity

As the Kazan story and other evidence suggest, the vision of Jews as a race that could be transformed into a modern nation through Jewish medicine and eugenics, under a democratized political regime and in a supranational polity, was deeply interiorized by hundreds of Jews, physicians or not. These people shared the common experience of participating in prerevolutionary Jewish politics and general cultural and scientific debates about self-liberation, self-improvement and self-organization, nation, race and class, autonomism, and Zionism. This generation lived through a moment when the most daring projects, such as the modernized all-imperial system of Jewish communal self-government or self-funded Jewish national medicine, both based on the most objective and extensive statistical knowledge, seemed real possibilities.

The experience of World War I and the Civil War had integrated this generation even deeper into a medicalized statistical culture of Jewish self-organization and self-help. Both its pre- and post-1917 versions treated relief and eugenics as inseparable, but after the 1917 revolution, the Jewish future—any future—appeared very literally predicated on elementary salvation and urgent statistical, medical, and eugenic measures in the present. Masses of simple "nondiscursive" Jews, who had lived through wartime calamities and pogroms, interacted with the biopolitical culture shared by the elites. Jewishness for them was becoming more and more carnal, defined by common physical experiences of violence, dislocation, hunger, generational break, and so on. The Jewish Bolshevik productivization discourse did not alter this carnality, but merely framed it differently. As was the case before 1917, cultural and ideological wars did not present an alternative to a racialized Jewishness but rather completed it. By rejecting a religious identification of Jewishness and in lacking a territorial substantiation of nationhood, Soviet Jewish political and cultural revolutionaries still needed to define the foundations of Jewishness as something objectively existing, albeit imperfect. The old, imperfect, Jewish nation had to be molded into a proletarian Jewish nation by its own proletarian elites.

That being said, today we know the story of the cultural molding of modern Jewishness under the Bolsheviks much better than the story of its biopolitical molding. The discursive eloquence of the cultural revolution and its presence in Soviet archives is the most obvious explanation. By contrast, it requires years of archival work to identify, reconstruct, and comprehend stories like the one that took place in Kazan. The gray zone then has become a gray zone now—a gray zone in our knowledge about Jewish biopolitics. Its participants, archives, and memory are gone, and the Kazan case study is very instructive in this sense. Most of the local personnel who worked for the ARA were arrested and disappeared in the Gulag as collaborators with foreigners, and the Soviet ARA archives were never fully accessible. Vishnevskii was arrested in Leningrad in 1937, and by then, his early work for the ARA and Jewish organizations was not even the main compromising episode. His personal file in the archive of the Russian Academy of Sciences includes a denunciation written by one of his younger anthropologist colleagues, who accused him of sabotaging the development of Soviet anthropology and presenting a distorted picture of it to foreign scholars.[54]

Many of the original OZE intellectual leaders happened to die in the 1920s and did not fall victim to Stalin's terror. Others such as Gran (who died in 1940) and Bramson (who died in 1939) were cautious enough in the 1930s to conceal their earlier activities. Solomon Vermel's autobiography, which opens Part II of the book, is a typical example of such self-censorship. Gran's and Bramson's official Soviet biographies add to the picture by showing how self-censorship was reinforced by state censorship.[55] Any mention of Jewish activism was purged from their biographies and other similar biographies of successful Soviet founders of sanitation policies, tuberculosis treatment and prevention, statistical institutions, and so on. Having been streamlined in this way, today their biographies seem to confirm the influential historiographic explanation of early Soviet modernity as the realization of the dreams of late imperial experts, unsatisfied with their role in the undermodernized imperial state and society. Presumably, they waited for the change of regime to become true "Foucauldian" experts working with the modern state or rather as a modern state. The Bolshevik Revolution enabled a long-awaited union of experts-professionals with the interventionist Soviet state, supportive of radical social engineering. Hence, they identified with this state and the Soviet version of modernity.[56]

Indeed, with the cleansing of Jewish activism from the biographies of the protagonists of this book—work that had already been accomplished for contemporary historians by the Soviet authorities in the 1930s—they became exemplary Soviet experts. But when the intellectual, political, and organizational contexts of the Jewish biopolitical project are reconstructed and recognized, its activists appear to be quite independent agents, advancing the vision of Jewish modernity and a social imagination that did not coincide with the Soviet project. Their nationalism had roots different from those of Soviet territorial nation-building, and their biopolitics stimulated the development of Soviet medicine, statistics, and eugenics rather than vice versa. The protagonists of this book were Jewish experts by choice and Soviet experts by chance and necessity, which does not make them anti-Soviet. Rather, we should reconsider their Soviet careers as having been founded on the knowledge, networks, and experience received in the world of Jewish biopolitics in an empire. They did not identify with the Soviet agenda and Soviet model of postimperial nation-building. Instead, their dual Soviet and Jewish professional and political belonging allowed them to pursue ideas and approaches that they considered objective, scientific, and useful to the Jews.

The OZE archives remained inaccessible throughout the entire Soviet period, when the dominant paradigm in European, Israeli, and American Jewish studies emphasized the suffering and suppression of Jewishness under the Soviets. This has changed only with a renewed interest in expressions of Soviet Jewishness and the opening of former Soviet archives. However, in the case of the OZE, a specific set of questions had to be formulated to make these archives speak about Jewish biopolitics as an original vision of postimperial modernity (in addition to traditional themes of Jewish national solidarity and institutions of communal charity). For obvious reasons, these questions, connecting Jews, especially Eastern European Jews, and race, were difficult to conceive in the wake of the Holocaust. As a result, the whole generation of Russian Jewish modernists and progressives, who articulated their vision of Jewish national particularity and postimperial integration in the language of mass Jewish medical statistics and eugenics, disappeared both physically and symbolically from the pages of Jewish and Russian/Soviet history.

Their unorthodox experience is not easy to narrate as part of the story of Soviet and even Soviet Jewish modernity. The main argument of this book has been that the racialized view of Jewishness that the Jewish partisans of self-racializing developed served as a middle ground for different political movements, intellectuals, and regular members of the Jewish mass society, who otherwise disagreed on many ideological issues. The racialization of Jewishness resolved for them the basic problem with objective foundations of nationality. Racialized Jewishness compensated for Jewish nonterritoriality and internal cultural heterogeneity, and allowed the Jewish physical body to be imagined as an authentic national space that could be studied, improved, and protected by national experts and activists. They could shelter the nation's body from the effects of colonial domination and external ideological manipulations. What happened with this middle ground when the old and new Jewish elites met in Soviet organizations such as Evobshchestkom, in which they had to embrace and share the responsibility for protecting the Jewish physical body under the new postcolonial condition?

As we have seen, the relationships between the old and the new elites were openly antagonistic. Both worked to protect Jews from anti-Semitic violence, but the new elite aspired to recover the Jewish bodies to mold them into a proletarian Soviet nation, using a general class blueprint. The productivization of physical labor was offered as a magical mechanism to reforge old Jews into new Soviet proletarian Jews. A conservative racial psyche as posited by Shternberg, or Jewish traditional ethnographic culture as conceptualized by Samuel Weissenberg or S. An-sky did not reflect the national identity of the proletarian Jews. Theirs was class consciousness expressed in national forms, as was the case in all other Soviet nations at the transitional evolutionary stage leading toward universal postnational communism. Jewish communists initiated the linguistic Yiddish revolution and the cultural antireligious revolution. They introduced Yiddish schools, theaters, journals, and workers' clubs. An important component of this cultural revolution was a radical assault on the old elites, especially those still popular with the masses.

A speaker at the First Jewish Communist Conference in Moscow in late October 1918 (the young former Bundist turned Bolshevik, Julius Shimeliovich) acknowledged that "Jewish workers, who fought so heroically against Tsarism and took an important place in the history of the Russian revolutionary movement until the February Revolution" met the October Revolution with complete passivity. It was necessary to draw Jewish workers "away from their false leaders and from the intelligentsia, who did not want the working class to become the master (*balebos*) of Russia."[57] Activists like this speaker belonged to the generation of Bundists and communists who had formed politically during the revolutions and the Civil War and did not share the imperial vision and the progressive expert discourse. They preferred economic, political, or cultural idioms of Jewishness to the language of biological race. However, Andrew Sloin's innovative analysis of the discussions at worker, party, and union meetings and in the Bolshevik (including the Jewish Bolshevik) press in the 1920s and 1930s, holds this generation responsible for "the slide from nationality to race" that reflected the failures of productivization magic.[58] Sloin argues that the "nonproductive" economic niches, such as *kustar'* (small craftsman) or *spekuliant* (speculator), were associated with innate qualities of Jews, and the party opposition (Trotskyism) was framed as a Jewish deviation. The clustering of Jews into a single category associated with parasitic economic activities, physical unfitness for industrial tasks, and deplorable moral vices produced a Soviet version of racialized Jewishness. The only politically satisfying responses to this iteration of the "Jewish race" were territorialization of the Jewish nation on Soviet terms, or assimilation—the complete obsolescence of all that was specifically Jewish in Soviet Jews.

As Sloin points out, assimilation was the answer embraced by the first properly Soviet generation of Jewish communists, the so-called generation of the Lenin Levy, admitted to the Bolshevik Party en masse following the death of Lenin on January 21, 1924. They came of age during the most turbulent times of wars and revolution and did not experience Jewishness as a stable social and cultural community. They also had no prior experience of any sort of Jewish national (including Bundist) politics. Hence, they more readily identified with proletarianization, internationalization, assimilation, and Soviet social mobility.[59] They rejected the grassroots culture that attributed to Jews a developed sense of Self that rested on old and immutable foundations such as racial psyche, specific physical constitution, characteristic pathology, and shared emotional and intellectual experience—the culture that proliferated in the gray zone in which Soviet Jewish communist biopolitics met OZE-type Jewish biopolitics. Instead, as Sloin shows, they accepted a new, Stalinist, negative codification of Jewishness in racial terms.

Conclusion

At the Finish Line

Люди роботы
 выглядят ровно:
взгляни
 на еврея
  землей полированного.
Здесь
 делом растут
  коммуны слова:
узнай
 хоть раз из семи—
который
 из этих двух
  из славян,
 который из них
  семит.
Не нам
 со зверьими сплетнями знаться.
И сердце
 и тощий бумажник свой
откроем
 во имя
 жизни без наций—
  грядущей жизни
   без нищих
   и войн!

—В. Маяковский, *Еврей*
 (*Товарищам из ОЗЕТа*), 1926

Human robots
 appear similar:
look
 at a Jew who is
  polished by land.
Here
 the words of the commune
  grow as deeds:
guess
 just once out of seven times—
which one
 of these two
  is from Slavs,
 which one of them
  is a Semite.
We refuse to accept
 wild gossip.
Let us open
 our hearts
 and our skinny wallets
in the name
 of a life without nations—
future life
 without paupers
 and wars!

—Vladimir Mayakovsky,
 Jew (To Comrades from OZET), 1926[1]

T HE RUSSIAN JEWISH "race for the future"—in the fields of science and biopolitics—was initiated in the postreform period by the enthusiasts of Jewish self-racializing, who relied on modern knowledge and self-help to resist imperial domination and what they perceived as its detrimental impact on imagined Jewish postimperial modernity. Since the beginning of the "race" in the 1860s–1880s, the postimperial future had remained a moving target, until the Soviet Union was constituted in December 1922 and the open future had effectively stabilized and become the accomplished postimperial present. It thus required of Jews a more total remaking than the advocates of Jewish self-racializing could ever have imagined—simultaneous and rapid revolutionary transformation into members of the proletarian class and a Soviet Jewish nation.

Bolsheviks, who once downplayed the connection between two principles—class and nationality—discovered the real problem of imperial diversity and learned to appreciate the mobilizing power of nationalism during the Civil War. In early 1919, their support for Bashkir national and territorial claims had decisively contributed to the Red Army's victory over the White forces of Admiral Alexander Kolchak, who refused to recognize the national autonomy of the Turkic-speaking Bashkirs of the southern Urals.[2] One year later, the national Tatar Autonomous Soviet Socialist Republic was created in the Middle Volga region, reflecting the strategic departure of the Bolsheviks from their earlier understanding of imperialism in purely class terms. In Central Asia, the local population's resistance to revolutionary—perceived as a new colonial—advancement from the former imperial center was soon subdued due to strategic alliances between the Bolsheviks and national-minded modern Muslim elites (Jadids). As Adeeb Khalid has shown, while serving as imperial intermediaries for the Soviet regime, Jadids used the postimperial situation to advance their own visions of the modernization and territorial nationalization of Central Asia.[3] Jadids turned the Bolsheviks—who came to the former imperial colonial possessions of Turkestan and the Bukhara and Khiva khanates with a program of proletarian revolution—into the enablers and facilitators of a nationalization that was never previously on the Bolshevik agenda. By 1924, the dynamic, multilayered social, linguistic, economic, and religious alliances in the region became nationalized and territorialized to the effect of forming initially two and eventually five national republics.[4] This version of "national communism" helped incorporate influential local elites into the Soviet project and became universally institutionalized in the 1920s as the policy of indigenization (*korenizatsiia*).

Bolshevik postimperial modernity thus appeared to be predicated on the affirmative action emancipation of the former victims of empire, who could implement their national visions within the institutional and ideological framework negotiated with the revolutionary regime. The new arrangement embracing the model of territorial nations helped not only to retain the former imperial space together. It also reflected a distinctive social evolutionism developed by the Bolsheviks, who had to deal with the challenges of the imperial situation inherited from the old regime.

The new regime accepted nation-building as a necessary step in preparing the diverse and unevenly developed population for full-scale participation in a socialist society and economy.[5] Although at the early stages of indigenization the Austrian cultural autonomy model still retained some influence in the Soviet Union, the centralization of production and distribution of economic resources as well as the one-party political system proved best accommodated by territorial nation-building, with local state institutions replicating those in the center, and a unified party structure guaranteeing ideological uniformity. The peoples who lacked any substantial proletarian layer (for example, the nomad Kazakhs), or cultural preconditions to immediately assume a nationality (for example, peoples of the Far North), or a territory they could claim for their titular republic (for example, Jews, also liable to the two other deficiencies) found themselves in a disadvantaged position with regard to negotiating their interests with the regime.[6] The protagonists of this book thus faced a twofold challenge: temporal and political. They had to learn to live in the present, accepting that the future had arrived and there was nothing else to wait for in terms of the postimperial transition. And they needed to find a strong position from which to negotiate the Jewish place in the achieved postimperial future.

In 1926, in Leningrad, the first collected volume was published in the series *Questions of Jewish Biology and Pathology* (*Voprosy biologii i patologii evreev, QJBP*). Its editors came from the ranks of leading scholars and organizers affiliated with the Society for Protection of the Health of the Jewish Population (Obshchestvo okhraneniia zdorov'ia evreiskogo naseleniia, OZE). They represented the tip of the iceberg, or, rather, the core of the network of Jewish scientists, physicians, and activists who played major roles in the late imperial and early Soviet culture of Jewish self-racializing. Veniamin Binshtok, Abram Bramson, Moisei Gran, and

Grigorii Dembo introduced their network as "a group of old public physicians [*obshchestvennykh vrachei*], who have been working for many years on problems of Jews' physical recovery." "Between 1912 and now," they wrote, "the idea of Jewish physical rejuvenation has assumed deeper and wider proportions."[7] The choice of the OZE founding year of 1912 as the key landmark challenged the Soviet canon in which the revolutionary year of 1917 signified the commencement of the new epoch. The stress on continuity also contrasted with the rhetoric of new class alignments, new Soviet elite, and new Soviet science. The series' only institutional affiliation was with the prerevolutionary center of modern Jewish scholarship—the Jewish Historical-Ethnographic Society (Evreiskoe istoriko-etnograficheskoe obshchestvo, EIEO). These choices together suggested that the *QJBP* collective was claiming a distinct genealogy and an autonomous vision of Jewishness—the genealogy and vision shorthanded as the Russian Jewish "race for the future."

As we have seen, the late nineteenth-century intellectual reconceptualization of the Jewish imperial condition in terms of modern sciences inspired in the early twentieth century not only a Jewish race science but also a social mobilization around notions of endangered Jewish national "authenticity" and problematic Jewish physicality. Uncovering the Jewish authentic racialized Self and protecting Jewish individual bodies and the collective national body from violence, degeneration, imperial hybridization, and assimilation meant preparing imperial Jews—a nonterritorial community—for the anticipated postimperial national-territorial modernity. In this postimperial modernity, Jews aspired to achieve both biomedical and political citizenship. As Warwick Anderson, Aro Velmet, and feminist scholars have argued, the denial of "biomedical citizenship" on grounds of racial or gender otherness effectively deferred political citizenship—a connection the Russian Jewish activists of self-racializing had been making all along.[8] While being disenfranchised politically and having interiorized a normative view of modernity as scientific, industrial, and nation-oriented, they deployed self-racializing to acquire discursive and biopolitical control over the imagined collective Jewish body. Scientific exploration of this body physical as body social, its medical protection, and its eugenic improvement (often, together with relief and rescue) became the main instruments of the Jewish "race for the future," and biopolitics became the preferred type of scientific politics.

The race argument posited common biological descent as an objective basis for nationality, thus overriding territorial and cultural decentralization, whereas the biopolitics of race substituted for the lack of political representation and served as a strategy of scientific nation-building and self-modernization. Jewish self-racializing grounded itself in Darwinian/Lamarckian epistemologies, and operated through comparative physical anthropology, medical statistics, and modern psychiatry and eugenics. In other words, it appealed to the authority of Western science. At the same time, the body language and body politics promoted within the culture of scientific self-racializing resonated with Jewish modernists' turn to the physical body as a language of the reflective self, and with "techniques of the body" in Judaism, connecting the reproduction of an identity to specific body-centered rituals.[9] The imperial policies of military conscription and official identification also helped to retain the centrality of the Jewish body in Jewish everyday life.[10] Finally and sadly, pogrom violence and wartime necropolitics performed a similar function. The success of the "race for the future" with the masses owed partially to this carnality of Jewish experiences, as well as to the popularity of slogans of self-emancipation and self-protection advanced by the ideologues of the culture of self-racializing. For this type of racialized science and politics, the exact parameters of a postimperial political arrangement held secondary significance, especially since the range of future alternatives for the Jews in Eurasia and Eastern Europe was not particularly broad. Pretty much any arrangement was deemed acceptable that would allow former imperial Jews to decolonize and join the emerging postimperial modernity on their own, scientifically defined, national terms.

The first *QJBP* collection (1926) offered a window on this prerevolutionary culture of self-racializing grounded in the notion that Jews "more than any other people have preserved the purity of their race."[11] At the same time, the collections clearly embraced the existing Soviet political framework—specifically, post–Civil War scientific population politics—for negotiating possible formats of Jewish nonterritorial national modernity. The first *QJBP* collection singled out two manifesto-like works by the two leading Soviet eugenicists, Nikolai Kol'tsov (*The Improvement of the Human Breed*, 1921) and Tikhon Iudin (*Eugenics: Teaching about the Perfection of Natural Characteristics of Humans*, 1925), as exemplifying the essence of Soviet transformation.[12] Many of the *QJBP* editors participated in the Russian Eugenic Society (established in 1920), specifically on its Committee for the Study of Jewish Race, of which the founder of the Moscow Institute for Experimental Biology and the Russian Eugenic Society, Kol'tsov, was also a member.[13]

When the society's Jewish Committee was organized in 1923, Leon Trotsky, then the second in command in Bolshevik leadership, was calling for the creation, by means of "the most complicated methods of artificial selection and psychophysical training," of a "higher sociobiological type, an *Übermensch,* if you will."[14]

A disturbing combination of the merciless "negative" class politics, which Trotsky restlessly promoted, with "positive" futuristic scientific biopolitics, received endorsement from many Bolshevik intellectuals. Kol'-tsov, who was briefly arrested in the early 1920s for alleged participation in an anti-Bolshevik conspiracy, survived only due to personal intervention by powerful patrons of eugenics such as Vladimir Lenin and Maxim Gorky. To his interrogators, Kol'tsov communicated his vision of eugenics as the engine of *real* change: "Contemporary eugenics—the science of ennoblement of humanity—teaches us that the fate of humanity depends not on changing the external conditions of people's lives but on changing the hereditary abilities of human nature."[15]

Thus, in the mid-1920s the enthusiasts of Jewish race science and biopolitics had reason to enter negotiations about the formats of Soviet Jewish modernity with some confidence. Gran, the former key OZE ideologist and organizer, championed a eugenic agenda in *QJBP.* While accusing negative eugenics, as practiced in England, the United States, and Italy, of validating the "biological leadership" of aristocratic classes, Gran endorsed Soviet eugenics as the most productive framework for exploring "the questions of race and nationality."[16] Like Gran, the *QJBP* contributors adapted the philosophy of general "human enhancement" to their own program of normalizing and cultivating the Jewish race for postimperial modernity. A lead eugenic article by Binshtok in the first collection covered the problem of Jewish giftedness (*odarennost'*). While reflecting a popular trend in Soviet eugenics of the time, it nonetheless heavily relied on Lev Shternberg's "in-house" concept of race as residing deep in the national psyche, protected from immediate political as well as geographic, social, and cultural influences. According to Binshtok (and Shternberg), this hidden atemporal race coexisted with physical race and social heritage, which were subjected to changes in varying degrees. The deep race, as per Shternberg and Binshtok, revealed itself in stable features of Jewishness such as "intellectualism and rationalism," "social emotionality and activity," and optimism, which had to be preserved, cultivated, and enhanced as the basis for national Jewish modernity.[17] Soviet Jewish eugenics was expected to scientifically guide this complex work of balancing the preservation of deep racial Jewishness with changes resulting from and required for Soviet transformation.

To this end, the *QJBP* contributors combined old and new methods. They continued developing the old-type population statistics that included demographic (registration of the annual numbers of deaths, births, marriages, morbidity, and migrations), economic, and more specialized anthropometric data.[18] The prerevolutionary and Civil War medicalization of Jewish childhood received a new impetus from the booming Soviet *paedology*—a biosocial and eugenic-oriented version of the science of the child (pedagogy and child psychology), widely popular in the early USSR.

Jewish paedologists insisted on testing Jewish children in Yiddish, not Russian or Ukrainian. They assessed their intellectual, psychological, and physical development separately from non-Jews and established special developmental standards for Jewish children based on their racial and psychological features and their cultural and social environment.[19] A growing number of articles popularized the then fashionable serological studies, substituting "race" with "blood group" as the main operational concept.[20] Here again the *QJBP* editors stressed the role of the Soviet state in creating and supporting labs and institutes for serological research, but insisted on the special role of such studies in establishing authentic Jewishness beyond the Soviet parameters of territoriality and proletarian culture.[21] Gran called on the *QJBP* contributors to embrace a eugenic genealogical approach in their studies of "the impact of agricultural work on the physical state of the Jewish population," thus balancing the reinvention of Jews as a productive class with a *longue durée* perspective on Jewishness as a lasting phenomenon.[22]

Gran personally experimented with collecting Jewish "genetic statistics" (family genealogies) in the oldest Russian Jewish agricultural colonies, established in the nineteenth century. In an article for the first *QJBP* collection (1926), coauthored with the Odessa sanitary doctor G. S. Matul'skii, Gran compared data for the two old agricultural colonies and two new, established under the Soviets in Pervomaiskii district in Ukraine. They took statistics collected in 1924–1925 by the two physicians working for the American Jewish Joint Distribution Committee (JDC)–OZE Medical Commission in Odessa, who examined 3,370 Jewish colonists—and along the way vaccinated them against diphtheria.[23] Similar to other *QJBP* contributors, Gran and Matul'skii treated these medical, anthropometric, and social statistics as reflective of Jewish racial stability and malleability, and approached Jewish agricultural colonies as an ongoing experiment with yet unclear results for collective Jewish "biology and pathology."[24] To provide a common structure for such studies, the *QJBP* editors suggested a range of questions:

> Does the Jewish mass possess the ability to toil the land and do industrial work? How can such a widespread change of occupation influence [the Jews'] mind and body ["psyche and physique" if translated literally]? What are the positive and negative sides of this transition? What are the ways and means of making it less painful? Solving these problems inevitably requires turning to race hygiene, eugenics, social biology, pathology, and to the studies of natural biological regrouping of masses and generations.[25]

The questions implied that biosocial sciences were to regulate the Soviet class-based productivization scenario, not just follow its path. Moreover, the scenario itself was to be tested against the assumed Jewish racial-psychological type and reconsidered as only one of the possible tracks for the Jews' "emigration" to modernity—a paradigm formulated in 1912 by Shternberg for Russian Jewish race science in the process of his reinterpretation of Franz Boas's famous anthropological study of Jewish immigrants in New York. This was not a conscious anti-Soviet gesture, just as the *QJBP* preoccupation with Jewish agriculturalists and industrial workers was not a protective strategy of artificially "speaking Bolshevik" to cover up for real subversion.[26] All the *QJBP* editors and contributors made livings and retained their public and academic status as Soviet professors, physicians, schoolteachers, and so on. Their "Soviet speech" reflected a possibility for this group to identify with Soviet scientific population politics and with Soviet modernity, which turned out to be the ultimate destination of their "race for the future." What the *QJBP* articles offered was not a radical alternative to Soviet class and national politics, but a form of its scientific national and, most importantly, nonterritorial Jewish appropriation.

This modestly independent stance vis-à-vis the Soviet ideological regime became harder to sustain with the beginning of the Cultural Revolution, the implementation of the First Five-Year Plan, and a decisive turn to the Birobidzhan territorial option for the Soviet Jewish nation in 1928.[27] The EIEO, which since 1923 had housed the Commission for the Study of Jewish Psychophysics and later served as a formal publisher of *QJBP*, was dissolved in 1929.[28] The political, institutional, scientific, and ideological space for experimentation with Soviet Jewishness was quickly shrinking. How, then, did the *QJBP* collections and the racialized, nonterritorial discourse represented in them persist into the 1930s? Perhaps, because even in its Stalinist version, the Soviet post–Civil War regime had fulfilled many of the Jewish "race for the future" fundamental objectives. It nationalized the political space of the former empire; integrated Jews into all spheres of civic life, economy, and culture; enabled normal Jewish science as part of Soviet academe; and promoted Jewish cultural nationalization. In the late 1920s, this was much more than the Jewish national minorities could have hoped for in the postimperial nation-states of Eastern and Central Europe. Whereas the archival documents that I was able to uncover reveal that the *QJBP* was, in fact, conceived in the early 1920s in the West and, as a transterritorial project, its implementation became possible in the USSR as a Jewish Soviet project, which was, of course, not accidental.

Indeed, the idea for the series was born in 1922, when the former OZE leadership in Petrograd and the new OZE leadership in Berlin began strategizing a regular journal that would be published in Germany and in German, and give voice to the activists of Jewish race science and biopolitics from Soviet Russia, the United States, Galicia, Romania, Lithuania, Poland, and other postimperial regions of Eastern and Central Europe.[29] From the very beginning, the implementation of this transterritorial project was hampered by a lack of funding, and letters from Berlin to the Petrograd/Leningrad contact person, Abram Bramson, reported few practical achievements between 1922 and 1924: "Dear Abram Moiseevich, . . . As you know, our goal is to launch a scientific journal. This project enjoys vast support. Several scholarly dignitaries from Germany, France, and England have agreed to participate. We encounter similar interest in the limitrophe states [*okrainnye gosudarstva*]. . . . We strongly count on receiving articles from you, and from the Russian-Jewish medical world in general (1924)."[30]

Both Berlin and Petrograd/Leningrad organizers worked on the lists of potential contributors.[31] The Russian list was ready in early 1924 and featured many names familiar to the readers of this book: Weissenberg, Dembo, Binshtok, and other participants of the old network of activists of Russian Jewish race science and biopolitics.[32] While Bramson was already collecting materials for the prospective journal, in Berlin the continuing lack of funds was greatly exacerbated by the October 1923 monetary reform that made available cash scarce and credit prohibitively expensive.[33] Responding to this problem, by the end of 1924 the organizers agreed to forgo the regular journal idea and instead launch a multilingual (German, French, and English) series of irregular thematic collections under the title "Materials for the Study of Jewish Biology and Pathology," and invite Jewish scholars residing in Berlin, Paris, and London to join the editorial board.[34] The exclusion of Soviet partners seems strange, but it might reflect a desire to protect them from the ire of the Soviet authorities while ensuring the publication's ideological independence and high status in Western academe.

The first almanac was expected to "establish, in a number of articles and essays, the status of Jewish biology and pathology as a special scientific discipline and a separate chapter of the general problem of race, racial hygiene and medicine, and social medicine and eugenics." The proposed list of rubrics included:

Jewish biology as a problem of science
Contemporary teaching about physical constitution in its application to
    Jews
Contemporary state of Jewish demography and sanitation statistics
Biology, pathology, and "nationality"
Sanitation consequences of World War I for the Jews
Sanitation medicine among Jews in different countries (Russia, Poland,
    Galicia, Romania, Lithuania, Palestine, and the United States)
Sanitation-biological aspects of Jewish emigration and colonization
The problem of health recovery of Jewish children and youth
Tuberculosis among Jews
Jewish anthropology and anthropometry
Jewish neuropsychic constitution
Jews as objects of eugenics, heredity, and internal secretion studies
Chronicle of scholarly life in the field of Jewish biology and pathology
Chronicle of the OZE and related organizations.[35]

There can be no doubt that the first *QJBP* volume published in Leningrad in 1926 followed this plan. Most probably, thanks to the partial restoration of private property and limited commercial publishing under the New Economic Policy, the financial resources for this venture had become available in the USSR before the Berlin committee could mobilize the required money and organizational will. In addition, as we have seen, by this time the Jewish activists of self-racializing, who had survived the Civil War and found a place for themselves among the Soviet educated class, felt confident enough to resume their influence on Soviet Jewish nation-building as scientific experts. In 1924–1925, one of these influential experts close to the Bolsheviks, the psychiatrist Aron Zalkind, defined the current agenda of Soviet science and social politics as reframing "biological, psychological, and social phenomena . . . as part of one and the same, fundamentally material, substance." Zalkind explained that "the objective was not just to *know* this substance in its various dialectically interconnected forms, but to acquire the ability to *mold* it at will in the desired direction."[36]

The two Soviet ministries that sporadically supported the OZE during the Civil War and later employed many of its activists in their institutions—the People's Commissariat for Education (Narkompros) and especially the People's Commissariat of Health (Narkomzdrav)—took this reframing and molding particularly seriously. In the 1920s, these two commissariats patronized eugenics, pedology, experimental psychology, "Freudo-Marxism," blood studies, and other sociobiological disciplines, all of which inspired the Jewish science of race.[37] Nikolai Semashko, the chair of Narkomzdrav and a personal acquaintance of Gran and other Jewish scholars connected to the *QJBP*, called himself a "Marxist-naturalist."[38] The "race for the future" culture could fit this definition, but there was little time left before the balance in Soviet human sciences had decidedly shifted toward a crude version of Marxism, and before Stalin's "Great Break" had delegitimized all claims for natural *longue durée* continuities that could hamper the First Five-Year Plan's radial economic and political constructivism.[39] The *QJBP* had to hit the brakes just as the long-conceived project was gaining momentum: the run of the second collection in 1928 was two thousand copies, twice that of the first collection.

Between the first and second collections, that is, in 1927, Semashko published a review of the first *QJBP* in the journal *Tribune of the Soviet Jewish Public* (*Tribuna evreiskoi sovetskoi obshchestvennosti;* 1927–1937), the official organ of the Society for Settling Toiling Jews on the Land (Obshchestvo zemleustroistva evreiskikh trudiashchikhsia, OZET). The recent supporter of eugenics and the "Marxist-naturalist" was apparently changing gears and aligning himself with a Jewish periodical that promoted a distinctly Soviet ideology of Jewish productivization and territorialization. Semashko's review was a carefully crafted endorsement of Soviet nation-building and the official ideology of the OZET. In the USSR, Semashko explained, "various nations could freely express their national traits," but his understanding of the "national autonomy" was now purely territorial and ethnographic. He therefore downplayed the collection's practical relevance for ongoing Soviet nation-building, instead emphasizing its contribution to "general ethnographic studies."[40] Likewise, Semashko welcomed Gran's interest in Jews' transition to agriculture, but disregarded his synthesis of a long-term genealogical approach with an analysis of short-term physical changes. Instead, the reviewer demanded that more attention be paid to economic and social conditions in the USSR—the only "normal conditions" for the productivization of Jews. Semashko sent the *QJBP* an important warning: "Incorrect methodology devalues the results of the most painstaking research. Reading this collection provokes questions about methodology."[41]

Gran received the message and offered a methodological defense in the opening article for the second collection, "On the Question of the Methodology of the Biological Exploration of Race and Nation." He agreed on the primacy of a materialist approach to race and nation but insisted that Soviet eugenics was a prime materialist science entitled to an expert role in Soviet nation-building ("A new scholarly discipline—eugenics—intertwines with the questions of race and nationality"). He endorsed Semashko's prioritization of "social and economic conditions under which race and nation develop," but noted that being "a pretty stable physical-biological type" race could evolve or persist under the influence of both social and biological factors.[42] Most interestingly, Gran warned against the tendency, apparent in Soviet race science, to focus on "underdeveloped" nations only. Protesting such selective racializing and extrapolating from the perspective of Jewish self-racializing, Gran called for the consideration of racial patterns among Russians, Ukrainians, or Tatars, that is, the peoples who championed nation-building after the revolution. Gran placed Jews among them—the peoples who exemplified not primitivism but development and progress.[43]

A year or two later, friendly negotiations with the old supporters of sociobiological sciences and biopolitics such as Semashko no longer made any sense. In 1929–1930, the leadership of practically all commissariats and governmental agencies was changed, and Semashko's generation of Soviet bosses interested in supporting experimental sociobiological sciences was gone.[44] In 1928, Gran, too, left his position in Semashko's Narkomzdrav and assumed a less politically visible professorship in social hygiene in Kazan. And just when the last *QJBP* collection came out in 1930, Semashko himself was removed from Narkomzdrav and became a regular professor at the First Moscow Medical Institute.[45] In 1930, Gran had to respond to explicitly political accusations from his colleagues in the Soviet medical community, suggesting that he personally and the *QJBP* collectively wanted to "supplement party politics" with anti-Soviet racial politics: "We do not want to question the 'purity' of the scholarly intentions of some of the collections' participants, but we cannot ignore that such quasi-scientific theories objectively pour water on the very suspicious 'mills'" (a reference to the popular Soviet idiom "pouring water on the mills of bourgeois propaganda").[46]

The last two *QJBP* collections, both published in 1930, produce a certain cognitive dissonance in a reader primarily because they continued to negotiate what had become nonnegotiable. Growing self-isolation of the Soviet Union that was building "socialism in one country" necessitated a more intense ideological opposition of Soviet science to "bourgeois" science, so that the global scale of comparison accepted in Jewish race science became a major compromising factor. An article by British physician Arthur George Hughes, which appeared in the first issue of *QJBP* in 1930 but must have been commissioned in 1928, gives a good illustration of the quickly evolving criteria of ideologically acceptability.

Hughes's article discussed the results of a study of "comparative intelligence" of Jewish and non-Jewish students in London elementary schools.[47] The data showed that Jews outperformed non-Jews, and the interpretation evoked the intellectualism and other implied positive qualities of the Jewish race. The article included the following observation: "Even the most ardent internationalist cannot deny that racial equality can be assumed only if the relative usefulness of each race is taken into account, because their abilities vary."[48] The Russian translation (which I have translated back into English) slightly distorted the English original: "Even the most ardent internationalist would admit that if races ought to be reckoned as equal, it is on a basis of diversity of usefulness and not of identity of capacity," making it sound less ambiguous and more racist.[49] Furthermore, the effect of mistranslation was greatly exacerbated by the lack of proper contextualization.

The sponsor of Hughes's study, the "bourgeois" Jewish Health Organization of Great Britain, was an OZE subdivision created in 1923 on the initiative of the OZE cofounder and now immigrant in Berlin Dr. Mikhail Shvartsman.[50] In the USSR of the 1930s, this was not the best pedigree, just as Hughes's ideological agenda to counter the widening opposition to Eastern European emigration in Britain was not a free pass for Soviet authorities who did not allow any emigration from the USSR. Hence, Soviet *QJBP* readers never learned that Hughes was responding to the 1925 study "The Problem of Alien Immigration into Great Britain, Illustrated by an Examination of Russian and Polish Jewish Children," by the eminent statistical demographers Karl Pearson and Margaret Moul. Their study substantiated the demand for "a cool measured statistical test" of physical and mental fitness as the main precondition for granting citizenship to immigrants. It posited the inferiority of the Jewish race, and hence the ineligibility of Jewish immigrants for citizenship.[51] Hughes responded with a "cool measure test" of his own that proved the opposite. Without these explanations and given the inaccuracies in translation, his article simply contributed to the discrediting of race science in the context of the Great Break.

The change of paradigms was sealed in 1932. That year, Moscow anthropologist Arkadii Iarkho summarized the zeitgeist of the day in an editorial for the new *Anthropological Journal,* which replaced the old *Russian Anthropological Journal* (*Russkii antropologicheskii zhurnal*), founded in 1900. Transmitting the official policy, Iarkho accused eugenics and physical anthropology of promoting biological determinism and harmful "racial and chauvinistic" recipes for human perfection, characteristic of bourgeois science. The title of Iarkho's editorial reflected the militant spirit of Cultural Revolution: "Against Idealistic Trends in Race Science in the USSR."[52] Under this slogan, major transformations of human sciences were initiated. Ethnography became largely a descriptive discipline, oriented toward registering progressive Soviet transformations of ethnographic cultures.[53] Physical anthropology was safely reduced to the study of prehistoric epochs, whereas eugenics was curtailed (the Russian Eugenic Society was closed in 1930).[54] The rise of Nazi race science delivered the final blow to the older race-based paradigms in the USSR.[55]

This major assault on "race" almost coincided with the rise in importance of the Birobidzhan district near the China-USSR border, which became the Jewish Autonomous Region in 1934. Ideologically, the Birobidzhan Soviet Jewish nation was constructed as young and underdeveloped, and hence deserving at most a region but not a national republic—a privilege reserved for fully developed Soviet nations. Gran sensed this nuance in 1928, when in his hidden polemic with Semashko he raised the issue of "primitivism" in the Soviet discourse and insisted that Soviet race science (and human sciences in general) would turn to the most developed Soviet nations as objects of study. Just a year later, in 1929, in order to salvage one of the sociobiological disciplines in crisis, paedology, Aron Zalkind decided not to fly against the wind and instead to apply sociological explanations to the developed nations and biological ones to the "primitive." He proposed reorienting paedology to the study of "the Bukhara Jews, the Moscow gypsies, Jewish kolkhoz in Siberia, and gypsy kolkhoz in the Northern Caucasus."[56] In this paradigm, Birobidzhan Jews were indeed symmetrical to Roma in that both groups were deemed in need of a territory and a proletarian national culture, and had to be collectivized and productivized.

The new primitivization of Jews in the Soviet discourse and the triumph of the territorial approach to nationality canceled major accomplishments of the Jewish "race for the future" in safeguarding an "authentic" Jewishness from being completely remolded or even reinvented to fit the Soviet form and content. But whatever the enthusiasts of Jewish self-racializing thought about the Birobidzhan project, in the 1930s it presented a viable postimperial arrangement for the Jews. Some saw it as a counterthesis to a Zionist solution, and others to European anti-Semitism and the Nazis' Jewish exclusion. The most controversial aspect of this catch-22 situation was that the Soviet state incorporated eugenic tropes and goals in its plan for the Birobidzhan Soviet Jewish nation, but alienated activists of Jewish self-racializing from defining their meaning and application. Eugenic elements in the Birobidzhan project reinforced Jews' primitivization and reinstalled a colonial framework against which the adepts of Jewish race science and biopolitics had struggled all along. In 1934, the chair of the Presidium of the USSR's Supreme Council (the de jure president of the USSR) and a great enthusiast of the Birobidzhan project, Mikhail Kalinin, described it as follows:

> Jews are the only nationality in the USSR numbering almost three million people and not having a statehood unit of their own. I think, under our conditions, the creation of an [autonomous] region is the only way for a normal state nationality to develop. . . . Many of those who had relocated later fell away, many have left [Birobidzhan]. Let the unfit go; not everybody can turn from a shtetl creature, a physically exhausted man into a courageous, resilient "colonizer." For this, one must undergo a rebirth. And what enables such a rebirth? These are the harsh, almost primordial nature of the land and the great and creative labor, which cannot be performed by a backward and weak man. The man should be strong—he needs to know how to resist and endure big hardships. If the first generation of "colonizers" will endure and stay, the next generation will be strong. These will be the true "Soviet" Jews like nowhere in the world. . . .

258

I think, the Birobidzhan Jewish nationality will bear no features of the shtetl Jews of Poland, Lithuania, Belorussia or even Ukraine. This is so because already now it is generating socialist "colonizers" of the free and rich land, [colonizers] with big fists and strong teeth. They will be forefathers of the regenerated and strong nation in the family of Soviet peoples. . . .

So, Birobidzhan as a statehood unit will simultaneously draw Jews to industry and agriculture. Then, despite the inevitable assimilation of some of the Jews, we will still be able to forge a healthy Soviet Jewish nation. I believe that this branch of Jewish nationality will grow strong.[57]

Kalinin's speech remarkably smacked of Rudyard Kipling's prose, although he did not expect Jews to carry out the "white man's" burden—they were only the imperfect tool of Soviet colonization of a neglected but strategic borderland. By carrying out their colonization mission, Jews could become a Soviet nation, and their eugenic transformation could occur along the way: a strong and healthy Soviet Jew would emerge from a feeble "shtetl creature." Kalinin excluded from this eugenic vision all those who did not fit or did not want to accept the new colonial mission and identity. He also rejected the possibility of Jews becoming a Soviet nation through engaging in nonterritorial self-decolonization in the epicenters of Soviet industrial and urban modernity in Ukraine, Belarus, or other places where they already lived. Kalinin bluntly compared Soviet Jewish settlers to "American cowboys," but cautioned that the latter were "predators with regard to nature, and enemies of the working people."[58] The Jews of Birobidzhan could not indulge themselves in any unrestrained colonial fantasies. Instead, they had to work hard and make of themselves a loyal proletarian nation. The eugenic goals were to be achieved through measures of rough social engineering bordering on "negative" eugenics—a "rebirth" through hard and hardly voluntary physical labor and a struggle with wild nature, strongly reminiscent of the Gulag concept for reforging the alien classes or "enemies of the people" into exemplary Soviet citizens.

The role of the state as the sole agent prescribing the criteria of nationality, so plainly evident in the case of Birobidzhan, received the fullest expression in the Soviet passportization campaign that started in 1932. Internal passports became mandatory documents for the urban population. They included an entry for nationality definable by "objective" criteria, that is, either mother's or father's origin/blood. Secret instructions for the police (*militsiia*) officials responsible for checking and revealing the true nationality of Soviet citizens made the racial nature of nationality abundantly clear to all carriers of internal Soviet passports. Someone born a Ukrainian, a Pole, a Korean, or a Jew could not choose to register as a Russian unless one of his or her parents was Russian by passport. Neither Russian language nor Russian education or self-identification would count. Mass deportations of entire peoples in the 1930s–1940s and the targeted arrests of hundreds of thousands under NKVD (Soviet security service) "national operations" only became possible on such a scale after the ascription and fixation of racialized nationalities in Soviet internal passports.[59]

Thus, under different names, "race" had assumed an important place in Soviet public discourse and practices. The post-1930s story of state-sponsored racializing can be traced further into the Soviet period and can include a new racialization of Jewishness in the postwar USSR, both by the regime and by the Jews themselves, a very different story from the one in this book.[60] Theoretically, a more direct link can be posited between the original Russian Jewish "race for the future" and the revival of racialized perceptions of Jewishness in Israel. I would, however, caution against such simplistic generalizations.[61] The imperial context that enabled the subaltern politics of race and the nonterritorial and grassroots nature of Russian Jewish self-racializing mark the Russian and Soviet cases as historically quite distinct and underscore the need for contextualizing the meaning, deployment, and implications of "race" in a radically different structural setting. In the Russian imperial Jewish story, race science and biopolitics functioned as a subaltern language and generated a social imagination that could be less teleological than were contemporary cultural or territorial nationalisms; they could inspire a radical developmental agenda and could mobilize for resistance and frame the politics of the future. Unlike the old imperial state, the new Soviet state had enough institutional and financial resources to control the production and dissemination of knowledge and hegemonic discourses, but the racializing of social reality persisted in new forms and shapes, reflecting the limits of the totalitarian utopia and exposing the internal incoherency of the official Soviet discourse.

The story told in this book suggests that race reemerges as a meaningful category at specific historical junctures that need to be carefully explored and contextualized, that racial discourses can speak in multiple languages and imply different developmental models, and, therefore, that ruptures in racial narratives matter as much if not more than structural continuities.

So, perhaps, the main message in this book should be the rupture argument. Jewish self-racializing in the colonial situation produced an epistemology and politics of race that differed from that in the Soviet "empire of nations" or the Jewish nation-state. In a similar vein, considering racialized Jewishness before and after the Holocaust would require integrating the rupture into our analytical framework. The Russian Jewish "race for the future" was embedded in a particular time and the imperial/postimperial historical juncture, and hence, in the words of Derek J. Penslar, had to be "historically and conceptually situated between colonial, anticolonial, and postcolonial discourse and practice."[62] Only such a complex contextualization uncovers "race" as a language of self-reflection and self-reinvention in the perpetual "imperial situation" of open-ended modernity.

ABBREVIATIONS

AMGUL   Archive of the Moscow State University named after M. V. Lomonosov, Moscow (Arkhiv Moskovskogo gosudarstvennogo universiteta imeni M. V. Lomonosova)

ARAN   Archive of the Russian Academy of Sciences, Moscow (Arkhiv Rossiiskoi akademii nauk)

DAKO   State Archive of the Kyiv Region, Kyiv (Derzhavnyi arkhiv Kyivs'koi oblasti)

DAOO   State Archive of the Odessa Region, Odessa (Derzhavnyi arkhiv Odes'koi oblasti)

GARF   State Archive of the Russian Federation, Moscow (Gosudarstvennyi arkhiv Rossiiskoi Federatsii)

IR NBUV   Manuscript Institute, Vernadskii National Library of Ukraine, Kyiv (Institut rukopysu Natsional'noi biblioteky Ukrainy imeni V. I. Vernads'kogo, Kyiv)

LSHA   Latvian State Historical Archive, Riga (Latvijas Valsts vēstures arhīva)

NART   National Archive of the Republic of Tatarstan, Kazan (Natsional'nyi arkhiv Respubliki Tatarstan)

RGALI   Russian State Archive of Literature and Art, Moscow (Rossiiskii gosudarstvennyi arkhiv literatury i iskusstva)

RGIA   Russian State Historical Archive, St. Petersburg (Rossiiskii gosudarstvennyi istoricheskii arkhiv)

SPbF ARAN | Archive of the Russian Academy of Sciences, St. Petersburg Branch (St. Peterburgskii filial arkhiva Rossiiskoi akademii nauk)

TsGAM | Central State Archive of the city of Moscow (Tsentral'nyi gosudarstvennyi arkhiv goroda Moskvy)

TsGA SPb | Central State Archive of St. Petersburg, St. Petersburg (Tsentral'nyi gosudarstvennyi arkhiv St. Peterburga)

TsGIA SPb | Central State Historical Archive of St. Petersburg, St. Petersburg (Tsentral'nyi gosudarstvennyi istoricheskii arkhiv St. Petersburga)

VULMC | Vilnius University Library's Manuscripts collection, Vilnius (Vilniaus valstubinis V. Kapsuko universitetas, Mokslinė biblioteka, Rankrasctu skyrius)

Organizations and Journals

ARA | American Relief Administration

EIEO | Jewish Historical-Ethnographic Society (Evreiskoe istoriko-etnograficheskoe obshchestvo)

EKO | Jewish Colonization Society (Evreiskoe kolonizatsionnoe obshchestvo)

EKOPO | Jewish Committee to Aid Victims of War (Evreiskii komitet dlia pomoshchi zhertvam voiny)

ENG | Jewish People's Group (Evreiskaia narodnaia gruppa)

Evobshchestkom | Jewish Public Committee for Relief of Pogrom Victims (Evreiskii obshchestvennyi komitet pomoshchi postradavshim ot pogromov)

Evsektsiia | Jewish Section of the Bolshevik Party (Evreiskaia sektsiia)

GODIVMO | *Annual Report on the Activities of the Imperial Vilna Medical Society (Godovoi otchet o deiatel'nosti Imperatorskogo Vilenskogo meditsinskogo obshchestva)*

IOLEAE | Imperial Society of Lovers of Natural Sciences, Anthropology and Ethnography (Imperatorskoe obshchestvo liubitelei estestvoznaniia, antropologii i etnografii)

IRGO | Imperial Russian Geographic Society (Imperatorskoe Russkoe Geograficheskoe Obshchestvo)

JDC | American Jewish Joint Distribution Committee

KD   Constitutional Democrats, members of the Constitutional-Democratic Party (Konstitutsionno-Demokraticheskaia Partiia)

MAE   Museum of Anthropology and Ethnography, St. Petersburg (Muzei antropologii i etnografii imeni Petra Velikogo)

Narkompros   People's Commissariat for Education (Narodnyi komissariat prosveshcheniia)

Narkomzdrav   People's Commissariat for Health (Narodnyi komissariat zdravookhraneniia)

OPE   Society for the Promotion of Culture among the Jews of Russia (Obshchestvo rasprostraneniia prosveshcheniia mezhdu evreiami v Rossii)

ORT   Society for Handicraft and Agricultural Work among the Jews of Russia (Obshchestvo remeslennogo i zemledel'cheskogo truda sredi evreev v Rossii)

OZE   Society for Protection of the Health of the Jewish Population (Obshchestvo okhraneniia zdorov'ia evreiskogo naseleniia)

OZET   Society for Settling Toiling Jews on the Land (Obshchestvo zemleustroistva evreiskikh trudiashchikhsia)

QJBP   *Questions of Jewish Biology and Pathology* (*Voprosy biologii i patologii evreev*)

RAZh   *Russian Anthropological Journal* (*Russkii antropologicheskii zhurnal*)

RSFSR   Russian Soviet Federative Socialist Republic (Rossiiskaia sovetskaia federativnaia sotsialisticheskaia respublika)

TASSR   Tatar Autonomous Soviet Socialist Republic (Tatarskaia avtonomnaia sovetskaia sotsialisticheskaia respublika)

Tatnarkomzdrav   People's Commissariat of Health of TASSR (Narodnyi komissariat zdravookhraneniia TASSR)

TOZ   Society for Protection of the Health of the Jewish Population, Poland (Towarzystwo ochrony zdrowia ludności żydowskiej)

NOTES

Introduction

Epigraph: Ann Laura Stoler, "Racial Regimes of Truth," in *Duress: Imperial Durabilities in Our Times* (Durham, NC: Duke University Press, 2016), 237–268, here 251.

1. Efron and Hart are the two authors who, while advancing somewhat differing approaches to Jewish race science, have greatly influenced this project: John M. Efron, *Defenders of the Race: Jewish Doctors and Race Science in Fin-de-Siècle Europe* (New Haven, CT: Yale University Press, 1994); Efron, *Medicine and the German Jews: A History* (New Haven, CT: Yale University Press, 2001); Efron, *German Jewry and the Allure of the Sephardic* (Princeton, NJ: Princeton University Press, 2016); Mitchell B. Hart, ed., *Jews and Race: Writings on Identity and Difference, 1880–1940* (Waltham, MA: Brandeis University Press, 2011); Hart, *Social Science and the Politics of Modern Jewish Identity* (Stanford, CA: Stanford University Press, 2000).

2. On empire as an epistemological condition and a context-setting category, see Alexander Semyonov, "Empire as a Context Setting Category," *Ab Imperio* 9, no. 1 (2008): 193–204.

3. Although the historiography of the Russian Empire is immense, "race" still presents a challenge to many historians who believe that it played no role in Russia's population politics or sciences. Sporadic attempts in the past to find a place for "race" in Russian and Soviet historical narratives got little traction within the field. Only recently has this tendency begun to reverse. See Discussion: Eric Weitz, "Racial Politics without the Concept of Race"; Francine Hirsch, "Race without the Practice of Racial Politics"; Amir Weiner, "Nothing but Certainty"; Alaina Lemon, "Without a Concept?"; and Weitz, "On Certainties and Ambivalences: Reply to My Critics," *Slavic Review* 61, no. 1 (2002): 1–29, 30–43, 44–53, 54–61, and 62–65; "Forum: The Multiethnic Soviet Union in Comparative Perspective": Adeeb Khalid, "Backwardness and the Quest for Civilization"; Adrienne Lynn Edgar, "Bolshevism, Patriarchy, and the Nation"; Peter A. Blitstein, "Cultural Diversity and the Interwar Conjuncture"; and Mark R. Beissinger, "Soviet Empire as 'Family Resemblance,'" *Slavic Review* 65, no. 2

(2006): 231–251, 252–272, 273–293, and 294–303; Eugene M. Avrutin, "Racial Categories and the Politics of (Jewish) Difference in Late Imperial Russia," *Kritika* 8, no. 1 (2007): 13–40; Avrutin, *Jews and the Imperial State: Identification Politics in Tsarist Russia* (Ithaca, NY: Cornell University Press, 2010); Maxim Matusevich, ed., *Africa in Russia, Russia in Africa: Three Centuries of Encounters* (Trenton, NJ: Africa World Press, 2007); Allison Blakely, *Russia and the Negro: Blacks in Russian History and Thought* (Washington, DC: Howard University Press, 1986); Vera Tolz, "Discourses of Race in Imperial Russia, 1830–1914," in *The Invention of Race: Scientific and Popular Representations,* ed. Nicolas Bancel, Thomas David, and Dominic Thomas (London: Routledge, 2014), 133–144; Tolz, *Russia's Own Orient: The Politics of Identity and Oriental Studies in the Late Imperial and Early Soviet Periods* (Oxford: Oxford University Press, 2011); Alaina Lemon, "'What Are They Writing about Us Blacks?' Roma and 'Race' in Russia," *Anthropology of East Europe Review* 13, no. 2 (1995): 34–40; David Rainbow, ed., *Ideologies of Race: Imperial Russia and the Soviet Union in Global Context* (Montreal: McGill–Queen's University Press, 2019); Andrew Sloin, *The Jewish Revolution in Belorussia: Economy, Race, and Bolshevik Power* (Bloomington: Indiana University Press, 2017); Edyta Bojanowska, *A World of Empires: The Russian Voyage of the Frigate Pallada* (Cambridge, MA: Harvard University Press, 2018); Marina Mogilner, *Homo Imperii: A History of Physical Anthropology in Russia* (Lincoln: University of Nebraska Press, 2013); Mogilner, "Racial Psychiatry and the Russian Imperial Dilemma of the 'Savage Within,'" *East Central Europe* 43 (2016): 99–133; Mogilner, "Classifying Hybridity in Nineteenth- and Early Twentieth-Century Russian Imperial Anthropology," in *National Races: Transnational Power Struggles in the Sciences and Politics of Human Diversity, 1840–1945,* ed. Richard McMahon (Lincoln: University of Nebraska Press, 2019), 205–240; Sergey Glebov, "Race and Politics: A History from an Imperial Borderland," in Marius Turda, ed., *A Cultural History of Race,* vol. 5, *A Cultural History of Race in the Age of Empire and Nation State (1760–1920),* ed. Marina Mogilner (London: Bloomsbury Academic, 2021), 93–110; Mogilner, "When Race Is a Language and Empire Is a Context," in "Critical Discussion Forum on Race and Bias," *Slavic Review* 80, no. 2 (2021): 207–215. On the Russian Empire as a field for imperial race science, see Mogilner, *Homo Imperii.* On the Jewish nationalists' political imagination, see Dmitry Shumsky, *Beyond the Nation-State: The Zionist Political Imagination from Pinsker to Ben-Gurion* (New Haven, CT: Yale University Press, 2018).

4. Ilya Gerasimov et al., "New Imperial History and the Challenges of Empire," in *Empire Speaks Out: Languages of Rationalization and Self-Description in the Russian Empire,* ed. Ilya Gerasimov, Jan Kusber, and Alexander Semyonov (Leiden: Brill, 2009), 3–32.

5. The term "strategic relativism" is coined as an opposite to Gayatri Chakravorty Spivak's characterization of the modern episteme of groupness as "strategic essentialism." Spivak, *In Other Worlds: Essays in Cultural Politics* (New York: Routledge, 1987), 205. Correspondingly, "strategic relativism" should be understood as a discourse and stance that relativizes the bounded and internally homogeneous nature of the constituent elements of the sociopolitical space and governance and produces a situation of uncertainty, incommensurability, and indistinction. For more, see Gerasimov et al., "New Imperial History," 20.

6. For a historiographic and methodological overview of new imperial history, see Ann Laura Stoler and Carole McGranahan, "Refiguring Imperial Terrains," in *Imperial Formations,* ed. Ann Laura Stoler, Carole McGranahan, and Peter C. Perdue (Santa Fe, NM: School for Advanced Research Press, 2007), 3–44; Stephen Howe, "Introduction: New Imperial Histories," in *The New Imperial Histories Reader,* ed. Stephen Howe (London: Routledge, 2010), 1–20; Gerasimov et al., "New Imperial History"; Frederick Cooper and Jane Burbank, *Empires in World History: Power and the Politics of Difference* (Princeton, NJ: Princeton University Press, 2010), esp. chap. 1 ("Imperial Trajectories"); and Dane Kennedy, *The Imperial History Wars: Debating the British Empire* (London: Bloomsbury, 2018).

7. Pieter Judson, "Finding Empire behind Multinationality in the Habsburg Case: Interview with Pieter Judson," by Alexander Semyonov, *Ab Imperio* 20, no. 1 (2019): 25–42, here 34.

8. Dipesh Chakrabarty, *Provincializing Europe: Postcolonial Thought and Historical Difference* (Princeton, NJ: Princeton University Press, 2007), 8.

9. Dipesh Chakrabarty, "Postcoloniality and the Artifice of History: Who Speaks for 'Indian' Pasts?," *Representations* 37 (Winter 1992): 1–26, here 5.

10. On the "flattening" and simplifying nature of biological versus cultural modes of constructing communities of origin, blood, or essentialized cultural proximity, see Paul Gilroy, *"There Ain't No Black in the Union Jack": The Cultural Politics of Race and Nation* (Chicago: University of Chicago Press, 1991); Ann Laura Stoler, *Carnal Knowledge and Imperial Power: Race and the Intimate in Colonial Rule* (Berkeley: University of California Press, 2002); Etienne Balibar, "Is There a 'Neo-Racism'?," in Etienne Balibar and Immanuel Wallerstein, *Race, Nation, Class: Ambiguous Identities,* trans. Chris Turner (1988; repr., London: Verso, 1991), 17–28.

11. Peter C. Perdue, "Erasing the Empire, Re-racing the Nation: Racialism and Culturalism in Imperial China," in Stoler, McGranahan, and Perdue, *Imperial Formations,* 145.

12. Terry Martin, *The Affirmative Action Empire: Nations and Nationalism in the Soviet Union, 1923–1939* (Ithaca, NY: Cornell University Press, 2001).

13. For an analysis of the rhetorical repertoire associated with this turn, see the thematic cluster in *Berichte zur Wissenschaftsgeschichte/History of Science and Humanities* 43 (2020): Marina Mogilner, "The Science of Empire: Darwinism, Human Diversity, and Russian Physical Anthropology," 96–118; Bruce Grant, "Missing Links: Indigenous Life and Evolutionary Thought in the History of Russian Ethnography," 119–140; and Riccardo Nicolosi, "The Darwinian Rhetoric of Science in Petr Kropotkin's *Mutual Aid. A Factor of Evolution* (1902)," 141–159.

14. Frantz Fanon, *The Wretched of the Earth* (New York: Grove Press, 1963), 10.

15. Derek J. Penslar, "Is Zionism a Colonial Movement?," in *Colonialism and the Jews,* ed. Ethan B. Katz, Lisa Moses Leff, and Maud S. Mandel (Bloomington: Indiana University Press, 2017), 275–300, here 285.

16. Svetlana Natkovich, "Questionable People: Inventing Modern Jewish Selves in the Writings of Russian-Jewish Authors, 1860s–1880s," in "The Substance of Jews: Discourses of Jewish Corporeal Performance in the Russian Imperial Setting" (unpublished manuscript, consulted in 2021), chap. 4. I am deeply grateful to Svetlana Natkovich for sharing this manuscript with me.

17. For an excellent scholarly biography of Shternberg, see Sergei Kan, *Lev Shternberg: Anthropologist, Russian Socialist, Jewish Activist* (Lincoln: University of Nebraska Press, 2009).

18. Ann Laura Stoler, "Colonial Aphasia: Disabled Histories and Race in France," in Stoler, *Duress,* 122–170, here 128.

19. Jeffrey Veidlinger, *Jewish Public Culture in the Late Russian Empire* (Bloomington: Indiana University Press, 2009), xii.

20. Sloin, *The Jewish Revolution in Belorussia,* 153.

1. The *Dawn* of the Jewish Race in the Late Nineteenth-Century Russian Empire

Epigraph: Ernest Renan, "Iudaizm kak rasa i kak religiia: Rech', chitannaia v kruzhke 'Saint-Simon' 27 ianvaria 1883 goda," *Voskhod* 4 (1883): 11–30, here 30. All translations are mine, unless otherwise specified.

1. Reinhart Koselleck, the premier theoretician of historical temporality, distinguished in particular the "saddle period" between 1750 and 1850, during which "the old experience of time was denaturalized" and the new horizon of the future had emerged, not having been directly determined by past experiences and knowledge. Progress acquired independent semantics and became an explanation of "the difference between the past so far and the coming future." New knowledge, ideas, concepts, and contemporary experiences began to form horizons of expectations. See Reinhart Koselleck, "Concepts of Historical Time and Social History," in *The Practice of Conceptual History: Timing History, Spacing Concepts*, trans. Todd Samuel Presner et al. (Stanford, CA: Stanford University Press, 2002), 115–130, here 120; and Koselleck, "The Eighteenth Century as the Beginning of Modernity," in Koselleck, *The Practice of Conceptual History*, 154–169.

2. Nicolas Bancel, Thomas David, and Dominic Thomas, eds., *The Invention of Race: Scientific and Popular Representations* (New York: Routledge, 2014); Marwa Elshakry, "When Science Became Western: Historiographical Reflections," *Isis* 101, no. 1 (2010): 98–110; Marina Mogilner, "Introduction: A Cultural History of Race in the Age of Empire and Nation State (1760–1920)," in Marius Turda, ed., *A Cultural History of Race*, vol. 5, *A Cultural History of Race in the Age of Empire and Nation State (1760–1920)*, ed. Marina Mogilner (London: Bloomsbury Academic, 2021), 1–18.

3. Urs App, *The Birth of Orientalism* (Philadelphia: University of Pennsylvania Press, 2009); Surekha Davies, *Renaissance Ethnography and the Invention of the Human: New Worlds, Maps, and Monsters* (Cambridge: Cambridge University Press, 2016); Richard McMahon, *The Races of Europe: Construction of National Identities in the Social Sciences, 1839–1939* (London: Palgrave Macmillan, 2016), esp. 67–81; Johann Friedrich Blumenbach, "On the Natural Varieties of Mankind, 1775," in *The Anthropological Treatises of Johann Friedrich Blumenbach* (London: Longman, Roberts and Green, 1865), 145–276. On sexual selection as a mechanism of race production, see Myrna Perez Sheldon, "Sexuality as a Secular Theory of Race," in Mogilner, *A Cultural History of Race*, vol. 5, 149–164; Evelleen Richards, *Darwin and the Making of Sexual Selection* (Chicago: University of Chicago Press, 2017).

4. For a summary of relevant historiography and important examples, see Mogilner, "Introduction," in *A Cultural History of Race*, vol. 5, 1–18.

5. V. M. Orel, ed., *Rossiiskaia akademiia nauk: 275 let sluzheniia Rossii* (Moscow: Ianus-K, 1999), 322–346; Michael Gordin, *A Well-Ordered Thing: Dmitrii Mendeleev and the Shadow of the Periodic Table* (Princeton, NJ: Princeton University Press, 2018); Gordin, *Scientific Babel: How Science Was Done before and after Global English* (Chicago: University of Chicago Press, 2015).

6. Jeffrey Veidlinger, *Jewish Public Culture in the Late Russian Empire* (Bloomington: Indiana University Press, 2009), 78.

7. N. A. Troinitskii, ed., *Kratkie obshchie svedeniia po imperii: Raspredelenie naseleniia po glavneishim sosloviiam, veroispovedaniiam, rodnomu iazyku i po nekotorym zaniatiiam* (St. Petersburg: parovaia tipo-lit. N. L. Nyrkina, 1905), 6–7, 14–15.

8. Shmuel Feiner, *The Jewish Enlightenment*, trans. Chaya Naor (Philadelphia: University of Pennsylvania Press, 2004); Immanuel Etkes, "On the Question of the Harbingers of East European Jewish Enlightenment" [in Hebrew], *Tarbiz* 57 (1988): 95–114; David E. Fishman, *Russia's First Modern Jews: The Jews of Shklov* (New York: New York University Press, 1995).

9. Mordechai Zalkin, "Scientific Thinking and Cultural Transformation in Nineteenth-Century East European Jewish Society," *Aleph* 5 (2005): 11–31, here 13; Tal Kogman, *Ha'maskilim be'madaim* [Maskilim in the sciences] (Jerusalem: Magness Press, 2013); Kogman, "The Emergence of Scientific Literature in Hebrew for Children and Youth in the

Nineteenth Century: Preliminary Directions of Research," *Jewish Culture and History* 17, no. 3 (2016): 249–263; Kogman, "Science and the Rabbis: Haskamot, Haskalah, and the Boundaries of Jewish Knowledge in Scientific Hebrew Literature and Textbooks," *Leo Baeck Institute Yearbook* 62 (2017): 135–149.

10. Some translations, however, had a wider influence. For example, Dr. Adolf Jellinek's book *Der jüdische Stamm* (1869), a precursor of modern "scientific" ethnic psychology that combined the racializing approach with cultural analysis, had tremendous influence both on Russian maskilim and on early proponents of scientific race. Chapters of the book were translated into Hebrew in 1869 by Peretz Smolenskin, who polemicized with Jellinek, and into Russian in 1870 by Adolf Landau, the future editor of *Voskhod*. See *Evreiskoe plemia: Etnograficheskie etiudy d-ra Adolfa Iellineka*, trans. from German, ed. A. E. Landau (St. Petersburg: Izdanie A. E. Landau, 1870).

11. Joakim Philipson, *The Purpose of Evolution: The "Struggle for Existence" in the Russian-Jewish Press 1860–1900* (Stockholm: Acta Universitatis Stockholmiensis, 2008), 166–167. See also Yaacov Shavit and Jehudah Reinharz, *Darvin ve'kama mi'bne mino: Evolutsia, geza, sviva ve'tarbut-yehudim kor'im et darvin, spenser, bakl ve'renan* [Darwin and some of his kind: Evolution, race, environment and culture—Jews read Darwin, Spencer, Buckle and Renan] (Tel Aviv: Hakibbutz Hameuchad, 2009).

12. The most elaborate analysis of conversions as experienced by ordinary Jews in the Pale is offered in Ellie R. Schainker, *Confessions of the Shtetl: Converts from Judaism in Imperial Russia, 1817–1906* (Stanford, CA: Stanford University Press, 2017).

13. Marwa Elshakry, *Reading Darwin in Arabic, 1860–1950* (Chicago: University of Chicago Press, 2014); Junaid Quadri, *Transformations of Tradition: Islamic Law in Colonial Modernity* (Oxford: Oxford University Press, 2021).

14. D. I. Sheinis, *Evreiskoe studenchestvo v Moskve (Po dannym ankety 1913 g.)* (Moscow: Izd. Moskovskogo komiteta OPE, 1913), 40.

15. Alexander Dmitriev, "Ukrainskaia nauka i ee imperskie konteksty (XIX–nachalo XX veka)," *Ab Imperio* 8, no. 4 (2007): 121–171, here 124; A. N. Dmitriev, "Philologists-Autonomists and Autonomy from Philology in Late Imperial Russia: Nikolai Marr, Jan Baudouin de Courtenay, and Ahatanhel Krymskii," *Ab Imperio* 7, no. 1 (2016): 125–167; A. N. Dmitriev, "Istoriia, sotsiologiia i poshuky 'natsional'noi nauky,'" *Krytyka* 17, no. 9–10 (2013): 17–20; Kira Il'ina, "Fridrikh Fater i Karl Gorman: Philology-klassiki iz nemetskikh universitetov v Rossii v seredine XIX veka," in *Sad uchenykh naslazhdenii*, ed. A. Dmitriev, N. Samutina, and E. Vishlenkova (Moscow: NIU VShE, 2017), 171–182; A. N. Dmitriev, "Langue ukrainienne et projet de science nationale: Les étapes d'une légitimation académique (fin des années 1880, début des années 1920)," in *Cacophonies d'empire: Le gouvernement des langues dans l'Empire russe et en Union soviétique*, ed. Juliette Cadiot, Dominique Arel, and Larissa Zakharova (Paris: CNRS Editions, 2010), 85–110; E. Vishlenkova, S. Malysheva, and A. Sal'nikova, *Terra Universitatis: Dva veka universitetskoi kul'tury v Kazani* (Kazan: KGU, 2005); Ramil' Valeev, *Orientalistika v Kazanskom universitete (1807-20-e gody XX veka)* (Mauritius: LAP LAMBERT Academic Publishing, 2019); Zenon Kohut, "The Development of Ukrainian National Historiography in Imperial Russia," in *Historiography of Imperial Russia: The Profession and Writing of History in a Multinational State*, ed. Thomas Sanders (Armonk, NY: M. E. Sharpe, 1999), 453–478; Thomas M. Prymak, *Mykola Kostomarov: A Biography* (Toronto: University of Toronto Press, 1996).

16. Vera Tolz, *Russia's Own Orient: The Politics of Identity and Oriental Studies in the Late Imperial and Early Soviet Periods* (Oxford: Oxford University Press, 2011).

17. Dmitry A. Elyashevich, *Pravitel'stvennaia politika i evreiskaia pechat' v Rossii, 1797–1917: Ocherki istorii tsenzury* (St. Petersburg–Jerusalem: Mosty kul'tury/Gesharim, 1999), 322.

18. S. M. Dubnov, *Pis'ma o starom i novom evreistve (1897–1907)* (St. Petersburg: Obshchestvennaia pol'za, 1907), 217.

19. Elyashevich, *Pravitel'stvennaia politika i evreiskaia pechat'*, 403.

20. On the reaction of the Russian Jewish press to this turn of Russian liberal discourse away from the ideal of Jewish emancipation, see John Doyle Klier, *Imperial Russia's Jewish Question, 1855–1881* (New York: Cambridge University Press, 1995), 350–369. On the connection between the emancipation of peasants in 1861 and the rise of Russian nationalism, see Mikhail Dolbilov, "The Emancipation Reform of 1861 in Russia and the Nationalism of the Imperial Bureaucracy," in *Construction and Deconstruction of National History in Slavic Eurasia,* ed. T. Hayashi (Sapporo: Slavic Research Center, 2003), 208–235.

21. Dubnov, *Pis'ma o starom i novom evreistve,* 223.

22. According to Elyashevich, Russian officials tolerated *Voskhod* because they "could not afford a complete disappearance of the Russian-Jewish press. They were afraid to deny the assimilated Russian-Jewish intelligentsia a public platform, thereby contributing to its growing radicalization. In addition, they needed a convenient and direct access to Jewish life (not through mediation of the specially hired Jewish censors or neophyte Christians)." Elyashevich, *Pravitel'stvennaia politika i evreiskaia pechat'*, 373, 408–409.

23. Os. Gruzenberg, "Literaturnaia letopis': Vzaimodeistvie ideinykh napravlenii," *Voskhod* 10 (1894): 15–50, here 16.

24. Goldenviezer quoted in Klier, *Imperial Russia's Jewish Question,* 103–104.

25. Iu. Goldendakh, "Odin iz sovremennykh evreiskikh voprosov," *Sion* 6 (1861): 93–96, here 96.

26. T. G. Geksli, *Mesto cheloveka v tsarstve zhivotnom,* trans. from the German by Iu. Goldendakh (Moscow: Izdatel'stvo doktora V. Karusa, 1862); *Lektsii sovremennykh nemetskikh klinitsistov,* comp. and trans. Iu. Goldendakh (Moscow: Tipografiia A. I. Mamontova, 1868–1882).

27. Goldendakh, "Odin iz sovremennykh evreiskikh voprosov," 95.

28. F. Gets, "Chto takoe evreistvo?," *Evreiskoe obozrenie* 6 (1884): 72–87, here 72.

29. Jonathan Frankel, *Prophecy and Politics: Socialism, Nationalism, and the Russian Jews, 1862–1917* (Cambridge: Cambridge University Press, 1981).

30. Among those who sympathetically discussed Renan's ideas in *Voskhod,* two names deserve special mention—Simon Dubnov and Iakov Rombro (known under the penname Philip Krantz). For example, see Ia. Rombro, "K voprosu o natsional'nostiakh: Po povodu dvukh chtenii Renana: 1. Qu'est-ce qu'une nation? Chitannaia v Sorbonne 18 marta 1882 goda. 2. Le judaïsme comme race at comme religion, chitannaia v Société historique (cercle Saint-Simon) 27 ianvaria 1883 g.," *Voskhod* 4 (1883): 1–18. Rombro/Krantz was a populist and later an activist of the Jewish labor movement in England and the United States. He moved to the United States to assume editorship of the socialist weekly *Arbeiter Zeitung* (Workers' newspaper) and became known as a Yiddish-language journalist and author of popular books on scholarly topics. On Krantz, see *The Cambridge History of American Literature,* ed. William Peterfield Trent et al. (New York: G. P. Putnam's Sons, 1921), bk. 3, pt. 3, 600–602.

31. "Renan o zadachakh evreiskoi natsii," *Den', organ russkikh evreev* 2 (January 10, 1870): 4.

32. Heather Bailey, *Orthodoxy, Modernity, and Authenticity: The Reception of Ernest Renan's "Life of Jesus" in Russia* (Newcastle, UK: Cambridge Scholars Publishing, 2008), 199.

33. G., "Ernst Renan i otzvuk ego zastupnichestva za evreev," *Russkii evrei* 4 (January 30, 1883): 10–13, here 11.

34. On the Tiszaeszlár blood libel, see Andrew Handler, *Blood Libel at Tiszaeszlár* (Boulder, CO: East European Monographs, 1980); Hillel J. Kieval, "The Importance of Place:

Comparative Aspects of the Ritual Murder Trial in Modern Central Europe," in *Comparing Jewish Societies*, ed. Todd M. Endelman (Ann Arbor: University of Michigan Press, 1997), 135–165; and Hillel J. Kieval, "Neighbors, Strangers, Readers: The Village and the City in Jewish-Gentile Conflict at the Turn of the Nineteenth Century," *Jewish Studies Quarterly* 12, no. 1 (2005): 61–79. On the reaction in the Russian press to the "affair," see Klier, *Imperial Russia's Jewish Question*, 434–436.

35. "Rech' Renana," *Russkii evrei* 22 (June 6, 1883): 24–31.

36. The last chapter of *Antichrist* appeared in Russian as a stand-alone piece under the title *The Destruction of Jerusalem*. Ernest Renan, *Razorenie Ierusalima*, trans. from the French by P. P. Nadezhdin (Moscow: V. N. Marakuev, 1886).

37. Kritikus [Dubnov], "Poslednie dni Ierusalima v izobrazhenii Renana," *Voskhod* 6 (1886): 7–18, here 18.

38. In the review, Dubnov cited Renan's speech to the Jewish Scholarly Society: "In a word, pure religion which will unite all humanity under its banner, will be only the realization of the religion of Isaiah; it will be the Jewish ideal religion freed from the dross accidentally alloyed to it." Kritikus [Dubnov], "Poslednie dni Ierusalima," 18. I am using the translation by Bailey, *Orthodoxy, Modernity, and Authenticity*, 207.

39. Shlomo Send, *The Invention of the Jewish People*, trans. Yael Lotan (London: Verso, 2009), 269.

40. Renan, "Iudaizm kak rasa i kak religiia," quoted from the English translation by Robert Pick, "Judaism: Race or Religion?," *Contemporary Jewish Record* 6, no. 4 (August 1, 1943): 435–448, here 446.

41. Pick, "Judaism," 438.

42. Pick, "Judaism," 448.

43. Bailey, *Orthodoxy, Modernity, and Authenticity*. Renan served as an authoritative source for both racists and anti-Semites, and humanists and philosemites, including Jews themselves. On this, see Shmuel Almog, "The Racial Motif in Renan's Attitude to Jews and Judaism," in *Antisemitism through the Ages*, ed. Shmuel Almog, trans. Nathan H. Reisner (Oxford: Pergamon, 1988), 255–278. Renan's views on race remain a subject of scholarly controversy today: Robert D. Priest, "Ernest Renan's Race Problem," *Historical Journal* 58, no. 1 (2015): 309–330; Zeev Sternhell, *The Anti-Enlightenment Tradition*, trans. David Maisel (New Haven, CT: Yale University Press, 2010). The list of Renan's works translated and published in turn-of-the-century Russia is long and includes most of the works by the French scholar. See Bailey, *Orthodoxy, Modernity, and Authenticity*, 292–297.

44. S. M. Dubnov, *Kniga zhizni: Vospominaniia i razmyshleniia. Materialy dlia istorii moego vremeni* (St. Petersburg: Peterburgskoe vostokovedenie, 1998), 132.

45. Bailey, *Orthodoxy, Modernity, and Authenticity*, 198.

46. Bailey, *Orthodoxy, Modernity, and Authenticity*, 140; Kritikus [Dubnov], "Drevniia istoriia evreev po Renanu," *Voskhod* 8 (1888): 11–24; 9 (1888): 18–33; Kritikus [Dubnov], "Period vtorogo Khrama v osveshchenii Renana," *Voskhod* 4 (1894): 31–48; 5 (1894): 14–25.

47. Priest, "Ernest Renan's Race Problem," 314–315.

48. All the letters first appeared in *Voskhod* between 1897 and 1906 (the only exception was the fifth letter, which appeared in *Futurity* [*Budushchnost'*]). The letters were revised and published under one cover in 1907, in Dubnov, *Pis'ma o starom i novom evreistve*. A comprehensive annotated bibliography of works by and about Dubnov can be found in Simon Rabinovitch, "Simon Dubnov," in *Oxford Bibliographies of Jewish Studies*, ed. David Biale (New York: Oxford University Press, 2014), http://www.oxfordbibliographies.com/abstract/document/obo-9780199840731/obo-9780199840731-0099.xml?rskey=2HTyGX&result=1&q=Simon+Dubnov#firstMatch.

49. Simon Rabinovitch, *Jewish Rights, National Rites: Nationalism and Autonomy in Late Imperial and Revolutionary Russia* (Stanford, CA: Stanford University Press, 2014).

50. The English translation is quoted from Simon Dubnov, "The Doctrine of Jewish Nationalism," in *Nationalism and History: Essays on Old and New Judaism*, ed. Koppel S. Pinson (Philadelphia: Jewish Publication Society, 1958), 80.

51. Frankel offered this conclusion in his original and thought-provoking introduction to the memoir of Dubnov's daughter, Sophie Dubnov-Erlich, about her father: Jonathan Frankel, "S. M. Dubnov: Historian and Ideologist," in *The Life and Work of S. M. Dubnov: Diaspora Nationalism and Jewish History*, by Sophie Dubnov-Erlich (Bloomington: Indiana University Press, 1991), 1–33, here 8.

52. Daniel P. Todes, *Darwin without Malthus: The Struggle for Existence in Russian Evolutionary Thought* (New York: Oxford University Press, 1989), 168; Alexander Vucinich, *Darwin in Russian Thought* (Berkeley: University of California Press, 1988).

53. Benjamin Nathans, "On Russian-Jewish Historiography," in Sanders, *Historiography of Imperial Russia*, 397–432, here 411.

54. S. Dubnov, "Neskol'ko momentov v istroii razvitiia evreiiskoi mysli," *Russkii evrei* 18 (May 1, 1881): 710.

55. The English edition of the letters is based primarily on this Hebrew edition: *Mikhtavim al ha-al ha-yahadut ha-yeshana veha-hadashah*, trans. Avraham Levinson (Tel Aviv: HaHoqer al Yad Devir, 1937); Pinson, *Nationalism and History*.

56. The translation here is mine. S. M. Dubnov, "Pis'mo pervoe: Teoriia evreiskogo natsionalizma," in Dubnov, *Pis'ma o starom i novom evreistve*, 1–28, here 1–2. This collection served as a source for the 1937 Hebrew edition.

57. I rely here on the English translation in Pinson, *Nationalism and History*, 75, but indicate inaccuracies in the translation in square brackets. Together, the instances of mistranslations seem to create a pattern: they aim at reducing the biological connotations of Dubnov's reasoning.

58. These discourses are hard to separate not only in Dubnov's case. Consider another quite typical example of the definition of the Jewish nation by Faifel Getz, a Russian Jewish journalist, graduate of Yur'ev and St. Petersburg Universities, an official "learned Jew" (*uchenyi evrei*, appointed in 1891) in the Vilna education district, and a contemporary of Dubnov. He wrote that nation is a "physiological-psychological and intellectual-moral" phenomenon that is formed by the "commonality of origin and historical past. . . . The commonality of each nation lies in its blood, bodily constitution, physiognomic traits, affinity of temperaments, in certain predispositions and qualities, customs, and habits." Gets, "Chto takoe evreistvo?," 78. Many more examples of this sort are to follow.

59. V.R., "Iz mira antropologicheskoi statistiki," *Voskhod* 8 (1881): 48–78, here 48, 77.

60. "Iz etiudov sravnitel'noi statistiki Legua: Biostatisticheskie preimushchestva evreiskoi rasy v Evrope," *Russkii evrei* 2 (September 9, 1879): 57–59.

61. "Iz etiudov sravnitel'noi statistiki Legua," 59, 58.

62. I. Tsederbaum, "O evreiskom narode," *Vestnik russkikh evreev* 20 (1871), cols. 625–629, here col. 625.

63. V.R., "Iz mira antropologicheskoi statistiki," 49.

64. V.R., "Iz mira antropologicheskoi statistiki," 51, 50.

65. On Jews' role in Russian colonialism, see a telling comparison in Israel Bartal, "Jews in the Crosshairs of Empire: A Franco-Russian Comparison," in *Colonialism and the Jews*, ed. Ethan B. Katz, Lisa Moses Leff, and Maud S. Mandel (Bloomington: Indiana University Press, 2017), 116–126.

66. Philipson, *The Purpose of Evolution*, 305–306.

67. Peter Holquist, "To Count, to Extract, and to Exterminate: Population Statistics and Population Politics in Late Imperial and Soviet Russia," in *A State of Nations: Empire and Nation-Making in the Age of Lenin and Stalin,* ed. Ronald Grigor Suny and Terry Martin (New York: Oxford University Press, 2001), 111–144.

68. On the German-type statistics and its flourishing in Otto von Bismarck's united Germany, see D. C. Coleman, *History and the Economic Past: An Account of the Rise and Decline of Economic History* (Oxford: Oxford University Press, 1987); Alon Kadish, *Historians, Economists, and Economic History* (New York: Routledge, 1989); Andrew Zimmerman, *Alabama in Africa: Booker T. Washington, the German Empire, and the Globalization of the New South* (Princeton, NJ: Princeton University Press, 2012).

69. Esther Kingston-Mann, "Statistics, Social Science, and Social Justice: The Zemstvo Statisticians of Pre-revolutionary Russia," in *Russia in the European Context, 1789–1914: A Member of the Family,* ed. Susan P. McCaffray and Michael Melancon (New York: Palgrave Macmillan, 2005), 113–140.

70. Ilya Gerasimov, *Modernism and Public Reform in Late Imperial Russia: Rural Professionals and Self-Organization, 1905–30* (New York: Palgrave Macmillan, 2009).

71. Max Weber, "On the Situation of Constitutional Democracy in Russia," in *Max Weber: Political Writings,* ed. Peter Lassman and Ronald Speirs (Cambridge: Cambridge University Press, 1994), 29–50, here 35. First appeared in *Archiv für Sozialwissenschaft und Sozialpolitik* 22, no. 1 (1906).

72. Zimmerman, *Alabama in Africa.*

73. On the national blindness of zemstvo statisticians and their discovery of nationalism, see Gerasimov, *Modernism and Public Reform,* esp. the chapter "Nation as the People," 156–167.

74. In this regard, historians even talk about the emergence of an alliance between men of politics and men of science in the Russian Empire by the mid-nineteenth century. Bruce W. Lincoln, *In the Vanguard of Reform: Russia's Enlightened Bureaucrats, 1825–1861* (DeKalb: Northern Illinois University Press, 1982); Marina Loskutova, "'Svedeniia o klimate, pochve, obraze khoziaistva i gospodstvuiushchikh rasteniiakh dolzhny byt' sobrany . . .': Prosveshchennaia biurokratiia, gumbol'dtovskaia nauka i mestnoe znanie v Rossiiskoi imperii vtoroi chetverti XIX v.," *Ab Imperio* 13, no. 4 (2012): 111–156.

75. Holquist, "To Count, to Extract and to Exterminate."

76. For a detailed discussion, see Marina Mogilner, *Homo Imperii: A History of Physical Anthropology in Russia* (Lincoln: University of Nebraska Press, 2013), chap. 15 ("The Discovery of Population Politics and Sociobiological Discourses in Russia"), 297–309.

77. Girsh M. Rabinovich, "Nevernye vyvody iz vernykh tsifr," *Russkii evrei* 7 (February 16, 1884): 25–28, here 27; 8 (February 23, 1884): 21–25.

78. For more on Gruzenberg's role in *Voskhod,* second only to its founder and executive editor Landau, and on his other engagements as a Jewish activist, see "Nekrolog: S. Gruzenberg," *Rassvet* 19 (1909): 16.

79. S. Gruzenberg, "O fizicheskom sostoianii evreev v sviazi s usloviiami ikh zhizni: Sanitarnyi ocherk," *Evreiskoe obozrenie* 2 (1884): 21–38; 3 (1884): 53–73; 4 (1884): 61–83.

80. Gruzenberg, "O fizicheskom sostoianii evreev," 31.

81. Gruzenberg, "O fizicheskom sostoianii evreev," 58.

82. Bernhard Blechmann, *Ein Beitrag zur Anthropologie der Juden* (Dorpat: W. Just's, 1882).

83. Dr. B. Greidenberg, "Ocherk antropologii evreev: Ein Beitrag zur Anthropologie der Juden, Inaug.-Dissert. Von D-r Bernhardt Blehmann. Dorpat)," *Russkii evrei* 1 (January 7, 1883): 31–34, here 31.

84. Gruzenberg, "O fizicheskom sostoianii evreev," 34, 37 ("kak-by slishkom mnogo dushi, no pri etom slishkom malo tela").

85. For a discussion of this problem, see Eugene M. Avrutin, *Jews and the Imperial State: Identification Politics in Tsarist Russia* (Ithaca, NY: Cornell University Press, 2010), 79.

86. S. Gruzenberg, "Glava iz antropologii evreev: O raspredelenii polov pri rozhdenii," *Voskhod* 11 (1887): 82–100, here 87–88.

87. Gruzenberg, "Glava iz antropologii evreev," 100.

88. ChaeRan Y. Freeze, *Jewish Marriage and Divorce in Imperial Russia* (Hanover, NH: Brandeis University Press, 2002), 272.

89. Gruzenberg, "O fizicheskom sostoianii evreev," 63.

90. Gruzenberg, "O fizicheskom sostoianii evreev," 67.

91. "Pis'ma S. O. Gruzenberga (1896–1899)," *Evreiskaia starina* 3–4 (1914): 385.

92. Klier, *Imperial Russia's Jewish Question*, xv.

93. "Za nedeliu," *Russkii evrei* 18 (May 6, 1883): 8–9, here 8.

94. M. O., "Za proshlyi god," *Voskhod*, March 1883, 31–48.

95. "Za nedeliu," *Russkii evrei* 8 (February 24, 1884): 9–11, here 9.

96. G. M. Rabinovich, "Fel'eton," *Russkii evrei* 26 (July 6, 1883): 28–34, here 29.

97. G. M. Rabinovich, "Fel'eton," *Russkii evrei* 27 (July 15, 1883): 30–37, here 31.

2. Samuel Abramovich Weissenberg

Epigraph: Samuel Weissenberg, "Die südrussischen Juden: Eine Anthropometrische Studie mit Berücksichtigung der Allgemeinen Entwickelungsgesetze," special issue, *Archiv für Anthropologie* 23, no. 3–4 (Braunschweig: Friedrich Vieweg und Sohn, 1895): 126.

1. Samuil Vaisenberg was the Russified version of his name he used in published works and private writings. The name on his birth certificate is Samuil-Jacub Vaisinberg. The document was recently located by a local historian in Kropyvnytskyi: Roman Liubarsky, accessed March 18, 2022, http://raiskiy.livejournal.com/81842.html.

2. Ephraim Fischoff, "Weissenberg, Samuel Abramovich," in *Encyclopaedia Judaica,* ed. Michael Berenbaum and Fred Skolnik, 2nd ed., vol. 20 (Detroit: Macmillan Reference, 2007), 738.

3. Amos Morris-Reich, "Jews between Volk and Rasse," in *National Races: Transnational Power Struggles in the Sciences and Politics of Human Diversity, 1840–1945,* ed. Richard McMahon (Lincoln: University of Nebraska Press, 2019), 175–203.

4. Today he is known by this name in just a few studies that draw exclusively on his German-language legacy: John M. Efron, *Defenders of the Race: Jewish Doctors and Race Science in Fin-de-Siècle Europe* (New Haven, CT: Yale University Press, 1994); Mitchell B. Hart, ed., *Jews and Race: Writings on Identity and Difference, 1880–1940* (Waltham, MA: Brandeis University Press, 2011).

5. Morris-Reich emphasizes that Weissenberg received his education in Germany and published in German, and does not consider his publications in other languages. He agrees with Efron that "the wider context in which Weissenberg developed as a physical anthropologist was the harassment of Eastern European immigrants . . . in Germany. . . . Weissenberg's turn to physical anthropology was inseparable from his recognition that the core of the anti-Semitic claim . . . was that the Jew bore physical and mental characteristics that prevented him from living peacefully with Aryans." Morris-Reich, "Jews between Volk and Rasse," 186.

6. Efron, *Defenders of the Race,* 91–122. Efron explains Weissenberg's turn to race science as a reaction to German anti-Semitism, which was "often shrouded in the supposedly

objective language of science. Weissenberg, with his training in German science and medicine, could see through the obfuscations of German scientific language to discern the antisemitic motives of the writers. Indeed, he ventured to counter the claims of a hostile scientific establishment by employing the language of German race science to the advantage of his own people" (94). At the same time, according to Efron, Weissenberg "was not preoccupied with the antisemitism of Russia, which was still based on the traditional Christian image of the religious Jew and did not so readily identify Jewish difference on the basis of race" (103). As my previous work shows, and this book confirms, this view simplifies and distorts the realities of the turn-of-the-century Russian Empire. See Marina Mogilner, *Homo Imperii: A History of Physical Anthropology in Russia* (Lincoln: University of Nebraska Press, 2013).

7. See Willard Sunderland, *Taming the Wild Field: Colonization and Empire on the Russian Steppe* (Ithaca, NY: Cornell University Press, 2004).

8. M. Polishchuk, *Evrei Odessy i Novorossii: Sotsial'no-politicheskaia istoriia evreev Odessy i drugikh gorodov Novorossii 1881–1904* (Moscow: Gesharim, 2002), 20–21.

9. I. A. Troinitskii, ed., *Pervaia vseobshchaia perepis' naseleniia Rossiiskoi imperii, 1897*, vol. 47, *Khersonskaia guberniia* (Moscow: Izdanie Tsentral'nogo statisticheskogo komiteta Ministerstva vnuternnikh del, 1904), 2–3.

10. Of the population, 2 percent (1,180) indicated Polish as their mother tongue, and 2.2 percent (1,383) identified themselves as Catholics (some of them must have been speaking Belorussian). There were also smaller communities of German speakers (425), Greeks (124), Moldavians (21), Armenians (16), and others. Troinitskii, *Pervaia vseobshchaia perepis'*, 47:2–3, table "Composition of the Available Population of Both Sexes."

11. The article "Yelisavetgrad (Elizavetgrad)" in *The Jewish Encyclopedia* volume published in 1906 gives a somewhat higher number of 24,340 Jews. I. Br. (Isaac Broyde, Office Editor, Doctor of the University of Paris, France) and S. J. (S. Janovsky, Counselor of Law, St. Petersburg, Russia), "Yelisavetgrad (Elizavetgrad)," in *The Jewish Encyclopedia*, vol. 12, *Talmud–Zweifel* (New York: Funk and Wagnalls, 1906), 592.

12. Mikhail Dolbilov, "The Emancipation Reform of 1861 in Russia and the Nationalism of the Imperial Bureaucracy," in *Construction and Deconstruction of National History in Slavic Eurasia,* ed. T. Hayashi (Sapporo: Slavic Research Center, 2003), 208–235.

13. *Ezhegodnik "Golosa Iuga": Adres-kalendar' i spravochnaia kniga po g. Elisavetgradu i uezdu* (Elisavetgrad: Izdanie redaktsii gazety Golos Iuga, 1913), 24, 76–77.

14. The correlation between patterns of industrialization and the probability of pogroms is explored in Darius Staliūnas, *Enemies for a Day: Antisemitism and Anti-Jewish Violence in Lithuania under the Tsars* (Budapest: Central European University Press, 2015).

15. I. Br. and S. J., "Yelisavetgrad (Elizavetgrad)," 592.

16. The city's *Address-Calendar* prepared by the editorial office of the liberal newspaper *Golos Iuga* (Voice of the South) specifically mentioned the pogroms of 1881 and 1905: "A sad page in Elisavetgrad history is the pogrom of 1881, which financially ruined many Jewish families. In 1905 Elisavetgrad, like many other towns in the Pale of Jewish settlement, became the arena of a Jewish pogrom that ended with a few human casualties and terrible devastation." *Ezhegodnik "Golosa Iuga,"* 63. On political views of the *Golos Iuga* editors, see V. M. Kryzhanivskii, "Elisavetgrads'ka gazeta 'Golos Iuga' ta Rosiis'ka tsenzura u svitli arkhivnykh documentiv," *Naukovi pratsi istorichnogo fakul'tetu Zaporiz'kogo natsional'nogo universitetu* 41 (2014): 265–269.

17. A. N. Pashutin, *Istoricheskii ocherk goroda Elisavetgrada* (Elisavetgrad: Tipo-litografiia Br. Shpolianskikh, 1897).

18. Samuel Weissenberg, *Das Wachstum des Menschen nach Alter, Geschlecht und Rasse* (Stuttgart: Verlag von Strecker und Schröder, 1911).

19. *Ezhegodnik "Golosa Iuga,"* 67, 254–268.

20. "Kirovograd," in *Kratkaia evreiskaia entsiklopediia*, vol. 4 (Jerusalem: Jewish University of Jerusalem, 1988), 304. By different estimates, from 1,300 to 3,000 Jews were murdered in Elisavetgrad during the pogroms on May 15–20, 1919. See O. V. Budnitskii, *Rossiiskie evrei mezhdu kransymi i belymi* (Moscow: ROSSPEN, 2005), 277. In English: Oleg Budnitskii, *Russian Jews between the Reds and the Whites, 1917–1920*, trans. Timothy Portice (Philadelphia: University of Pennsylvania Press, 2012).

21. By November 1915, the Jewish Distribution Committee already registered 4,496 Jewish refugees in Elisavetgrad (which was about 15 percent of the prewar number of Jews in the city): "Report of November, 1915," in *Reports Received by the Joint Distribution Committee of Funds for Jewish War Sufferers Joint Distribution Committee of the American Funds for Jewish War Sufferers*, ed. Felix Moritz Warburg and Albert Luca (New York: Clarence S. Nathan, 1916), 20.

22. Just as the share of Jews among the inhabitants of Elisavetgrad had declined by one-third during the early 1920s, their proportion in the population of Moscow and Leningrad had increased severalfold by 1926 (to 6.5 percent and 5.2 percent, respectively). *Vsesoiuznaia perepis' naseleniia 17 dekabria 1926 g.: Kratkie svodki*, vol. 4, *Narodnost' i rodnoi iazyk naseleniia SSSR* (Moscow: TsSU SSSR, 1928), 44, 51.

23. After Zinovyev's arrest in 1934, the city was renamed once again, this time after Sergei Kirov, the first secretary of the Leningrad City Committee of the All-Union Communist Party, a loyal supporter of Joseph Stalin and one of the most popular Bolshevik leaders. He was assassinated based on a secret order from Stalin in 1934. Zinovyevsk became Kirovo first and, in 1939, Kirovograd. In 2016, this Ukrainian provincial town was given a new name—Kropyvnytskyi.

24. Efron, *Defenders of the Race*, 93.

25. See Weissenberg's biography in V. I. Binshtok, "Nekrolog: S. A. Weissenberg," in *Voprosy biologii i patologii evreev* [Questions of Jewish biology and pathology (QJBP)], vol. 3, no. 1 (Leningrad: Izd. Evreiskogo Istoriko-Etnograficheskogo Obshchestva, 1930), 165–167. Weissenberg's private clinic was located in the building that he owned at Ivanovskaia Street. See *Ezhegodnik "Golosa Iuga,"* 257. In 1913, the Elisavetgrad *Address-Calendar* listed fifty-seven physicians (working for the local zemstvo, city or communal hospitals, the military hospital, or in private practice), excluding dentists. Judging by their names, at least twenty-six of these physicians were Jewish. *Ezhegodnik "Golosa Iuga,"* 257–258.

26. Weissenberg, "Dvizhenie evreiskogo naseleniia Zinov'evska (Elizavetgrada) za 1901–1925," in *QJBP*, vol. 2 (Leningrad: Izd. Evreiskogo istoriko-etnograficheskogo obshchestva, 1928), 189–204.

27. The documents were recently published by the local Kirovograd historian, Vladimir Bos'ko, accessed March 18, 2022, http://raiskiy.livejournal.com/81842.html.

28. Bos'ko, http://raiskiy.livejournal.com/81842.html.

29. Weissenberg, *Die südrussischen Juden*, 126.

30. Efron, *Defenders of the Race*, 96.

31. During his expedition to Palestine, Weissenberg collected measurements on Arabs, Samaritans, and, most importantly for his project, Palestine's Jews who never left the primordial land. He declared their type as the *Urtypus* and designated them as *Judaeus primigenius*. Weissenberg, "Die autochthone Bevölkerung Palästinas in Anthropologischer Beziehung," *Zeitschrift für Demographie und Statistik der Juden* 5, no. 9 (1909): 129–139, esp. 131.

32. Weissenberg, *Die südrussischen Juden*, 578–579. A good summary in Russian: Weissenberg, "Evreiskii tip," *Rassvet*, no. 14–15 (April 5, 1913): 57–61.

33. As Celia Brickman has noted, "The Jews of Europe . . . were variously described as 'Oriental,' 'primitive,' 'barbarian,' 'white Negroes,' 'mulatto,' and 'a mongrel race.' . . . The Eastern European Jew, in particular, could be described 'as the exemplary member of the 'dark-

skinned' races." Brickman, *Aboriginal Populations in the Mind: Race and Primitivity in Psychoanalysis* (New York: Columbia University Press, 2003), 163.

34. Johann Friedrich Blumenbach, *On the Natural Varieties of Mankind* (New York: Berman, 1969). Johann Friedrich Blumenbach, "On the Natural Varieties of Mankind, 1775," in *The Anthropological Treatises of Johann Friedrich Blumenbach* (London: Longman, Roberts and Green, 1865), 145–276.

35. Efron, *Defenders of the Race*, 107.

36. This fact is mentioned in Binshtok's obituary of Weissenberg: Binshtok, "Nekrolog," 166.

37. The name appears elsewhere in this volume as Imperial Society of Lovers of Natural Sciences, Anthropology and Ethnography (IOLEAE), but the society itself often omitted "Imperial" in its official title.

38. The International Congress of Anthropology and Prehistoric Archaeology took place in Moscow in 1892.

39. "A Letter by Dr. of Medicine, S. Weissenberg," 1892, ARAN, f. 446, op. 2, d. 109, ll. 1-1 rev. On Anuchin, see Mogilner, *Homo Imperii*, 17–100, 133–166.

40. Nikolai Maliev, *Antropologicheskii ocherk bashkir* (Kazan: Tipografiia IKU, 1876); Nikolai Zograf, *Antropologicheskii ocherk meshcheriakov Zaural'skoi chasti, Permskoi gubernii* (Moscow: Tipografiia M.N. Lavrova and K, 1878).

41. "A Letter by Dr. of Medicine, S. Weissenberg, 1889," ARAN, f. 446, op. 2, d. 109, l. 2.

42. On the school and its philosophy and methodology, see Mogilner, *Homo Imperii*.

43. Weissenberg, *Die südrussischen Juden*, 568–573; "Protokol publichnogo zasedeniia 25-go oktiabria 1895 g.," *Izvestiia IOLEAE* 15 [*Trudy Antropologicheskogo otdela 8, no. 1–3*] (1895): 491–493, here 492.

44. For a discussion of the Moscow's school philosemitism, see Mogilner, *Homo Imperii*, 217–250.

45. S. Weissenberg, "Karaimy," *RAZh* 17–18, no. 1–2 (1904): 66–75, here 74. The article was republished in the Russian-language Karaim journal: S. Weissenberg, "Antropologiia karaimov," *Karaimskaia zhizn'* 1 (1911): 17–24. The German version was published as "Die Karäer der Krim," *Globus* 84, no. 9 (1903): 139–143.

46. On the Bashkirs, see Charles Steinwedel, *Threads of Empire: Loyalty and Tsarist Authority in Bashkiria, 1552–1917* (Bloomington: Indiana University Press, 2016).

47. Weissenberg, "Karaimy," 74, 75. He confirmed the mixed nature of Karaims in an onomatological study: S. Weissenberg, "Familii karaimov i krymchakov," *Evreiskaia starina* 6 (1913): 384–399.

48. For example, *RAZh*, vol. 23–24, no. 3–4, in 1905 featured Dzhavakhov's article "Toward the Anthropology of Georgia" ("K antropologii Gruzii," 1–46) and Weissenberg's article "On Body Proportions of a Newborn and a Three-Month-Old Child," based exclusively on Jewish anthropometric statistics ("O proportsiiakh tela u novorozhdennogo i trekhmesiachnogo rebenka," 106–126). *RAZh*, vol. 27–28, no. 3–4, in 1907 featured Weissenberg's article "Growth of Head and Face," also based on the Jewish statistics ("Rost golovy i litsa," 67–85), and Dzhavakhov's "Toward the Anthropology of Georgia: Georgians of Kakheti" ("K antropologii Gruzii: Gruziny kakhetii," 127–167). In 1912, *RAZh*, vol. 32, no. 4, published Weissenberg's "Karaims and Krymchaks from the Anthropological Point of View" ("Karaimy i krymchaki s antropologicheskoi tochki zreniia," 35–56) and Dzhavakhov's own "Caucasian Jews" ("Kavkazskie evrei," 57–74).

49. A. N. Dzhavakhov, "Antropologiia Gruzii. I. Gruziny kartalinii i kakhetii," *Izvestiia IOLEAE* 116 [*Trudy Antropologicheskogo otdela 36*] (1908): 1–242.

50. For a detailed analysis of Dzhavakhov's Georgian race science, see Mogilner, *Homo Imperii*, 208–213.

51. S. Weissenberg, "Kavkazskie evrei v antropologicheskom otnoshenii," *RAZh* 30–31, no. 2–3 (1912): 137–163, here 138. Tables with anthropometric statistics and indexes made up to twelve pages of the article (151–163). For the earlier German version, see Weissenberg, "Die Kaukasischen Juden in Anthropologischer Beziehung," *Archiv für Anthropologie* 8, no. 4 (1909): 237–245.

52. S. Weissenberg, "Die Jemenitischen Juden," *Zeitschrift für Ethnologie* 41, no. 8 (1909): 309–327.

53. Weissenberg, "Kavkazskie evrei v antropologicheskom otnoshenii," 148.

54. Earlier, Weissenberg measured eighteen Jews from Central Asia living in the Bukharan quarter of Jerusalem. They came from different regions of Central Asia but spoke a common dialect of Persian. The only physical features they had in common were their medium height and dark complexion; otherwise, all their indexes differed. Weissenberg concluded that Central Asian Jews were a racially mixed group. Weissenberg, "Die Zentralasiatischen Juden in Anthropologischer Beziehung," *Zeitschrift für Demographie und Statistik der Juden* 5, no. 7 (1909): 103–106. He finalized his research project on the Jews of the Caucasus and Central Asia with the comparison of local Jews and Armenians: "The foundation of the similarity between Armenians and Jews is thus to be sought in the actual blending of both peoples. But this did not take place in the prehistoric period on the soil of Palestine, but rather in the historical period on the soil of Armenia." S. Weissenberg, "Armenier und Juden," *Archiv für Anthropologie* 14 (1915): 383–387, here 387. For the Russian version of the same article, see Weissenberg, "K Antropologii Armian," *RAZh* 37–38, no. 1–2 (1916): 61–65.

55. A. N. Dzhavakhov, "Kavkazskie evrei (Po povodu stat'i S. A. Weissenberga)," *RAZh* 32, no. 4 (1912): 57–74.

56. Dzhavakhov, "Kavkazskie evrei," 57. For the original, see Weissenberg, "Kavkazskie evrei v antropologicheskom otnoshenii," 137.

57. S. Weissenberg, "Istoricheskie gnezda Kavkaza i Kryma (iz otcheta o letnei poezdke 1912 goda)," *Evreiskaia starina* 6, no. 1 (1913): 51–69, here 53. Similar descriptions can be found in his article on Turkestani Jews: Weissenberg, "Evrei v Turkestane (otchet o letnei poezdke 1912 goda)," *Evreiskaia starina* 4, no. 4 (1912): 390–405. Writing about Jews in Turkestan, Weissenberg also began with the history of persecutions by ancient indigenous populations and continued with the description of the present-day cultural degradation of the local Jews. "It is extremely strange that indigenous Jews do not strive toward European education. . . . At least I met no one among them who would have completed a vocational school [*kurs srednego uchilishcha*], not to mention higher educational establishments. . . . It should also be noted that Jewish literacy does not particularly flourish among Bukharan Jews." Weissenberg, "Evrei v Turkestane," 398–399.

58. Weissenberg, "Istoricheskie gnezda Kavkaza i Kryma," 60.

59. Compare this to a later article by Weissenberg on Kurdish Jews, where he writes about the medieval Kurdish Jewish community as an important cultural and demographic factor in the region. Since that time, he notes, Asia "has more and more delved into darkness, which also completely enfolded local Jews. . . . Endless prosecutions lasting for entire centuries resulted in a substantial drop of their number. . . . Long-lasting separation from the rest of the Jewry became a cause of their mental backwardness. . . . The only silver lining illuminating their otherwise gloomy life—the hope for a better future, for redemption. . . . My memory retains in particular an image of a good-natured, hefty young man with an expression of deep sadness in his widely open dark eyes. As I was taking his anthropological measurements, he repeated nonstop: 'Maybe, *geula* will come? If God wants, He can give us *geula*!'" Weissenberg, "Kurdistanskie evrei," *Novyi voskhod*, no. 10–11 (March 10, 1915): 79–84, here 81–82.

60. Dzhavakhov, "Kavkazskie evrei," 76, 75.

61. Dzhavakhov, "Kavkazskie evrei," 60. This argument was not unique to Dzhavakhov. It came to the fore of political discussions in the Caucasus and the empire in general after the enactment of the so-called May 3 laws of 1882, stipulating that all the Jews living in villages in the Pale of Settlement had to resettle in towns. They also could not receive mortgages, hold leases, or manage any land outside of the towns. More restrictions followed. Theoretically, the laws were applicable only to the Jews of the Pale, however, de facto there were attempts to apply them to Jews living in the Caucasus as well. This triggered the discussion in the local society, in which Georgian intellectuals often sided with Jewish leaders insisting on a complete assimilation of Jews, Georgian Jews in particular, into the Georgian culture and way of life. The Russian Jewish press followed the debate. For example, in 1904 the Jewish journal *Futurity* (*Budushchnost'*) republished one characteristic article from the newspaper *Bakinskie izvestiia* (Baku news), which argued that in the course of many centuries the Jews of the Caucasus "have completely merged with the rest of the population; they have nationalized. . . . The type of the local Jew and the type of the indigenous peasant do not present the same difference as exists between the Russian peasant and the Jew." "O pravakh gruzinskikh evreev," *Budushchnost'* 14 (April 9, 1904): 263–266.

62. Dzhavakhov, "Kavkazskie evrei," 72.

63. On the concept of "Jewish physiognomy" in race science, see Mogilner, *Homo Imperii*, 217–250.

64. Weissenberg, *Die südrussischen Juden*, 30.

65. "Izvestiia i zametki," *Izvestiia IOLEAE* 90 [*Trudy Antropologicheskogo otdela* 18, no. 1–3] (1897): 501.

66. "Otzyv prof. D. N. Anuchina o trude d-ra S. A. Weissenberg *Die südrussischen Juden*," *Izvestiia IOLEAE* 90 [*Trudy Antropologicheskogo otdela* 18, no. 1–3] (1897): 502–504, here 503.

67. "Otzyv prof. D. N. Anuchina," 504.

68. "Izvestiia i zametki," *RAZh* 30–31, no. 2–3 (1912): 188–189, here 189. See also a draft of this introduction in El'kind's own handwriting: A. El'kind, "Ob antropologicheskikh rabotakh d-ra S. A. Weissenberga za poslednie gody," AMGUL, f. 61, op. 1, d. 319 (Kratkie otchety o nauchnykh issledovaniiakh, provedennykh chlenami Obshchestva po zadaniiam otdelov i otdelenii za 1912 g.), ll. 7–9.

69. "Noveishie issledovaniia evreev d-ra S. A. Weissenberga," *RAZh* 30–31, no. 2–3 (1912): 190.

70. Weissenberg, *Das Wachstum des Menschen*.

71. The influence of this work on textbooks was specifically mentioned in the obituary for Weissenberg, published in *QJBP* in 1930: Binshtok, "Nekrolog," 166.

72. D. Ts., "Bibliography: *Das Wachstum des Menschen nach Alter, Geschlecht und Rasse* von D-r S. Weissenberg. Stuttgart, 1911," *Evreiskii meditsinskii golos* 1–2 (1911): 68–70, here 68, 70.

73. Weissenberg, most probably, would justify his selection of case studies by using the data on *Russian* Jews, hence comparing it to the corresponding statistics for Russians and by selecting a nation known as having many tall individuals—the English. The rest should be relegated to the level of the subconscious.

74. Weissenberg, "Kavkazskie evrei v antropologicheskom otnoshenii," 151.

75. S. Weissenberg, "The Jewish Racial Problem," in Hart, *Jews and Race*, 78.

76. S. Weissenberg, "Evrei kak rasa," *Evreiskii meditsinskii golos* 2–4 (1908): 157–162.

77. He later published a separate article describing this experiment: S. Weissenberg, "Koganity i levity v istorii i sovremennosti," *Evreiskaia starina* 10 (1918): 112–131.

78. S. Weissenberg, "Beiträge zur Volkskunde der Juden," *Globus* 77, no. 8 (1900): 130–131, here 130. I am using the translation provided by Efron, *Defenders of the Race*, 111.

79. "Pis'mo Weissenberga S. A.—An-skomu S. A. June 10, 1912," IR NBUV, f. 339, d. 213, l. 1.

80. "Pis'mo Weissenberga S. A.—An-skomu S. A. March 4, 1912," IR NBUV, f. 339, d. 211, l. 1. Judging by An-sky's report on the meeting to the principal financial sponsor of the expedition, Vladimir Ginsburg, Weissenberg was among the twenty participants of the meeting in Petersburg on March 24–25, 1912: "Twenty participants came to the meeting, including three from Moscow (Dr. Vermel, Marek, and Engel) and Dr. S. A. Weissenberg from Elisavet-grad. Papers on anthropology, ethnography, folklore, and popular music were presented; the research framework was discussed, especially issues such as whether economic studies should be included, [what] the methods of work [should be], the itinerary, and so on." "Pis'mo S. Rappoporta V. G. Gintsburgu, baronu. April 13, 1912," IR NBUV, f. 339, d. 339, l. 1.

81. Simon Dubnov, *Ob izuchenii istorii russkikh evreev i ob uchrezhdenii russko-evreiskogo istoricheskogo obshchestva* (St. Petersburg: A. E. Landau, 1891), 36–37. Quoted in Nathaniel Deutsch, *The Jewish Dark Continent: Life and Death in the Russian Pale of Settlement* (Cambridge, MA: Harvard University Press, 2011), 7.

82. "Pis'mo S. Rappoporta V. G. Gintsburgu, baronu. June 30, 1912," IR NBUV, f. 339, d. 211, l. 1.

83. As examples of his ethnographic writings, see Weissenberg, "Imena iuzhnorusskikh evreev," *Etnograficheskoe obozrenie* 96–97, no. 1–2 (1913): 76–109; "Evrei v russkikh poslovitsakh," *Evreiskaia starina* 8, no. 2 (1915): 228–231; "Evrei v velikorusskoi chastushke," *Evreiskaia starina* 8, no. 1 (1915): 119–120; and "Miniatiura iz evreiskogo martirologa," in *Perezhitoe. Sbornik, posviashchennyi obshchestvennoi i kul'turnoi istorii evreev v Rossii*, vol. 4 (St. Petersburg: Tipogr. I. Fleishmana, 1913), 333–335.

84. Deutsch, *The Jewish Dark Continent*, 44.

85. "Khronika: Ob antropologicheskom tipe evreev, doklad S. A. Weissenberga," *Etnograficheskoe obozrenie* 101–102, no. 1–2 (1914): 283–284, here 284.

86. Morris-Reich, "Jews between Volk and Rasse," 189.

87. S. Weissenberg, "Nachalo razlozheniia v russko-evreiskoi zhizni," *Rassvet*, no. 28 (July 13, 1912): 38–41, here 41.

88. See a discussion of his conservative views on sexual relationships and female emancipation in Efron, *Defenders of the Race*, 120–121.

89. "Pis'mo Weissenberga S. A.—An-skomu S. A. April 6, 1912," IR NBUV, f. 339, d. 209, l. 1.

90. Samuel Weissenberg, "Jüdische Museen und Jüdisches in Museen," *Mitteilungen der Gesellschaft für Jüdische Volkskunde* 23, no. 3 (1907): 77–88; Weissenberg, "Zur Sozialbiologie und Sozialhygiene der Juden," *Archiv für Rassen- und Gesellschaftsbiologie* 19, no. 4 (1927): 402–418.

91. S. Rappaport, "Plan evreiskoi etnograficheskoi ekspeditsii imeni barona G. O. Gintsburga i predvaritel'naia smeta raskhodov," TsGIA SPb, f. 2049 (Sheftel' M. I.), op. 1, d. 84, l. 2.

92. When the program was eventually published and Weissenberg had not received his copy, he wrote to its editor, Lev Shternberg, with some resentment: "I think that as a member of the Committee, I also hold some rights to it, so I hope that you would not refuse to make a call on my behalf to order one copy to be mailed to me. I would be very grateful for this" (May 21, 1915). "Pis'ma Weissenberga S. A. Shternbergu L. Ia, 1912–25," SPbF ARAN, f. 282, op. 2, d. 46, ll. 2–2rev, here l. 2.

93. "Pis'mo Weissenberga S. A.—An-skomu S. A. March 4, 1912," l. 1rev; "Pis'mo Weissenberga S. A.—An-skomu S. A. June 10, 1912," l. 1.

94. In a handwritten report for the Anthropological Division on his expedition to the Caucasus and Crimea in 1912, Weissenberg explained that his main purpose was anthropo-

logical research. However, he also tried to collect observations on the way of life and traditions of local Jews. S. Weissenberg, "Otchet o poezdke v Zakaspiiskii krai, na Kavkaz i v Krym," AMGUL, f. 61, op. 1, d. 319, l. 50.

95. Weissenberg, "Die Kaukasischen Juden," 237–245, here 238.

96. S. Weissenberg, "Istoricheskie gnezda Kavkaza i Kryma (so snimkami tipov i drevnikh pamiatnikov)," *Evreiskaia starina* 5, no. 1 (1913): 51–69, here 69.

97. Weissenberg, "Istoricheskie gnezda Kavkaza i Kryma (so snimkami tipov)," 52.

98. Weissenberg, "Na okrainakh," *Rassvet*, no. 11 (March 15, 1913): 17–18, here 18.

3. Arkadii Danilovich El'kind

Epigraph: A. D. El'kind, "Antropologicheskoe izuchenie evreev za poslednie desiat' let," *Russkii antropologicheskii zhurnal* [Russian anthropological journal (*RAZh*)] 30–31, no. 2–3 (1912): 1–50, here 1.

1. A. D. El'kind, "Evrei (Sravnitel'no-antropologicheskoe issledovanie, preimushchestvenno po nabliudeniiam nad pol'skimi evreiami)," *Izvestiia IOLEAE* 104 [*Trudy Antropologicheskogo otdela* 21] (1903): 1–458.

2. "El'kind, Arkadii Danilovich," in *Evreiskaia entsiklopediia: Svod znanii o evreistve i ego kul'ture v proshlom i nastoiashchem,* ed. Dr. L. Katsenel'son, vol. 16 (St. Petersburg: Obshchestvo dlia nauchnykh evreiskikh izdanii i izd-vo Brokgauz i Efron, 1913), col. 249.

3. "Diplom," TsIAM, f. 418, op. 90, d. 709, l. 8. The exact quotation from the certificate reads: "Aron-Girsh Donov El'kind, son of a *meshchanin,* of Judaic faith." *Meshchanin* means registered in the *meshchanskii* estate; the closest equivalent is "burgher." At the turn of the century, people registered as *meshchane* could be peasants, representatives of intellectual professions, craftsmen, and so on.

4. B. N. Vishnevskii, "Pamiati vrachei-antropologov D. P. Nikol'skogo i A. D. El'kinda," *Zhurnal Kazanskogo mediko-antropologicheskogo obshchestva* 1 (1921): 179–185, here 183.

5. See materials of the first general population census of the Russian Empire (1897) online in "Demoskop Weekly" by the Institute of Demography at the National Research University "Higher School of Economics": for Novgorodskii uezd (region) and Novgorod, accessed March 18, 2022, see www.demoscope.ru/weekly/ssp/rus_lan_97_uezd.php?reg=912; for Mogilevskii uezd and Mogilev, accessed March 18, 2022, see www.demoscope.ru/weekly/ssp/rus_lan_97_uezd.php?reg=793.

6. Vishnevskii, "Pamiati vrachei-antropologov," 183.

7. Expeditions are mentioned in the report: "Noveishie antropologicheskie issledovaniia v Rossii," *RAZh* 1, no. 1 (1900): 124–126, here 125.

8. "Protokoly publichnogo zasedaniia 25-go oktiabria 1895 g.," *Trudy Antropologicheskogo otdela* 8, no. 1–3 (1895): 491–493.

9. A. D. El'kind, "Privislianskie poliaki (Antropologicheskii i kraniologicheskii ocherk)," *Izvestiia IOLEAE* 90 [*Trudy Antropologicheskogo otdela* 18, no. 1–3] (1897): 1–392.

10. "El'kind, Arkadii Danilovich," in *Evreiskaia Entsiklopediia,* vol. 16, col. 249.

11. Vishnevskii, "Pamiati vrachei-antropologov," 184.

12. Sergi undermined the validity of the widely used cephalic index (the percentage of breadth to length of a skull), introduced by the Swedish anatomist and anthropologist Anders Retzius in 1844, as a substantial basis for human racial classification. Instead, he developed cranial morphology, which he believed held the key to the persistence of primitive biological traits in some populations. Peter D'Agostino, "Craniums, Criminals, and the 'Cursed Race': Italian Anthropology in American Racial Thought, 1861–1924," *Comparative Studies*

in Society and History 44, no. 1 (2002): 319–343; Fedra A. Pizzato, "How Landscapes Make Science: Italian National Narrative, The Great Mediterranean, and Giuseppe Sergi's Biological Myth," in *Mediterranean Identities: Environment, Society, Culture*, ed. Borna Fuerst-Bjeliš (Rijeka, Croatia: IntechOpen, 2017), 79–89, open access book, http://www.intechopen.com/books/mediterranean-identities-environment-society-culture.

13. Presumably, Sergi developed his method after having lined up four hundred Melanesian skulls on several tables and carefully observed each of them and the whole collection with the unaided eye of a museum visitor. Without taking any measurements, he began grouping skulls together in categories and subcategories. Each skull type designated a separate race as a species of the hominid family. For a more detailed account and analysis, see Aaron Gillette, *Racial Theories in Fascist Italy* (London: Routledge, 2002), 25–32.

14. "Izvestiia i zametki: Antropologicheskii otdel Obshchestva liubitelei estestvoznaniia: A. D. El'kind, 'O cherepnykh tipakh prof. Sergi sredi nekotorykh kraniologicheskikh kollektsii Munkhenskogo antropologicheskogo instituta,'" *RAZh* 1, no. 2 (1900): 114.

15. The catacombs, which are on two levels between 5 and 16.3 meters below the surface, were used from the second to the fourth century, with the maximum number of interments at the end of that period. They were accidentally unearthed in 1859. For details, see Jessica Dello Russo, "The Discovery and Exploration of the Jewish Catacomb of the Vigna Randanini in Rome: Records, Research, and Excavations through 1895," *Roma Subterranea Judaica* 5 (2011): 1–24. Dello Russo underlines that the discovery was accidental, and the excavation privately conducted; the sites themselves were all eventually abandoned or even destroyed. This confirms El'kind's description of the site and explains the easy access to the skulls that he had obtained.

16. A. D. El'kind, "Zametka o cherepakh iz evreiskikh katakomb v Rime," *Izvestia IOLEAE* 95 [*Trudy Antropologicheskogo otdela* 19] (1899): 230–231, here 230.

17. El'kind, "Zametka o cherepakh," 231.

18. Cesare Lombroso, *L'antisemitismo e le scienze moderne* (Torino: L. Roux, 1894), appendix.

19. Lombroso, *L'antisemitismo*; Felix von Luschan, "Die anthropologische Stellung der Juden," *Correspondenz-Blatt der deutschen Gesellschaft für Anthropologie, Ethnologie und Urgeschichte* 23 (1892): 94–102. Luschan believed that the racial basis of modern Jews had been formed by ancient Hittites, with some admixture of Amorites ("Aryans") and Semites.

20. In 1880, there were 16,000 Jews in Moscow (of whom 8,025 were officially registered). As a result of the anti-Jewish measures of the city governor, Grand Duke Sergei Alexandrovich, who attempted the expulsion of Jews in 1891–1894, this number declined. According to the imperial census of 1897, Moscow's Jewish population did not exceed 8,000, 40 percent of whom claimed Russian as their native tongue. Leonid Praisman, "Moscow," in *The YIVO Encyclopedia of Jews in Eastern Europe*, last consulted March 19, 2022, http://www.yivoencyclopedia.org/article.aspx/Moscow.

21. El'kind, "Evrei (Sravnitel'no-antropologicheskoe issledovanie)."

22. El'kind, "Antropologicheskoe izuchenie evreev."

23. "Ob'iavlenie dlia napechataniia v gazetakh," TsIAM, f. 418, op. 90, d. 709, l. 3. On the defense, see ll. 4–4 rev.

24. "Antropologicheskoe izuchenie evreev i disput A. D. El'kinda," *Zemlevedenie* 1–2 (1913): 229–234, here 234.

25. David Schimmelpenninck van der Oye, *Toward the Rising Sun: Russian Ideologies of Empire on the Path to War with Japan* (DeKalb: Northern Illinois University Press, 2001).

26. For the view of the first Russian Revolution as the peak of mass politics and its ethnicization, see Ilya Gerasimov, ed., *Novaia imperskaia istoriia Severnoi Evrazii*, vol. 2, *Balansirovanie imperskoi situatsii, XVIII–XX vv.* (Moscow: Ab Imperio, 2017), 420–540.

27. Laura Engelstein, *The Keys to Happiness: Sex and the Search for Modernity in Fin-de-Siècle Russia* (Ithaca, NY: Cornell University Press, 1992); Marina Mogilner, *Mifologiia "podpol'nogo cheloveka": Radial'nyi mikrokosm v Rossii nachala XX veka kak predmet semioticheskogo analiza* (Moscow: NLO, 1999), 61–100; Daniel Beer, *Renovating Russia: The Human Sciences and the Fate of Liberal Modernity, 1880–1930* (Ithaca, NY: Cornell University Press, 2008).

28. Alexander Semyonov, "'The Real and Live Ethnographic Map of Russia': The Russian Empire in the Mirror of the State Duma," in *Empire Speaks Out: Languages of Rationalization and Self-Description in the Russian Empire*, ed. Ilya Gerasimov, Jan Kusber, and Alexander Semyonov (Leiden: Brill, 2009), 191–228; Vladimir Levin, "Die jüdischen Wähler und die Reichsduma," in *Von Duma zu Duma: Hundert Jahre russischer Parlamentarismus*, ed. Dittmar Dahlmann and Pascal Trees (Bonn: Bonn University Press and V&R Unipress, 2009), 155–172; Levin, "Russian Jewry and the Duma Elections, 1906–1907," in *Jews and Slavs*, vol. 7, ed. W. Moskovich, L. Finberg, and M. Feller (Jerusalem: Hebrew University of Jerusalem, 2000), 233–264.

29. Marina Vitukhnovskaia-Kaupalla, *Finskii sud vs "Chornaia sotnia": Rassledovanie ubiistva Mikhaila Gertsenshteina i sud nad ego ubiitsami (1906–1909)* (St. Petersburg: Izd-vo Evropeiskogo universiteta, 2015).

30. Robert Weinberg, *Blood Libel in Late Imperial Russia: The Ritual Murder Trial of Mendel Beilis* (Bloomington: Indiana University Press, 2013). Throughout the book, I use "Kiev" to reflect the usage in the sources, and I use "Kyiv" when speaking in my own voice and in the notes to reflect the contemporary Ukrainian norm.

31. "V Sovet imperatorskogo Moskovskogo universiteta. 15 dekabria 1912," TsIAM, f. 418, op. 90, d. 709, ll. 5–5 rev.

32. "Diplom." The archival copy features El'kind's handwriting: "I have received the original certificate."

33. For more on formal procedures involved in the process of preparation for academic service in Russian imperial universities, see A. E. Ivanov, *Vysshaia shkola Rossii v kontse XIX–nachale XX veka* (Moscow: Institut istorii AN SSSR, 1991), 210–217.

34. A. D. El'kind, "O zadachakh vozobnovliaemogo zhurnala," *RAZh* 37–38, no. 1–2 (1916): 1–3, here 1–2.

35. "Anthropological Notes: Russian Anthropologists," *American Anthropologist*, n.s., 26, no. 1 (1924): 135.

36. "Noveishie antropologicheskie issledovaniia v Rossii," 125.

37. A. D. El'kind, "Evrei (s 25 ris. i 4 diagr.)," *RAZh* 11, no. 3 (1902): 1–44.

38. "Prisuzhdenie premii po antropologii imeni ego imperatorskogo vysochestva velikogo kniazia Sergeia Aleksandrovicha A. D. El'kindu za ego issledovanie 'Evrei,'" *RAZh* 11, no. 3 (1902): 117–119.

39. El'kind, "Evrei (Sravnitel'no-antropologicheskoe issledovanie)."

40. See, for example, the introductory part of his 1902 article: El'kind, "Evrei (s 25 ris. i 4 diagr.)," 1–9. Most importantly, see his addendum to the dissertation: El'kind, "Antropologicheskoe izuchenie evreev."

41. "Prisuzhdenie premii," 117.

42. Marina Mogilner, *Homo Imperii: A History of Physical Anthropology in Russia* (Lincoln: University of Nebraska Press, 2013), 133–140.

43. V. V. Bunak, "Deiatel'nost' Anuchina v oblasti antropologii," *RAZh* 3–4 (1924): 5.

44. V. Bunak, "Morfologicheskoe opisanie mozga D. N. Anuchina," *RAZh* 3–4 (1926): 28, 11.

45. Vistula Land or Vistula Country (Privislenskii krai) was the name applied to the former lands of Congress Poland from the 1880s, following the defeat of the January Uprising

(1863–1864) and the general nationalization of politics in the empire. It was meant to annihilate any possible association with the idea of a historical "Poland." M. Dolbilov and A. Miller, *Zapadnye okrainy Rossiiskoi imperii* (Moscow: NLO, 2007).

46. El'kind, "Privislianskie poliaki."

47. Julian Tal'ko-Hryntsevich, "K antropologii ukrainskikh i litovskikh evreev," *Protokoly Russkogo antropologicheskogo obshchestva* 3 (1892): 3–57; Julian Talko-Hryncewicz, "Charakterystyka fizyczna Ludności zydowskieij Litwy i Rusi," *Zbiór wiadomości do antropologii krajowej* 16 (1892): 1–62; Józef Mayer and Izydor Kopernicki, "Charakterystyka fizyczna Ludności galicyjskiej," in *Zbiór wiadomości do antropologii krajowej,* vols. 1 and 2 (Kraków: Drukarnia Uniwersytetu Jagiellońskiego, 1877, 1885).

48. I. I. Pantiukhov, *Antropologicheskie nabliudeniia na Kavkaze: Chitannye v zasedanii Kavkazskogo otdela IRGO* (Tiflis: Tipogr. K. P. Kozlovskogo, 1893), 35–38.

49. El'kind, "Evrei (s 25 ris. i 4 diagr.)," 12.

50. "Antropologicheskoe izuchenie evreev i disput," 230.

51. El'kind, "Evrei (s 25 ris. i 4 diagr.)," 18, 19.

52. El'kind, "Evrei (s 25 ris. i 4 diagr.)," 42.

53. "Prisuzhdenie premii," 119.

54. "Antropologicheskoe izuchenie evreev i disput," 233.

55. El'kind, "Evrei (Sravnitel'no-antropologicheskoe issledovanie)," "Predislovie," iii.

56. El'kind, "Evrei (Sravnitel'no-antropologicheskoe issledovanie)," "Posviashchenie," iii.

57. D. A. (Dmitrii Anuchin), "Evrei (v antropologicheskom otnoshenii)," in *Entsiklopedicheskii slovar' F. A. Brokgauza and I. E. Efrona,* vol. 11 (St. Petersburg: Tipo-litogr. I. A. Efrona, 1893), 426–428.

58. D. A., "Evrei (v antropologicheskom otnoshenii)," 426.

59. "Protokol publichnogo zasedaniia 25 oktiabria 1895 g.," *Izvestiia IOLEAE* 90 [*Trudy Antropologicheskogo otdela* 18, no. 1–3] (1897): 491–493, here 492.

60. "Protokol publichnogo zasedaniia 25 oktiabria 1895 g.," 428.

61. Although Anuchin also used statistical data on the height of Jewish conscripts in his famous study of height as a race indicator, based on the anthropometric data of Bavarian, Hungarian, and Russian imperial conscripts, he never personally measured Jewish skulls, collected anthropometric statistics from the living Jewish population, or debated in the press with other students of Jewish race. D. N. Anuchin, "O geograficheskom raspredelenii rosta muzhskogo naseleniia Rossii (po dannym o vseobshchei voinskoi povinnosti v imperii za 1874–1883 gg.) sravnitel'no s raspredeleniem rosta v drugikh stranakh," *Zapiski Geograficheskogo obshchestva* 8 (1889): 1–184.

62. Joseph Jacobs, *The Jewish Race: A Study in National Character* (London: privately printed, 1899); Jacobs, *Jewish Statistics: Social, Vital and Anthropometric* (London: D. Nutt, 1891); Luschan, "Die anthropologische Stellung der Juden." Useful analysis of the views of all the scholars mentioned by El'kind: John M. Efron, *Defenders of the Race: Jewish Doctors and Race Science in Fin-de-Siècle Europe* (New Haven, CT: Yale University Press, 1994), 20–28, 58–90.

63. El'kind, "Evrei (s 25 ris. i 4 diagr.)," 4.

64. A. A. Ivanovskii, *Ob antropologicheskom sostave naseleniia Rossii* [*Izvestia IOLEAE* 105; *Trudy Antropologicheskogo otdela* 22] (Moscow: Otdel. tipografii t-va I. D. Sytina, 1904), 7, 30.

65. El'kind, "Evrei (Sravnitel'no-antropologicheskoe issledovanie)," iii.

66. Ivanovskii, *Ob antropologicheskom sostave,* 196.

67. Ivanovskii, *Ob antropologicheskom sostave,* 142–143.

68. Ivanovskii, *Ob antropologicheskom sostave,* 152.

69. See one of his earlier articles: A. A. Ivanovskii, "Evrei (Ripley über die Anthropologie der Juden [*Globus* 76, no. 2])," *RAZh* 1, no. 2 (1900): 87.

70. Mogilner, *Homo Imperii*, 136–140.

71. Rudolf Virchow, "Gesamtbericht über die von der deutschen anthropologischen Gesellschaft veranlassten Erhebungen über die Farbe der Haut, der Haare und der Augen der Schulkinder in Deutschland," *Archiv für Anthropologie* 16 (1886): 275–475. For a useful analysis of Virchow's survey of schoolchildren, see Andrew Zimmerman, *Anthropology and Antihumanism in Imperial Germany* (Chicago: University of Chicago Press, 2001), 136–141.

72. On Virchow's political and anthropological liberalism, see Andrew Evans, "A Liberal Paradigm? Race and Ideology in Late-Nineteenth-Century German Physical Anthropology," *Ab Imperio* 8, no. 1 (2007): 113–138.

73. Andrew Zimmerman, "Race and World Politics: Germany in the Age of Imperialism, 1878–1914," in *The Oxford Handbook of Modern German History*, ed. Helmut Walser Smith (Oxford: Oxford University Press, 2011), 359–377, here 367.

74. El'kind, "Privislianskie poliaki," introduction.

75. El'kind, "Privislianskie poliaki," introduction.

76. El'kind, "Evrei (Sravnitel'no-antropologicheskoe issledovanie)," 42.

77. A. D. El'kind, "M. Fishberg. *The Comparative Pathology of the Jews.* 'N.Y. Med. Jour.,' 1901, March–April," *RAZh* 11, no. 3 (1902): 107–111; El'kind, "M. Fishberg. *The Relative Infrequency of Tuberculosis among Jews* ('American Medicine,' 1901, November)," *RAZh* 11, no. 3 (1902): 11; El'kind, "M. Fishberg: Physical Anthropology of the Jews. Washington, 1902," *RAZh* 13, no. 1 (1903): 154–155; El'kind, "M. Fishberg. Physical Anthropology of the Jews. 2. Pigmentation. Washington, 1903," *RAZh* 14, no. 2 (1903): 90–91. Here El'kind provided incorrect information; both parts of Fishberg's *Physical Anthropology of the Jews* appeared in the journal *American Anthropologist*, n.s., 4 (1902): 684–706, and 5 (1903): 89–106. The most detailed treatment of Fishberg's works published after 1903 could be found in El'kind's dissertation addendum: El'kind, "Antropologicheskoe izuchenie evreev," 2–18.

78. Maurice Fishberg, *The Jews: A Study of Race and Environment* (London: Walter Scott, 1911). The book was republished in 2006 by Transaction Publishers (New Brunswick, NJ). Fishberg, "Preface from *Jews: A Study of Race and Environment*," in *Jews and Race: Writings on Identity and Difference, 1880–1940*, ed. Mitchell B. Hart (Waltham, MA: Brandeis University Press, 2011), 21–23, here 22–23.

79. Maurice Fishberg, "Physical Anthropology of the Jews. I.—The Cephalic Index," *American Anthropologist*, n.s., 4, no. 4 (1902): 701.

80. Fishberg, *The Jews*, 470, 472.

81. There are also reasons to believe that El'kind at least partially shared the arguments often advanced by Zionists about the future of Eastern Europe Jewry: "With whom is the Jew of Eastern Europe to assimilate if he is to assimilate at all? Clearly with the Russian muzhik or the Galician or Polish peasant. But this is a proposal that a superior race shall become absorbed by a greatly inferior, a stronger by a weaker, a sober by a particularly unsober one, and is altogether contrary to the course of race absorption." *Zionism: A Jewish Statement to the Christian World* (New York: Federation of American Zionists, 1907), 4.

82. A. El'kind, "Anthropologische Untersuchungen über die russisch-polnischen Juden und der Wert Dieser Untersuchungen für die Anthropologie der Juden im allgemeinen," *Zeitschrift für Demographie und Statistik der Juden* 2, no. 4 (1906): 49–54, and 5 (1906): 65–69. Here I am quoting the English translation from Hart, *Jews and Race*, 81.

83. El'kind, "Dr. M. Fischberg. *Zur Frage der Herkunft des blonden Elements im Judentum.* 'Zeitschrift für Demographie und Statistik der Juden,' 1907, no. 1 and 2," *RAZh* 25–26, no. 1–2 (1907): 250–255, here 255.

84. El'kind, "Dr. M. Fischberg. *Zur Frage der Herkunft,*" 255. Haidamacks were the discontented peasants and Cossacks of the Greek Orthodox faith in eighteenth-century Poland, who conducted organized attacks on their Catholic masters—the Polish nobles, and the Jews. Bands composed of runaway serfs, Zaporozhians, and Cossacks from Ukrainian lands under Muscovy's control in 1734 and again in 1750 robbed and destroyed many towns, villages, and estates in Kyev, Volhynia, and Podolia, killing a great number of Jews and Polish nobles.

85. "Antropologicheskoe izuchenie evreev i disput," 232.

86. Joseph Jacobs, "On the Racial Characteristics of Modern Jews," *Journal of the Anthropological Institute of Great Britain and Ireland* 15 (1886): 23–62, here 32.

87. "Antropologicheskoe izuchenie evreev i disput," 232.

88. El'kind, "Evrei (Sravnitel'no-antropologicheskoe issledovanie)," 42.

89. "Antropologicheskoe izuchenie evreev i disput," 232. In a different case, when Anuchin criticized the work by his St. Petersburg colleague Fiodor Volkov, who in the study "The Anthropological Features of the Ukrainian People" (1916) posited the existence of the pure Ukrainian type and explained all deviations from it by marginal ethnic admixtures, he elaborated on the theme of ethnicity: "Volkov constantly speaks about 'ethnic' influences, 'ethnic' admixtures etc., but the Greek word ethnos—the people [*narod*] implies the spiritual essence of the people, and not its bodily characteristics. Ethnic influence can be felt in language, way of life, folklore, customs, costume, ornaments and so on, but not in the height, the length of legs or the shape of noses." D. Anuchin, "K antropologii ukraintsev," *RAZh* 1–2 (1918): 48–60, here 54.

90. El'kind, "Evrei (Sravnitel'no-antropologicheskoe issledovanie)," 42; emphasis added.

91. "Antropologicheskoe izuchenie evreev i disput," 232.

92. Mogilner, *Homo Imperii,* 17–23.

93. Zimmerman, "Race and World Politics," 361.

94. A. D. El'kind, "Die sozialen Verhältnisse der Juden in Preussen und Deutschland. Statistisch dargestellt von Arthur Ruppin. *Jahrbücher für Nationalökonomie und Statistik,* herausgegeb. von I. Conrad, März 1902," *RAZh* 11, no. 3 (1902): 111–112; El'kind, "Dr. J. M. Judt. *Zydzi jako rasa fizyczna. Analiza z dzidziny antropologii. Z 24 rysunkami, mapa i tablicami w tekse.* Warsawa, 1902," *RAZh* 11, no. 3 (1902): 105–107; El'kind, "K sotsial'noi bor'be s vyrozhdeniem," *Evreiskii meditsinskii golos* 2–4 (1908): 141–156, here 144.

95. El'kind, "K sotsial'noi bor'be s vyrozhdeniem," 156.

96. A. D. El'kind, "K antropologii negrov: Dagomeitsy (s 6 ris.)," *RAZh* 29, no. 1 (1912): 20–35.

97. El'kind, "K antropologii negrov," 34–35.

4. Lev Iakovlevich Shternberg

Epigraphs: Etienne Balibar, "Is There a 'Neo-Racism'?," in Etienne Balibar and Immanuel Wallerstein, *Race, Nation, Class: Ambiguous Identities,* trans. Chris Turner (1988; repr., London: Verso, 1991), 17–28, here 22; Shternberg quoted in Bruce Grant, "Foreword," in L. Ia. Shternberg, *The Social Organization of the Gilyak* (New York: American Museum of Natural History, 1999), xxiii–lvi, here lv.

1. Shternberg first enrolled in the St. Petersburg University Department of Physics and Mathematics but was arrested and expelled in November 1882 for clandestine revolutionary activities. In 1883, the minister of education allowed the expelled students to reapply, and

this time Shternberg chose Odessa. For more, see Sergei Kan, *Lev Shternberg: Anthropologist, Russian Socialist, Jewish Activist* (Lincoln: University of Nebraska Press, 2009), 8–13.

2. The story of Shternberg's legalization in Petersburg is carefully reconstructed in Kan, *Lev Shternberg*, 117–121. It took a special appeal from a number of academicians to gain permission for Shternberg, then already a recognized specialist on the Gilyak (Nivkh) ethnography and language, to take his exams.

3. Grant, "Foreword," xlvi–xlvii.

4. S. Ratner-Shternberg, "L. Ia. Shternberg i leningradskaia etnograficheskaia shkola v 1904–1927 gg," *Sovetskaia etnografiia* 2 (1935): 143–154; T. R. Roon, "Lev Iakovlevich Shternberg: U istokov sovetskoi etnografii," in *Vydaiushchiesia otechestvennye etnologi i antropologi XX veka,* ed. V. Tishkov and D. Tumarkin (Moscow: Nauka, 2004), 49–94.

5. Ratner-Shternberg, "L. Ia. Shternberg i leningradskaia etnograficheskaia shkola."

6. In prison, Shternberg learned English and Italian, read and wrote poetry and fiction, and kept a diary and a journal with reflections on the scholarly works of authors such as Charles Darwin, Herbert Spencer, Pierre-Joseph Proudhon, and Karl Marx. Kan, *Lev Shternberg*, 13, 17.

7. Kan, *Lev Shternberg*, 43–52, 59–63.

8. Kan, *Lev Shternberg*, 45, xix, 49.

9. Grant, "Foreword"; Kan, *Lev Shternberg*.

10. Vl. Bogoraz, "Shternberg kak chelovek i uchenyi," *Etnografia* 2 (1927): 267–282; Bogoraz, "Shternberg kak etnograf," *Sbornik MAE* 7 (1928): 4–30; M. Krol, "Vospominaniia o L. Ia. Shternberge," *Katorga i ssylka* 57–58 (1929): 214–236; Nina Gagen-Torn, *Lev Iakovlevich Shternberg* (Leningrad: Vostochnaia literatura, 1975); T. Staniukovich, "L. Ia. Shternberg i Muzei antropologii i etnografii," *Sovetskaia etnografiia* 5 (1986): 81–91.

11. Kan, *Lev Shternberg*, xviii.

12. Khaim Lev was the oldest son of Iankel Moishe Shternberg, a businessman from Zhitomir. Kan describes him as more open-minded than other representatives of his milieu: he tried farming at some point in his life, enjoyed the outdoors, and had a decent command of Russian alongside his native Yiddish. After his sons had graduated from heder, he enrolled them in the local Russian gymnasium. Shternberg's mother was much more pious and traditional in her outlook. Kan, *Lev Shternberg*, 1–3.

13. Kan, *Lev Shternberg*, 15, 19–20.

14. Erich Haberer, *Jews and Revolution in Nineteenth-Century Russia* (Cambridge: Cambridge University Press, 1995); Jonathan Frankel, *Prophecy and Politics: Socialism, Nationalism, and the Russian Jews, 1862–1917* (Cambridge: Cambridge University Press, 1981).

15. Daniel Beer, *Renovating Russia: The Human Sciences and the Fate of Liberal Modernity, 1880–1930* (Ithaca, NY: Cornell University Press, 2008).

16. The piece was first published in April 1905 in the newspaper *Syn otechestva* (Son of the Fatherland). Later it was reprinted four times in different editions.

17. The newspaper was published in St. Petersburg in Russian from January to August 1907, when it was closed by the Censorship Committee. See Christoph Gassenschmidt, *Jewish Liberal Politics in Tsarist Russia, 1900–1914: The Modernization of Russian Jewry* (New York: New York University Press, 1995), 57.

18. See a nuanced analysis in V. E. Ke'lner, "Ikh tseli mogut byt' vysoki, no oni—ne nashi tseli (M. M. Vinaver—antisionist)," *Judaica Petropolitana* 1 (2013): 114–132.

19. *Programmnye polozheniia Evreiskoi narodnoi gruppy,* a brochure in SPbF ARAN, f. 282 (Shternberg Lev Iakovlevich), op. 1, d. 177, ll. 40–41 rev.

20. *Programmnye polozheniia*, 41.

21. L. Shternberg, "Zadachi russkogo evreistva," in *Pervyi uchreditel'nyi s'ezd Evreiskoi narodnoi gruppy* (St. Petersburg: Tsentral'naia tipo-litografiia M. Ia. Minkova, 1907), 11–32.

22. M. B. Ratner, "Sovremennaia postanovka evreiskogo voprosa," in I. V. Gessen, Ratner, and L. Ia. Shternberg, *Nakanune probuzhdeniia: Sbornik statei po evreiskomu voprosu* (St. Petersburg: Izd-vo A. G. Rozen, 1906), 99–148, here 130–131.

23. Ratner, "Sovremennaia postanovka evreiskogo voprosa," 147.

24. L. Ia. Shternberg, "Tragediia shestimillionnogo naroda," in Gessen, Ratner, and Shternberg, *Nakanune probuzhdeniia*, 163–187, here 187.

25. "Sekuliarizatsiia natsional'nosti," *Svoboda i ravenstvo* 7 (February 1, 1907): 1–3, here 2.

26. L. Sh-g (Lev Shternberg), "Besedy s chitatelem. XV," *Novyi voskhod* 26 (June 30, 1911): 7–14, here 7–8.

27. L. Sh-g, "Besedy s chitatelem. XV," 8–9.

28. Jeffrey Veidlinger, "Introduction," in *Going to the People: Jews and the Ethnographic Impulse*, ed. Jeffrey Veidlinger (Bloomington: Indiana University Press, 2016), 1–26, here 5.

29. Robert M. Seltzer, *Simon Dubnow's "New Judaism": Diaspora Nationalism and the World History of the Jews* (Leiden: Brill, 2014).

30. On Sikorskii's arguments during the trial, see Marina Mogilner, "Human Sacrifice in the Name of a Nation: The Religion of Common Blood," in *Ritual Murder in Russia, Eastern Europe, and Beyond: New Histories of an Old Accusation*, ed. Eugene M. Avrutin, Jonathan Dekel-Chen, and Robert Weinberg (Bloomington: Indiana University Press, 2017), 130–150.

31. L. Sh. (Lev Shternberg), "Chetvertaia nedelia protsessa," *Novyi voskhod* 43 (October 24, 1911): 3–6, here 4.

32. Marina Mogilner, *Homo Imperii: A History of Physical Anthropology in Russia* (Lincoln: University of Nebraska Press, 2013), 34–53.

33. L. Ia. Shternberg, "Sakhalinskie giliaki," *Etnograficheskoe obozrenie*, no. 60 (1904): 1–42; no. 61 (1904): 19–55; no. 63 (1904): 66–119; L. Ia. Shternberg, "V. F. Miller kak etnograf," *Zhivaia starina* 22, no. 3–4 (1913): 417–425.

34. With only one exception: A. Gren, "Zakavkazskie iudei: Istoriko-etnograficheskii eskiz," *Etnograficheskoe obozrenie* 1 (1893): 15–21.

35. Veidlinger, "Introduction," 5.

36. Kan, *Lev Shternberg*, 174–177.

37. "Protokol zasedaniia etnograficheskogo otdela RGO ot 16 dekabria 1911 goda," *Zhivaia starina* 20, no. 1 (1911): xlviii–xlix.

38. Grant, "Foreword," xxxvii.

39. Quoted in Kan, *Lev Shternberg*, 51.

40. It was reprinted under the title "A Recently Discovered Case of Group Marriage," *Die Neue Zeit* 11, bund 1, no. 12 (1892): 373–375.

41. Kan, *Lev Shternberg*, 55, and elsewhere.

42. Grant, "Foreword."

43. Both Radlov and Shternberg shared the view that "the goal of an Academic Museum is to build an exhibition illustrating the evolution of human culture from the prehistoric period to the highest cultures of the modern day": 1903 memo written by Radlov with input from Shternberg, quoted in Kan, *Lev Shternberg*, 135–136.

44. Kan, *Lev Shternberg*, 204.

45. S. An-sky, "Natsionalizm tvorcheskii i natsionalizm razgovornyi," *Evreiskii mir* 19–20 (September 10, 1910): 12–20, here 12; L. Sh-g (Lev Shternberg), "Besedy s chitateliami," *Novyi voskhod* 10 (March 10, 1911): 6–10, here 6.

46. "Ob organizatsii evreiskoi etnograficheskoi ekspeditsii imeni barona Goratsia Osipovicha Gintsburga," *Novyi voskhod* 23 (June 7, 1913): 24–25, here 24.

47. A letter from the Dneprovsk factory workers in Dubrovna, Mogilev province, explained that the factory employed "exclusively Jews," who set up a mutual aid fund and a library. Their poor financial situation was exacerbated by a strike "that happened after Passover" and did not allow them to subscribe to *Jewish Antiquity*. The workers asked to donate it to their library: "Among us, Jewish workers, there are many who would read the journal 'Jewish Antiquity' for 1909 and 1910 with great appetite." See "Pis'mo 'Gospodam chlenam Komiteta EIEO,' polucheno 2 dek.[abria] 1910," TsGIA SPb, f. 2134 ("Redaktsia zhurnala 'Evreiskaia starina'"), d. 1, ll. 69–69 rev; "Pis'mo 'Redaktsia "Karaimskoi zhizni,"' obrashchaetsia 11 iunia 1911," TsGIA SPb, f. 2134, d. 1, l. 200.

48. L. Ia. Shternberg, "Chelovechestvo," in *Entsiklopedicheskii slovar' F. A. Brokgauza i I. A. Efrona*, vol. 38, *Chelovek-Chuguevskii polk* (St. Petersburg: Tipografiia aktsionernogo obshchestva Brokgauz-Efron, 1903), 486–488, here 487. See also Shternberg's article "Tailor ili Teilor," in *Entsiklopedicheskii slovar' F. A. Brokgauza i I. A. Efrona*, vol. 37, *Tai-Termity* (St. Petersburg: Tipografiia aktsionernogo obshchestva "Izdatel'skoe delo," 1901), 485–487.

49. See, for example, L. Sh-g (Lev Shternberg), "Besedy s chitateliami," *Novyi voskhod* 6 (1910): 4–7; 9 (1910): 3–7; 3 (1911): 6–9.

50. Shternberg quoted in Sergei Kan, "'To Study Our Past, Make Sense of Our Present and Develop Our National Consciousness': Lev Shternberg's Comprehensive Program for Jewish Ethnography in the USSR," in Veidlinger, *Going to the People*, 64–84, here 71.

51. Gabriella Safran, "Jews as Siberian Natives: Primitivism and S. An-sky's *Dybbuk*," *Modernism/Modernity* 13, no. 4 (2006): 635–655, here 644.

52. On primitivism and the problematization of "instinct" in the early twentieth-century psychiatry and medicine, see Marina Mogilner, "Racial Psychiatry and the Russian Imperial Dilemma of the 'Savage Within,'" *East Central Europe* 43 (2016): 99–133.

53. Nathaniel Deutsch, *The Jewish Dark Continent: Life and Death in the Russian Pale of Settlement* (Cambridge, MA: Harvard University Press, 2011), 5.

54. As Eugene M. Avrutin and Harriet Murav have noted, many of the photographs taken during the expedition still "reflect the underlying conceptual models of physical anthropology." Avrutin and Murav, "Photographing the Jewish Nation," in Avrutin, Valerii Dymshits, Alexander Ivanov, Alexander Lvov, Harriet Murav, and Alla Sokolova, eds., *Photographing the Jewish Nation: Pictures from S. An-sky's Ethnographic Expeditions*, ed. Eugene M. Avrutin, Valerii Dymshits, Alexander Ivanov, Alexander Lvov, Harriet Murav, and Alla Sokolova (Waltham, MA: Brandeis University Press, 2014), 1–26, here 12. At the beginning of 1914, Shternberg edited the first volume of the 2,087-question *The Jewish Ethnographic Program: The Person*: S. An-sky, *Dos yidishe etnografishe program: Der mentsh*, vol. 1, ed. Lev Shternberg (Petrograd: Drukerei fun Iosef Lurie et comp., 1914). The second volume was never published. It radically differed from Shternberg's Sakhalin questionnaires, which were almost exclusively focused on kinship terms and relations. Unlike them, *The Jewish Ethnographic Program* was not practical, came out late, and was never actually used in the field. According to unpublished correspondence, it satisfied An-sky much more than it did Shternberg, whose degree of actual involvement in its development, besides giving it his authoritative scholarly name, remains unclear. For more, see Nathaniel Deutsch, "Thrice Born; or, Between Two Worlds: Reflexivity and Performance in An-sky's Jewish Ethnographic Expedition and Beyond," in Veidlinger, *Going to the People*, 27–44.

55. Deutsch, "Thrice Born," 33.

56. Gustav Spiller, ed., *Papers on Inter-racial Problems, Communicated to the First Universal Races Congress, Held at the University of London, July 26–29, 1911* (London: P. S. King and Son, 1911). On the congress, see Elliot M. Rudwick, "W. E. B. Du Bois and the Universal Races Congress of 1911," *Phylon* 20 (1959): 372–378; Michael D. Biddiss, "The Universal Races Congress of 1911," *Race* 13 (1971): 37–46; Paul Rich, "'The Baptism

of a New Era': The 1911 Universal Races Congress and the Liberal Ideology of Race," *Ethnic and Racial Studies* 7 (1984): 534–550; Paul Gilroy, *The Black Atlantic: Modernity and Double Consciousness* (Cambridge, MA: Harvard University Press, 1993), 144–145, 214; Robert John Holton, "Cosmopolitanism or Cosmopolitanisms? The Universal Races Congress of 1911," *Global Networks: A Journal of Transnational Affairs* 2 (2002): 153–170; Ian Christopher Fletcher, "Introduction: New Historical Perspectives on the First Universal Races Congress of 1911," *Radical History Review* 92 (Spring 2005): 99–102; and Susan D. Pennybacker, "The Universal Races Congress, London Political Culture, and Imperial Dissent, 1900–1939," *Radical History Review* 92 (Spring 2005): 103–117.

57. Spiller, *Papers on Inter-racial Problems*, xiii.

58. Alexander Yastchenko, "The Role of Russia in the Mutual Approach of the West and the East," in Spiller, *Papers on Inter-racial Problems*, 195–207, here 195, 203.

59. L. Sh-g, "Besedy s chitateliami," *Novyi voskhod* 31 (August 4, 1911): 4–11, here 5.

60. Spiller, *Papers on Inter-racial Problems*, 13–24 (Luschan); 25–39 (Fouillée); 57–61 (Margoliouth); 453–461 (Tevfik).

61. Israel Zangwill, "The Jewish Race," in Spiller, *Papers on Inter-racial Problems*, 268–279.

62. L. Sh-g, "Besedy s chitateliami," *Novyi voskhod* 31 (August 4, 1911): 5.

63. Edna Nahshon, ed., *From the Ghetto to the Melting Pot: Israel Zangwill's Jewish Plays* (Detroit: Wayne State University Press, 2005); Meri-Jane Rochelson, *A Jew in the Public Arena: The Career of Israel Zangwill* (Detroit: Wayne State University Press, 2008).

64. John M. Efron, *Defenders of the Race: Jewish Doctors and Race Science in Fin-de-Siècle Europe* (New Haven, CT: Yale University Press, 1994), 126.

65. Zangwill, "The Jewish Race," 270–271.

66. Zangwill, "The Jewish Race," 277, 279.

67. L. Sh-g, "Besedy s chitateliami," *Novyi voskhod* 10 (March 10, 1911): 6–10, here 7.

68. Ignaz Zollschan, *Das Rassenproblem unter besonderer Beruecksichtigung der theoretischen Grundlagen der juedischen Rassenfrage* (Vienna: W. Braumüller, 1910).

69. L. Sh-g, "Besedy s chitateliami," *Novyi Voskhod* 10 (March 10, 1911), 8–10.

70. For an analysis of Zollschan's anthropology in the context of Jewish politics in Austria and Germany, see Efron, *Defenders of the Race*, 153–166.

71. Notes on "Jews. Heredity," SPbF ARAN, f. 282, op. 1, d. 176, ll. 95–96.

72. L. Ia. Shternberg, "Problema evreiskoi natsional'noi psikhologii," *Evreiskaia starina* 11 (1924): 5–44.

73. Kan, *Lev Shternberg*, 155.

74. L. Shternberg, "Sakhalinskie evrei (Iz vospominanii politicheskogo ssyl'nogo)," *Evreiskii mir* 6 (1912): 99–114, here 103.

75. Kan, *Lev Shternberg*, 155.

76. He departed on July 26, 1905. Kan, *Lev Shternberg*, 156–157.

77. Bruce Grant, "Appendix B: An Interview with Zakharii Efimovich Cherniakov. Moscow, June 1996," in Shternberg, *The Social Organization of the Gilyak*, 245–255, here 247.

78. Student quoted in Kan, *Lev Shternberg*, 279.

79. L. Ia. Shternberg, "Notes and Extracts on the Problem of Race," SPbF ARAN, f. 282, op. 1, d. 130, l. 153.

80. Shternberg, "Notes and Extracts," l. 155.

81. Shternberg, "Notes and Extracts," l. 159.

82. Shternberg, "Notes and Extracts," ll. 67–68, 37.

83. Shternberg, "Notes and Extracts," l. 50.

84. Shternberg, "Notes and Extracts," ll. 34–36, 95, 217–228.

85. Shternberg, "Notes and Extracts," l. 95.
86. Shternberg, "Notes and Extracts," l. 117.
87. "Nauchnye sobraniia obshchestva," *Evreiskaia starina* 6, no. 2 (1913): 6.
88. "Doklad L. Ia. Shternberga 'O rasovom voprose v noveishikh rabotakh po antropologii evreev,'" *Novyi voskhod* 18 (May 4, 1912): 45–48.
89. L. Shternberg, "Noveishie raboty po antropologii evreev (Doklad, chitannyi v Evreiskom istoriko-etnograficheskom obshchestve, 23 aprelia 1912 g.)," *Evreiskaia starina* 5, no. 3 (1912): 302–329.
90. "Nauchnye sobraniia obshchestva," 6.
91. In 1912 the two men actively corresponded. Weissenberg in particular was eager to report on his ethnographic observations to Shternberg, who played an important role in the EIEO and was a professional ethnographer. For example, he wrote to Shternberg from Elisavetgrad on August 12, 1912:

> Deeply respected Lev Iakovlevich!
> I just received your letter and I rush to express to you personally and to the whole Committee my sincere gratitude for the opportunity granted [to me] to visit Crimea. I will try to spend the money and time in the most effective way; moreover, I am thinking about traveling to the Caucasus as well—this, obviously, on my own account. I plan to depart in late April, and therefore I would be very grateful if you can order that the money be sent to me *asap*. I would be glad to take a phonograph with me, in case you can send me one. . . . I hope that my camera will serve a little longer, although it has been severely battered in Palestine. Please allow me one more request. If I am correct, you have connections to the Big Encyclopedia. I want to submit an article for it on human height based on my book *Des Wochrhum & Meuschen*. When you have a chance, find out about this and let me know . . .
> With deep respect,
> S. Vaisenberg (Weissenberg).

"Pis'ma Weissenberga S. A. Shternbergu L. Ia. 1912–1925," SPbF ARAN, f. 28, op. 2, d. 46, ll. 1–1 rev. See also a letter written in May 1915, in which Weissenberg asks Shternberg to send him a few books for review and the program of An-sky's expedition. SPbF ARAN, f. 28, op. 2, d. 46, l. 2.
92. Shternberg, "Noveishie raboty," 302.
93. Shternberg, "Noveishie raboty," 302. Shternberg referred to a 1912 edition of Zollschan, *Das Rassenproblem.*
94. Maurice Fishberg, *The Jews: A Study of Race and Environment* (London: Walter Scott, 1911); *Reports of the Immigration Commission,* vol. 38, *Changes in Bodily Form of Descendants of Immigrants (Final Report)* (Washington, DC: Government Printing Office, 1911); Franz Boas, "Changes in Bodily Form of Descendants of Immigrants," *American Anthropologist* 14 (1912): 530–562.
95. Shternberg, "Noveishie raboty," 303.
96. Shternberg, "Noveishie raboty," 305.
97. Shternberg, "Noveishie raboty," 308.
98. Ignacy Maurycy Judt, *Die Juden als Rasse* (Berlin: Jüdischer Verlag, 1903).
99. Shternberg, "Noveishie raboty," 311.
100. Efron, *Defenders of the Race,* 160.
101. Shternberg, "Noveishie raboty," 309.
102. Shternberg, "Noveishie raboty," 313–314.
103. Shternberg, "Noveishie raboty," 315.
104. Shternberg, "Noveishie raboty," 316.
105. Shternberg, "Noveishie raboty," 316.

106. The commission was known by the name of its chair, William P. Dillingham, a senator from Vermont. For a useful overview, see Dorothee Schneider, "The United States Government and the Investigation of European Emigration in the Open Door Era," in *Citizenship and Those Who Leave: The Politics of Emigration and Expatriation,* ed. Nancy L. Green and François Weil (Urbana: University of Illinois Press, 2007), 195–210.

107. For a detailed analysis of multiple influences on Progressives of different leanings and generations, see James S. Pula, "The Progressives, the Immigrant, and the Workplace: Defining Public Perceptions, 1900–1914," *Polish American Studies* 52, no. 2 (1995): 57–69.

108. *Reports of the Immigration Commission,* 38:5.

109. *Reports of the Immigration Commission,* 38:5.

110. *Reports of the Immigration Commission,* 38:2.

111. Shternberg, "Noveishie raboty," 318.

112. Shternberg, "Noveishie raboty," 318.

113. Shternberg, "Noveishie raboty," 319, 320, 321, 325.

114. *Reports of the Immigration Commission,* 38:2.

115. Shternberg, "Noveishie raboty," 325.

116. Shternberg, "Noveishie raboty," 327.

117. Shternberg, "Noveishie raboty," 328.

118. Leonard B. Glick, "Types Distinct from Our Own: Franz Boas on Jewish Identity and Assimilation," *American Anthropologist* 84 (1982): 545–565, here 557.

119. George W. Stocking Jr., "Anthropology as Kulturkampf: Science and Politics in the Career of Franz Boas," in *The Ethnographer's Magic, and Other Essays in the History of Anthropology* (Madison: University of Wisconsin Press, 1992), 92–113, here 113.

120. Ezra Mendelsohn, "Should We Take Notice of Berthe Weill? Reflections on the Domain of Jewish History," *Jewish Social Studies* 1 (1994): 22–39, here 32.

121. Franz Boas, *Anthropology and Modern Life,* new and rev. ed. (New York: Transaction, 1932); Boas, *Race and Democratic Society* (New York: J. J. Augustin, 1945), 38–42.

122. Mendelsohn, "Should We Take Notice?," 30.

123. John Murray Cuddihy, *The Ordeal of Civility: Freud, Marx, Levy-Strauss, and the Jewish Struggle with Modernity* (New York: Dell, 1974).

124. George W. Stocking Jr., "The Scientific Reaction against Cultural Anthropology," in *Race, Culture, and Evolution: Essays in the History of Anthropology* (New York: Free Press, 1968), 270–307, here 276. See also Gelya Frank, "Jews, Multiculturalism, and Boasian Anthropology," *American Anthropologist,* n.s., 99, no. 4 (1997): 731–745.

125. Alexander Lesser, "Franz Boas," in *Totems and Teachers: Perspectives on the History of Anthropology,* ed. Sydel Silverman (New York: Columbia University Press, 1981), 1–3; Frank, "Jews, Multiculturalism, and Boasian Anthropology."

126. Andrew Sloin, *The Jewish Revolution in Belorussia: Economy, Race, and Bolshevik Power* (Bloomington: Indiana University Press, 2017), 1.

127. L. Ia. Shternberg, "Problema evreiskoi etnografii," *Evreiskaia starina* 12 (1928): 11–16.

128. Sloin, *The Jewish Revolution in Belorussia,* 18.

129. Mark B. Adams, "Eugenics in Russia: 1900–1940," in Mark B. Adams, ed., *The Wellborn Science: Eugenics in Germany, France, Brazil, and Russia* (New York: Oxford University Press, 1990), 153–216, here 184.

130. The head of the EIEO, Dubnov, edited ten volumes, 1909–1916: four issues per year and one issue of volume 10 in 1918. After Dubnov's emigration from Soviet Russia in 1922, Shternberg assumed his responsibilities and edited volumes 11 (1924) and 12 (1928). S. L. Tsinberg edited the final, thirteenth volume, which came out in 1930. The society was officially closed in December 1929 for "conducting activities that are ideologically alien to Soviet society

and Jewish toiling masses." V. Lukin, "K stoletiiu obrazovaniia peterburgskoi nauchnoi shkoly evreiskoi istorii," in *Istoriia evreev v Rossii: Problemy istochnikovedeniia i istoriografii* (St. Petersburg: Peterburgskii evreiskii universitet, 1993), 23.

131. L. Ia. Shternberg, "Problema evreiskoi natsional'noi psikhologii," *Evreiskaia starina* 11 (1924): 5–44, here 7.

132. Shternberg, "Problema evreiskoi natsional'noi psikhologii," 8.

133. Shternberg, "Problema evreiskoi natsional'noi psikhologii," 10.

134. Shternberg, "Problema evreiskoi natsional'noi psikhologii," 13–15.

135. Shternberg, "Problema evreiskoi natsional'noi psikhologii," 27–32.

136. Shternberg, "Problema evreiskoi natsional'noi psikhologii," 29.

137. Shternberg, "Problema evreiskoi natsional'noi psikhologii," 10.

138. Shternberg, "Problema evreiskoi natsional'noi psikhologii," 11.

139. Shternberg, "Problema evreiskoi natsional'noi psikhologii," 11, 12.

140. Given Shternberg's openness in the 1920s to different unconventional theories of human nature and culture, from eugenics to Marrism (on the latter, see Kan, *Lev Shternberg*, 320–322), his rejection of the mainstream psychological theory of the day, Freudianism (as a speculative and ethnographically unproven explanation of primitive psychology as a neurosis), and his lack of interest in Carl Gustav Jung's theory of archetypes, on which he wrote next to nothing, are telling. He seemed to reject outright theories that did not embrace positivist epistemology and downplayed the biological and racial aspects of psychological phenomena. For Shternberg's views on Freudianism, see Shternberg, "Sovremennaia etnologiia," *Etnografiia* 1, no. 1–2 (1926): 15–43.

141. The "academic" section of the plan included the title "On the Anthropology of Jews (with regard to the Changeability of Major Anthropological Traits)." See the handwritten, unsigned, and undated proposal in SPbF ARAN, f. 282, op. 1, d. 176 (Materialy po "evreiskomu voprosu"), ll. 746–746 rev.

142. "Antropologiia evreev," SPbF ARAN, f. 282, op. 1, d. 176, ll. 725–727.

143. Instructions are analyzed in Kan, "'To Study Our Past,'" 72–76.

144. N. M. Mogilianskii, "Predmet i zadachi etnografii," *Zhivaia starina* 26 (1916): 1–22, here 11.

145. David G. Anderson, Dmitry V. Arzyutov, and Sergei S. Alymov, eds., *Life Histories of* Etnos *Theory in Russia and Beyond* (Cambridge: Open Book Publishers, 2019).

146. David G. Anderson, "Notes from His 'Snail's Shell': Shirokogoroff's Fieldwork and the Groundwork for *Etnos* Thinking," in Anderson, Arzyutov, and Alymov, *Life Histories of* Etnos *Theory,* 212. "Shikorogoroffs" refers to Sergei and his wife and collaborator, Elisaveta.

147. S. M. Shirokogorov, *Mesto etnografii sredi nauk i klassifikatsiia etnosov* (Vladivostok: Svobodnaia Rossiia, 1922); Shirokogorov, *Etnos* (Shanghai: Izvestiia Vostochnogo fakul'teta gosudarstvennogo Dal'nevostochnogo universiteta, 1923); and others.

148. David G. Anderson and Dmitry V. Arzyutov, "The *Etnos* Archipelago: Sergei M. Shirokogorov and the Life History of a Controversial Anthropological Concept," *Current Anthropology* 60, no. 6 (2019): 741–753, here 745.

149. Anderson and Arzyutov, "The *Etnos* Archipelago," 746.

150. Anderson and Arzyutov, "The *Etnos* Archipelago," 746.

151. Kan, *Lev Shternberg*, 343. The book was Sergei Shirokogorov, *Etnos: Issledovanie osnovnykh printsipov izmeneniia etnicheskikh i etnograficheskikh iavlenii* (Shanghai: Sibpress, 1923).

152. Shirokogoroff quoted in Dmitry V. Arzyutov, "Order out of Chaos: Anthropology and Politics of Sergei M. Shirokogoroff," in Anderson, Arzyutov, and Alymov, *Life Histories of* Etnos *Theory,* 260.

153. In the 1969 article "Ethnos and Endogamy" in the journal *Soviet Ethnography* (*Sovetskaia etnografiia*), which initially stirred controversy but eventually became instrumental in reestablishing the category of *etnos* as a sociobiological entity in Soviet scholarly discourse, Bromlei wrote, as if evoking Shternberg: "*Etnoses* as specific entities exist objectively . . . what is the objective mechanism of ethnic integration? A development of the hypothesis according to which ethnic communities in a stable state, that is, typical for them, are necessarily characterized by endogamy, can shed some light on this question. . . . However, if the assertion that relatively small ethnic communities, especially tribes, are endogamous sounds quite familiar, the claim about endogamy of large *etnoses,* such as, for example, modern nations, can seem quite unexpected. Hence, it needs to be tested." Ju. V. Bromlei, "Etnos i endogamiia," *Sovetskaia etnografiia* 6 (1969): 84–91, here 84–85.

154. Anderson and Arzyutov, "The *Etnos* Archipelago," 741.

155. Frédéric Bertrand, *L'anthropologie soviétique des années 20–30: Configuration d'une rupture* (Bordeaux: Presses universitaires de Bordeaux, 2002); D. V. Arziutov, S. S. Alymov, et al., *Ot klassikov k marksizmu* (St. Petersburg: MAE RAN, 2014).

156. Anderson and Arzyutov, "The *Etnos* Archipelago," 750.

157. V. A. Tishkov, *Rekviem po etnosu* (Moscow: Nauka, 2003).

A Necessary Introduction

1. Solomon Samuilovich Vermel, *Iz patologii evreev (Rozhdaemost', smertnost', zabolevaemost')* (Moscow: T-vo tipografii A. I. Mamontova, 1911), 1–2; Vermel, *Iz patologii evreev,* 21. Sergei Petrovich Botkin was a famous Russian clinician, therapist, and activist, chief surgeon of the emperor, and president of the Medical Association of St. Petersburg, one of the founders of modern Russian medical science and education. Mark Antokol'skii was the founder of the realist school in Russian sculpture. Solomon Vermel, *Dushevnye bolezni u evreev (Iz Kazanskoi okruzhnoi lechebnitsy)* (Kazan: Tipo-litografiia universiteta, 1917), 16.

2. Vermel, *Dushevnye bolezni u evreev,* 15.

3. For example, he dedicated many pages to exposing the anti-Semitism of the Warsaw-based psychiatrist and amateur anthropologist E. V. Erikson: Marina Mogilner, "Racial Psychiatry and the Russian Imperial Dilemma of the 'Savage Within,'" *East Central Europe* 43 (2016): 99–133.

4. In a letter to Vermel written on the occasion of the EIEO's revival (May 1924), Lev Shternberg addressed him as "one of the oldest and most active members of our society and an old contributor to *Evreskaia starina.*" "Pis'mo Evreiskogo istoriko-etnograficheskogo obshchestva Vermeliu S. S.," RGALI, f. 119, op. 2, d. 14, ll. 1–2.

5. Vermel, *Iz patologii evreev;* Vermel, *Dushevnye bolezni u evreev.*

6. S. S. Vermel, *Istoricheskii ocherk deiatel'nosti Moskovkskogo otdeleniia Obshchestva dlia rasprostraneniia prosveshcheniia mezhdu evreiami v Rossii, 1864–1914* (Moscow: OPE, 1917); Vermel, *V. G. Korolenko i evrei (Vospominaniia, pis'ma)* (Moscow: Pechatnyi stanok, 1924); Vermel, *Moskovskoe izgnanie (1891–1892): Vpechatleniia, vospominaniia* (Moscow: Der emes, 1924). Also see Vermel, "Prestupnost' evreia (Po dannym Nauchnogo kabineta po izucheniiu prestupnikov," *Russkii evgenicheskii zhurnal* 2, no. 2–3 (1924): 153–158.

7. S. S. Vermel, *Isaak Il'ich Levitan i ego tvorchestvo* (St. Petersburg: Tipo-litografiia A. E. Laudau, 1902), 8–9.

8. "Zaiavlenie Vermelia Solomona Samoilovicha v professional'nyi komitet Sanitarnogo otdela o naznachenii personal'noi pensii s prilozheniem svedenii o sluzhebnom stazhe i nauchno-literaturnoi rabote. 16 marta 1934," manuscript, RGALI, f. 119, op. 1, d. 1, ll. 1–6.

9. S. S. Vermel, *Golovnye boli, sushchnost', klassifikatsiia i lechenie* (Moscow: Gosizdat, 1927); Vermel, "Prestuplenie, ego priroda i sushchnost'," *Zhurnal nevropatologii i psikhiatrii im. Korsakova* 3 (1926): 6866–6877.

10. Vermel, "Prestupnost' evreia," 156.

11. "Temy, namechennye gigienicheskoi komissiei Moskovskogo otdeleniia OZE, 1919," RGIA, f. 1545, op. 1, d. 168 ("Protokoly zasedanii Komiteta OZE za 1918–1919 gody"), l. 80.

12. "Deiatel'nost' obshchestva v 1920–1921 gg.," RGIA, f. 1545, op. 1, d. 96 ("Otchet o rabote Komiteta OZE za period 1912–1921. T. 2."), l. 60.

13. Ann Laura Stoler, "Colonial Aphasia: Disabled Histories and Race in France," in *Duress: Imperial Durabilities in Our Times* (Durham, NC: Duke University Press, 2016), 122–170, here 128.

5. Russian Jewish Physicians and the Politics of Jewish Biological Normalization

1. Daniel Brower, "Social Stratification in Russian Higher Education," in *The Transformation of Higher Learning, 1860–1930*, ed. Konrad H. Jarausch (Chicago: University of Chicago Press, 1983), 255, 215; Lisa Rae Epstein, "Caring for the Soul's House: The Jews of Russia and Health Care, 1860–1914" (PhD diss., Yale University, 1995), 68n45.

2. John F. Hutchinson, "Society, Corporation or Union? Russian Physicians and the Struggle for Professional Unity (1890–1913)," *Jahrbücher für Geschichte Osteuropas* 30 (1982): 37–53; Hutchinson, "Politics and Medical Professionalization after 1905," in *Russia's Missing Middle Class: The Professions in Russian History*, ed. Harley D. Balzer (Armonk, NY: M. E. Sharpe, 1996), 89–116.

3. Epstein, "Caring for the Soul's House," 248.

4. Veronika Lipphardt, *Biologie der Juden: Jüdische Wissenschaftler über "Rasse" und Vererbung, 1900–1935* (Göttingen: Vandenhoeck & Ruprecht, 2008), 66, 133, 305.

5. Nancy Frieden, *Russian Physicians in an Era of Reform and Revolution, 1856–1905* (Princeton, NJ: Princeton University Press, 1981); John F. Hutchinson, *Politics and Public Health in Revolutionary Russia, 1890–1918* (Baltimore: Johns Hopkins University Press, 1990).

6. For a general review of its activities in the nineteenth century, see *Devianostoletie Imperatorskogo Vilenskogo meditsinskogo obshchestva: 12 dekabria 1895 goda* (Vilna: Tipografiia A. Minskera, 1895).

7. Calculated on the basis of the lists of names of the society's members that are preserved in its archive in the Vilnius University Library's Manuscripts collection (Vilniaus valstubinis V. Kapsuko universitetas, Mokslinė biblioteka, Rankrasctu skyrius): VULMC, "Vilniaus medicinos draugija," f. 26-158, ll. 1–2, 4; f. 26-159, ll. 1–3, etc.

8. Thus, the Jewish Society's members composed monthly reports on patients and illnesses in the Vilna Jewish Hospital and presented these data at the society's meetings. Other physicians did not work with ethnic medical statistics. For example, see *Godovoi otchet o deiatel'nosti Imperatorskogo Vilenskogo meditsinskogo obshchestva za 1875 g.* [Annual report on the activities of the Imperial Vilna Medical Society for 1875] (Vilna: Pechatnia A. G. Syrkina, 1875), 16 (hereafter cited as *GODIVMO* and year).

9. See, for example, such lists from 1876: VULMC, "Vilniaus medicinos draugija," f. 26-156, added to l. 1. In 1866, the society reported 40 members (VULMC, "Vilniaus medicinos draugija," f. 26-151, l. 2); in 1886, 233 members (*GODIVMO za 1886 g.* [Vilna: Izdanie

IVMO, 1887], 3); in 1891, 196 members (*GODIVMO za 1891 g.* [Vilna: Izdanie IVMO, 1892], 3; VULMC, "Vilniaus medicinos draugija," f. 26-159, l. 4 rev.); and in 1905, 280 members (*GODIVMO za 1905 g.* [Vilna: Izdanie IVMO, 1908], 4).

10. In the early 1890s, the treasurer was Dr. L. I. Zeidler: *GODIVMO za 1891 g.; GODIVMO za 1892 g.* (Vilna: Izdanie IVMO, 1893); *GODIVMO za 1894 g.* (Vilna: Izdanie IVMO, 1894). He was succeeded by another Jewish doctor, Ven'iamin Ioselevich Fin: *GODIVMO za 1893 g.* (Vilna: Izdanie IVMO, 1895); *GODIVMO za 1895 g.* (Vilna: Izdanie IVMO, 1896); and others. After Fin, Grigorii Iakovlevich Gershun was elected treasurer: *GODIVMO za 1900 g.* (Vilna: Izdanie IVMO, 1901). In 1910, the treasurer was Il'ia Mordukhovich Blokh: *Protokoly zasedanii Imperatorskogo Vilenskogo meditsinskogo obshchestva za pervoe polugodie 1911 g.* (Vilna: Promen', 1911).

11. I found only one exception: in 1910, Abram L'vovich Kogan was elected the vice-chair. *Protokoly zasedanii Imperatorskogo Vilenskogo meditsinskogo obshchestva za pervoe polugodie 1912 g.* (Vilna: Pormen', 1912), *Protokol 1143 ocherednogo zasedaniia IVMO, god CVI, #8; Protokoly zasedanii Imperatorskogo Vilenskogo meditsinskogo obshchestva za 1913 g.* (Vilna: Promen', 1913), *Protokol 1163 ocherednogo zasedaniia IVMO, god CVIII, #9*; and so on.

12. For example, the well-known doctor Tsemakh Iosifovich Shabad was a regular presenter: Shabad, "O Vserossiiskoi lige dlia bor'by s tuberkulezom: Ob uchrezhdenii Vselenskogo otdela Vserossiiskoi ligi dlia bor'by s bugorchatkoi." See *Doklady, chitannye v zasedanii Imperatorskogo Vilenskogo meditsinskogo obshchestva 16 aprelia 1911* (Vilna: Vilenskoe otdelenie Vserossiiskoi ligi dlia bor'by s tuberkulezom, 1911). In 1897, Shabad made at least five presentations. Mentioned in VULMC, "Vilniaus medicinos draugija," f. 26-3361, ll. 1–6, 7–11. He made the same number in 1895: f. 26-3360, ll. 1–23, 24–31.

13. See, for example, *GODIVMO za 1875 g., 16*.

14. *GODIVMO za 1905 g.,5*, and later annual reports.

15. "V Vilenskoe Imperatorskoe meditsinskoe obshchestvo naslednikov vracha Ven'iamina Samuilovicha Fina, kuptsov I. A. Gol'berga i A. M. Naishulia," VULMC, "Vilniaus medicinos draugija," f. 26-540, ll. 1–1 rev. The Ministry of National Education confirmed the draft of the stipend's regulations on August 22, 1905 (l. 8).

16. VULMC, "Vilniaus medicinos draugija," f. 26-540, l. 8 rev.

17. *Nedel'naia khronika Voskhoda* 18 (1899): 541–542.

18. The author of the comment that is quoted in the text, the journalist of the newspaper *Odesskie novosti* (Odessa news), used this expression as an epigraph to his column. "Iorik. Obo vsem. Dlia zhida ia paradnykh komnat ne otkroiu. Izrechenie nachal'nika stantsii," *Odesskie novosti*, no. 6439 (October 5, 1904): 5.

19. "Dikaia vykhodka," *Budushchnost'*, no. 40 (October 8, 1904): 3.

20. EJWiki.org, "Belostok—evreiskaia obshchina," accessed August 21, 2021, http://www.ejwiki.org/wiki/%D0%91%D0%B5%D0%BB%D0%BE%D1%81%D1%82%D0%BE%D0%BA_(%D0%B5%D0%B2%D1%80%D0%B5%D0%B9%D1%81%D0%BA%D0%B0%D1%8F_%D0%BE%D0%B1%D1%89%D0%B8%D0%BD%D0%B0).

21. "Protokol #1, Godichnogo zasedaniia Obshchestva vrachei Mogilevskoi gubernii, 12 ianvaria 1907," in *Protokoly zasedanii Obshchestva vrachei Mogilevskoi gubernii s prilozheniem trudov chlenov ego za 1906, 1907 i 1908 gg.* (Mogilev: Tipo-litografiia Ia. N. Podzemskogo, 1910), 145–146.

22. *Ustav Vilenskogo kruzhka vrachei* (Vilna: Pechatnia D. Kreisnesa i Sh. Koval'skogo, 1911).

23. *Ustav Vilenskogo kruzhka vrachei, 1, 8*.

24. Epstein, "Caring for the Soul's House," 89. Doctor Aybolit is a fictional character from the children's poems *Aybolit* and *Barmaley* by Korney Chukovsky, based on the adaptation of *Doctor Dolittle* by Hugh Lofting.

25. Epstein, "Caring for the Soul's House," 255–256.

26. For more on Gran in the context of the OZE, see Gari Pozin, *Obshchestvo okhraneniia zdorov'ia evreiskogo naseleniia: Dokumenty, fakty, imena* (St. Petersburg: Tetra, 2007), 25–32, 85, and 142–152 and, in this book, in chapters 7 and 8 and in the Conclusion.

27. DAKO, f. 10, op. 1, d. 402, 1l. 1–22, here l. 18.

28. DAKO, f. 10, op. 1, d. 583, ll. 1–20, here l. 3, l. 13.

29. DAKO, f. 10, op. 1, d. 583, ll. 1–20, here l. 3, l. 13.

30. DAKO, f. 10, op. 1, d. 583, l. 20.

31. DAKO, f. 10, op. 1, d. 402, ll. 1–22; DAKO, f. 10, op. 1, d. 583, ll. 1–20.

32. "Vozzvanie 'O-va okhraneniia zdorov'ia evreiskogo naseleniia,'" *Novyi voskhod* 41 (October 11, 1912): 2.

33. DAKO, f. 10, op. 1, d. 514, ll. 1–2.

34. In 1870, more than 50 percent of the society's members were Jews. *Protokoly zasedanii Obshchestva odesskikh vrachei*, January 10–May 2, 1870; *Protokoly zasedanii Obshchestva odesskikh vrachei*, September 5–December 12, 1870 (Odessa: OOV, 1870).

35. For more on the activities of the Society, see Julia Leskova, "Obshchestvo odesskikh vrachei," *Migdal Times*, no. 54 (February 28, 2011): 2, https://www.migdal.org.ua/times/54 /4971/.

36. John P. Davis, *Russia in the Time of Cholera: Disease under Romanovs and Soviets* (New York: Bloomsbury Academic, 2018), esp. 84–89.

37. Leskova, "Obshchestvo odesskikh vrachei."

38. DAOO, f. 9 (Otdel'nyi tsenzor po vnutrennei tsenzure v g. Odesse), op. 1 [1887–1888], d. 100 (*Protokoly zasedanii Obschestva odesskikh vrachei*, 52 ll.), l. 31.

39. See, for example, the discussion of Rudolf Virchow's and other anthropologists' works at the meeting on October 17, 1887: DAOO, f. 9, op. 1 [1887–1888], d. 100, l. 4.

40. On Virchow's popularity in Russia, see Marina Mogilner, *Homo Imperii: A History of Physical Anthropology in Russia* (Lincoln: University of Nebraska Press, 2013), esp. 187–198.

41. Virchow's visit to Odessa is described in Konstantin Vasil'ev, *Meditsina Iuzhnoi Pal'miry: Istoriia zdravookhraneniia v Odesse, 1794–1920* (Odessa: Opium, 2009), 210–212, quotation on 212.

42. Iakov (Jacob) Bardakh (1857–1929) was a microbiologist and one of the first directors of Odessa's bacteriological station. His father, Iuly Meyer Bardakh, was a Hebraic scholar, a teacher at the Odessa Talmud Torah (religious school for boys of modest economic standing), preparing students for yeshiva; he was also a local censor of Jewish books.

43. DAOO, f. 9, op. 1 [1887–1888], d. 100 (*Protokoly zasedanii Obshchestva odesskikh vrachei*, 52 ll.), ll. 6–8.

44. Epstein, "Caring for the Soul's House," 185n28.

45. E. B. Zlodeeva, "Pirogovskie s'ezdy vrachei i ikh rol' v stanovlenii sistemy gosudarstvennogo zdravookhraneniia v Rossii (konets XIX–nachalo XX v.)" (PhD diss., State Russian Presidential Academy of State Service, Moscow, 2004), 102.

46. "K redaktoru 'Russkikh vedomostei,'" *Russkie vedomosti*, no. 135 (May 22, 1905): 4.

47. DAOO, f. 2, d. 260 ("Kantseliariia odesskogo gradonachal'nika MVD"), l. 18 rev.

48. DAOO, f. 2, d. 260, ll. 23 rev., 23 rev.–24, 24 rev.

49. See the list of Society of Russian Physicians' members in 1906—the turning point in the society's history. This list included the old members who contributed to its plans for self-liquidation and the new members who joined the society in 1905. All together there were seventy members, most of whom never attended the society's meetings and did not make presentations. Among them were three honorary members and six member-correspondents. "Spisok chlenov Obshchestva russkikh vrachei v Odesse (Po svedeniiam 1906 g.)," in *Trudy*

Obshchestva russkikh vrachei g. Odessy, vol. 5 (Odessa: Slavianskaia tipografiia E. Khriso-gelos, 1907), 1–3.

50. "Godovoi otchet o deiatel'nosti Obshchestva russkikh vrachei za 1905–1906 gg., sostavlennyi sekretarem V. Rudnevym," in *Trudy Obshchestva russkikh vrachei g. Odessy,* vol. 6 (Odessa: Tipografiia russkoi rechi, 1908), 25–28.

51. This letter of the board to the ordinary members is quoted in "Godovoi otchet o deiatel'nosti Obshchestva russkikh vrachei za 1905–1906 gg.," 27.

52. "Otchet sekretaria, Dr. L. M. Rozenfelda, o deiatel'nosti Obshchestva odesskikh vrachei za 1905–1907 gg.," in *Trudy Obshchestva odesskikh vrachei,* vol. 7 (Odessa: Takhnik, 1908), 69.

53. Robert Weinberg, *The Revolution of 1905 in Odessa: Blood on the Steps* (Bloomington: Indiana University Press, 1993).

54. Gokhman exemplifies the rare case of a person of the Judaic faith who received personal permission from the minister of education, I. D. Delianov, to assume a permanent position at an imperial university. As a strong mathematician with a doctoral degree, he was appreciated as a unique specialist. In 1887–1906, Gokhman taught practical mechanics, descriptive geography, and the theory of gravity, as well as hydrostatics and hydrodynamics at Imperial Novorossiisk University. DAOO, f. 45, op. 19, d. 18, l. 15. But in 1913, when Gokhman applied for the renewal of his contract and the university's president (rector) petitioned, on behalf of the Department of Physics and Mathematics, the overseer of the Odessa educational district, Vladimir Smol'ianinov, the decision was negative. Smol'ianinov warned the university's president that, "For the future, I will ban not only individuals of non-Christian faiths but also *inorodtsy* in general, even Christians among them, from serving at the Imperial Novorossiisk University. Exceptions are possible only in the rarest cases, when candidates of Russian origin are completely absent, and only with regard to those candidates from the ranks of *inorodtsy,* whose dedication to Russian state values I do not doubt." DAOO, f. 45, op. 19, d. 18, l. 4–4 rev.

55. DAOO, f. 45, op. 19, d. 18, l. 10 rev.

56. "Rech' professora Levasheva," in *Trudy Obshchestva russkikh vrachei g. Odessy,* vol. 6, 6; "Spisok chlenov Obshchestva russkikh vrachei v Odesse (Po svedeniiam 1907 g.)," in *Trudy Obshchestva russkikh vrachei g. Odessy,* vol. 6, 1–3. The list indicates the dates of admissions into the society.

57. DAOO, f. 2, op. 7, d. 260, l. 18; d. 142, ll. 32–32 rev.

58. Iakov Iul'evich Bardakh, chair; Grigorii Il'ich Gimel'farb, Iakov Vladimirovich Zil'berberg, [unreadable] Krantsfel'd, secretaries; Samuil Grigor'evich Shteinfinkel', treasurer; Leon Borisovich Bilik, librarian. DAOO, f. 2, op. 7, d. 142, ll. 29–29 rev. Manuscript.

59. "Spisok chlenov Obshchestva russkikh vrachei v Odesse (Po svedeniiam 1907 goda)," 1–3; "Otchet sekretaria, Dr. L. M. Rozenfelda," 70.

60. "Otchet sekretaria, Dr. L. M. Rozenfelda," 75.

61. DAOO, f. 45, op. 19, d. 1429, ll. 1–8.

62. DAOO, f. 10, op. 1 [1907–1911], d. 2, l. 26.

63. M. Sherman, "Evreiskii meditsinskii zhurnal," *Rassvet* 3 (January 18, 1909): 13–15, here 14; see also his review of the second volume: M. Sherman, "Bibliografiia: *Evreiskii meditsinskii golos," Rassvet* 14 (April 4, 1910): 41–42.

64. See, for example, L. Sheinis, "Psevdo-antropologicheskie osnovy antisemitizma," *Russkoe bogatstvo* 5 (May 1900): 85–110; Dr. L. I. Sheinis, "Obladaet li evreiskaia rasa immunitetom po otnosheniiu k alkogolizmu?," *Evreiskii meditsinskii golos* 1 (1910): 37–47, here 39.

65. M. S. Shvartsman, "Evreiskaia obshchestvennaia meditsina," *Evreiskii meditsinskii golos* 1–2 (1910): 49–64.

66. "Uchreditel'noe sobranie 'O-va okhraneniia zdorov'ia evreiskogo naseleniia,'" *Novyi voskhod* 44 (November 1, 1912): 24–26, here 24. For another detailed report on the meeting,

see "Sobranie O-va okhraneniia zdorov'ia evreiskogo naseleniia," *Rassvet* 44 (November 2, 1912): 19–21.

67. Given the scale and unique importance of OZE work, the ridiculously small number of studies dedicated to the society's history is surprising. The earliest publications were primarily personal recollections: L. Gourevitch, *Twenty-Five Years of Ose (1912–1937)* (Paris: OSE, 1937); Leon Wulman, ed., *In Fight for the Health of the Jewish People (50 Years of OSE)* (New York: World Union OSE and American Committee of OSE, 1968). The Russian scholar Gari Pozin was the first one to comprehensively work with the OZE archives. Although his research is a valuable contribution to this understudied topic, his publications retranslate the OZE's own self-narrative and provide "facts" without engaging modern historiography and critical analysis: Pozin, *Obshchestvo okhraneniia zdorov'ia evreiskogo naseleniia;* G. P. Pozin, "Istoriia Obshchestva okhraneniia zdorov'ia evreiskogo naseleniia (1912–1921)," in *Materialy XVI Ezhegodnoi mezhdunarodnoi mezhdistsiplinarnoi konferentsii po Iudaike,* vol. 27 (Moscow: Sefer, 2009), pt. 3, 78–98. Another scholar from Leningrad, Michael Beizer, authored more analytical studies, but they also reproduce the official narrative of the OZE: Beizer, "The Society for the Protection of the Health of the Jewish Population," in *The YIVO Encyclopedia of Jews in Eastern Europe,* accessed March 20, 2022, https://yivoencyclopedia .org/article.aspx/OZE; Beizer, "The Emergence of the Society for the Protection of the Health of the Jewish Population (OZE): Historical Circumstances and Ideology of the Founders" (paper presented at the OZE [OSE–Union] Centennial Conference, Paris, June 2012), http:// pluto.huji.ac.il/~beizer/files/OZE%20100%20Paris.pdf; Beizer, *Relief in Time of Need: Russian Jewry and the Joint, 1914–24* (Bloomington, IN: Slavica, 2015). Other publications focus on OZE history in interwar Eastern Europe: Iris Borowy and Wolf Gruner, eds., *Facing Illness in Troubled Times: Health in Europe in the Interwar Years, 1918–1939* (Frankfurt am Main: Peter Lang, 2005); Nadav Davidovitch and Rakefet Zalashik, "'Air, Sun, Water': Ideology and Activities of OZE (Society for the Preservation of the Health of the Jewish Population) during the Interwar Period," *Dynamis* 28 (2008): 127–149; Zalashik and Davidovitch, "Taking and Giving: The Case of the JDC and OZE in Lithuania 1919–26," *Eastern European Jewish Affairs* 39, no. 1 (2009): 57–68.

68. "Ustav Obshchestva okhraneniia zdorov'ia evreiskogo naseleniia (Izvlecheniie)," in *Zadachi Okhraneniia zdorov'ia russkikh evreev* (St. Petersburg: Gramotnost', 1913), 17.

69. "'Organizatsiia vrachebno-sanitarnoi pomoshchi evreiskomu naseleniiu': Tezisy k dokladu M. M. Grana," GARF, f. 9458, op. 1, d. 153, ll. 5–6, here l. 4.

70. Vrach-zemets, "Vrachebno-sanitarnaia organizatsiia sredi evreiskogo naseleniia," *Novyi voskhod* 19 (May 15, 1914): 27–28, here 28.

71. "Uchreditel'noe sobranie 'O-va okhraneniia zdorov'ia evreiskogo naseleniia,'" 25.

72. Advertisement in the newspaper *Den',* no. 11 (453) (January 12, 1914), 1.

73. Roman Liubars'kii, "'Bogom dannyi', lud'my—zabytyi . . ." [in Ukrainian], accessed August 21, 2021, http://poet.inf.ua/proza/bogom-dannyiy-lyudmi-zabityiy/.

74. "Protokol zasedaniia Biuro Komiteta obshchestva 6 iunia 1920," RGIA, f. 1545, op. 1, d. 168 ("Protokoly zasedanii Komiteta OZE za 1918–1919 gody"), l. 124.

75. Other members of the Central Committee: Doctors Naum (Nahum) Botvinnik (1872–1939), S. Frumkin (1886–1918), Aron Zalkind (1866–1931), Yehuda Leib Katzenelson (1846–1917), Mikhail Shvartsman (1880–1937), Abram Bramson (1871–1939), Iakov Eiger (1862–1932), Veniamin Binshtok (?–1933), and Grigorii I. Dembo (1872–1939). Attorney Grigorii Gol'dberg (1869–1922) and a few others represented other professions. In 1914, the committee's composition was the following: S. A. Kaufman, chair; A. Gol'dberg, vice-chair. Members: N. R. Botvinnik, V. I. Binshtok, A. M. Bramson, S. E. Vaisblit, Ia. E. Bavla, G. I. Dembo, M. M. Gran, A. V. Zalkind, Ia. B. Eiger, M. S. Shvartsman. "Otchet obshchestva OZEN za 1914 g.," RGIA, f. 1545, op. 1, d. 9a ("Otchet o rabote Komiteta OZE za period 1914–1921. T. 1"), l. 14.

76. *Izvestiia Komiteta OZE* 2 (1917): 1–2; RGIA, f. 1545, op. 1, d. 165, l. 13.

77. "Iz istorii OZE," RGIA, f. 1545, op. 1. d. 9a ("Otchet o rabote Komiteta OZE za period 1914–1921. T. 1"), l. 142.

78. Epstein, "Caring for the Soul's House," 260.

79. "K istorii OZE," RGIA, f. 1545, op. 1, d. 96 ("Otchet o rabote Komiteta OZE za period 1914–1921. T. 2"), ll. 173–74.

80. By early 1918, the OZE reported 125 kindergartens in 100 localities for 12,000 children. "Otchet o rabote Komiteta OZE za period 1914–1921. T. 2," RGIA, f. 1545, op. 1, d. 9b, l. 28. On playgrounds: Nadav Davidovitch and Rakefet Zalashik stress in their study of the interwar OZE in East-Central Europe that "the idea of playgrounds for children and the attention given to their plays as part of their development was a totally new concept within Jewish communities. In 1914, the OZE published the first book in Yiddish on children's playing." Davidovitch and Zalashik, "'Air, Sun, Water,'" 133. In 1917, the OZE reported 32 playgrounds with 32,000 registered users. "Otchet o rabote Komiteta OZE za period 1914–1921. T. 2," RGIA, f. 1545, op. 1, d. 9b, l. 28. On summer camps: another OZE report lists 26 summer colonies in 1916 (for 2,515 children) and 16 in 1917 (for 1,305 children). "Otchet o rabote Komiteta OZE za period 1914–1921. T. 2," RGIA, f. 1545, op. 1, d. 9b, l. 28.

81. During the first year of OZE operations, its chapters were officially registered in Odessa, Vitebsk, Brest-Litovsk, Vol'sk, Kovna, Minsk, and Feodosia, although de facto OZE cells operated in more localities, including Moscow outside of the Pale. "K istorii OZE," RGIA, f. 1545, op. 1, d. 96, ll. 166–172.

82. In 1917–1919, the OZE publication program declined due to a lack of funds, but its regular news publication continued in Russian (*Izvestiia Petrogradskogo tsentral'nogo komiteta OZE*). The bulk of scientific publications in the fields of Jewish medicine and eugenics, which resumed in the 1920s, was also primarily in Russian. But a gradual transition toward making Yiddish a language of Jewish science in addition to Russian was also taking place within OZE circles. For example, the branch of the OZE that was formally established in Vilna in 1918 (and informally the OZE had been active there since 1914 under the leadership of the chair of the Vilna Jewish Physicians Society, Dr. Tsemakh Shabad), published its journal in Yiddish. *Fulksgrezunt* (People's health), founded in 1923, was the official organ of the OZE and the TOZ (Towarzystwo ochrony zdrowia ludności żydowskiej, Society for Protection of the Health of the Jewish Population), which combined the mission of disseminating knowledge with presenting scientific research.

83. "Komitet Obshchestva okhraneniia zdorov'ia evreiskogo naseleniia (otchet)," GARF, f. 9458, op. 1, d. 143, ll. 11–11 ob.

84. The Jewish organizations that formed this consortium were the EKOPO, the OZE, the ORT, the OPE, the EIEO, and the EKO (Evreiskoe kolonizatsionnoe obshchestvo—Jewish Colonization Society). The EKOPO's executive leadership included the director, Genrikh Sliozberg (1863–1937), who had previously worked for the OPE; the secretary, Leontii Bramson (1869–1941), from the OZE; the OZE chairman, S. A. Kaufman (1839/40–1918); and the lawyer Maksim M. Vinaver (1862–1926), representing the EIEO. For more, see Polly Zavadivker, "Fighting 'On Our Own Territory': The Relief, Rescue, and Representation of Jews in Russia during World War I," in *Russia's Home Front in War and Revolution, 1914–22*, bk. 2, *The Experience of War and Revolution*, ed. Adele Lindenmeyr, Christopher Read, and Peter Waldron (Bloomington, IN: Slavica, 2015), 86–90. Within the EKOPO-led arrangement, the OZE received money from the Princess Tatiana Committee, a philanthropic organization founded and sponsored by the Romanov family, providing support to refugees instead of the failed state. Thus, in 1916 the Tatiana Committee allocated 340,000 rubles for OZE projects: RGIA, f. 1545, op. 1, d. 9b, ll. 46–55.

85. "'Organizatsiia vrachebno-sanitarnoi pomoshchi evreiskomu naseleniiu': Tezisy k dokladu M. M. Grana," GARF, f. 9458, op. 1, d. 153, l. 3, l. 4.

86. "'Organizatsiia vrachebno-sanitarnoi pomoshchi evreiskomu naseleniiu': Tezisy k dokladu M. M. Grana," l. 4.

87. "'Organizatsiia vrachebno-sanitarnoi pomoshchi evreiskomu naseleniiu': Tezisy k dokladu M. M. Grana," ll. 5, 6.

88. "Komitet Obshchestva okhraneniia zdorov'ia evreiskogo naseleniia (otchet)," GARF, f. 9458, op. 1, d. 143, l. 11 ob.

89. "Polozheniia doklada d-ra M. S. Shvartsmana 'Nervno-psikhicheskoe zdorov'e evreev i zadachi ego okhraneniia," GARF, f. 9458, op. 1, d. 153, ll. 12–15, here ll. 12 rev., 13 rev., 12 rev.

90. According to Eric Lohr, the number of Jewish refugees and deportees during the war reached one million. Lohr, "The Russian Army and the Jews: Mass Deportation, Hostages, and Violence during World War I," *Russian Review* 60 (2001): 404–419, here 404. The OZE offered a different estimate. By the time mass relocations stopped in 1916, Jewish relief committees had registered 211,691 refugees and deportees in 1,907 localities, who had appealed for help. The OZE estimated that a similar number must have survived without charitable relief. Counting Jews who stayed in Warsaw after the retreat of the Russian army, and Jews who stayed in Lithuania, the approximate number of dislocated Jews was 600,000. "Otchet o rabote Komiteta OZE za period 1914–1921. T. 2," RGIA, f. 1545, op. 1, d. 9b, l. 20.

91. "Polozheniia doklada d-ra M. S. Shvartsmana," ll. 14, 15. See another paper on the same topic presented by Shvartsman at the First OZE All-Ukrainian Conference in 1918: DAKO, f. P-4018, op. 1, d. 8, l. 6.

92. *Die Entstehung der Gesellschaft OSE und ihre ersten Massnahmen* (Berlin: Verband für Gesundheitsschutz der Juden OSE, 1925), 4.

93. An OZE report in 1915 gives the number of employees as 714. "Otchet o rabote Komiteta OZE za period 1914–1921. T. 1," RGIA, f. 1545, op. 1, d. 9a, l. 14.

94. E. N. Levina, "V vikhre mirovykh sobytii ('O.Z.E.' v Vostochnoi i Tsentral'noi Evrope)" [In the whirlwind of world events (O.Z.E. in Eastern and Central Europe)], in Wulman, *In Fight for the Health of the Jewish People,* 13–31, here 13.

95. Steven Zipperstein, "The Politics of Relief: The Transformation of Russian Jewish Communal Life during the First World War," in *Studies in Contemporary Jewry,* vol. 4, *The Jews and the European Crisis, 1914–1921,* ed. Jonathan Frankel, Peter Y. Medding, and Ezra Mendelsohn (New York: Oxford University Press, 1988), 22–40.

96. Zavadivker, "Fighting 'On Our Own Territory,'" 84.

97. "Protokol zasedaniia Komiteta OZE ot marta 21 dnia 1918 goda," RGIA, f. 1545, op. 1, d. 168 ("Protokoly zasedanii Komiteta OZE za 1918–1919 gody"), l. 24.

98. "Iz istorii OZE," RGIA, f. 1545, op. 1, d. 9a, l. 145.

99. "Iz istorii OZE," RGIA, f. 1545, op. 1, d. 9a, l. 141.

100. "Iz istorii OZE," RGIA, f. 1545, op. 1, d. 9a, ll. 142, 143.

101. "Iz istorii OZE," RGIA, f. 1545, op. 1, d. 9a, ll. 142, 143.

102. "Deiatel'nost' obshchestva v 1920–1921 gg.," RGIA, f. 1545, op. 1, d. 96, l. 58.

103. "Deiatel'nost' obshchestva v 1920–1921 gg.," RGIA, f. 1545, op. 1, d. 96, l. 60.

6. Medical Statistics and National Eugenics

1. B. Gol'dberg, "O fizicheskom vyrozhdenii u evreev," *Rassvet* 48 (December 14, 1908): 17–19, here 17. The author uses relevant data from the 1897 imperial population census and concludes that they do not support the perception of Jews as a degenerate nation.

2. I. Sapir, "Eshche o vyrozhdenii evreev," *Rassvet* 1 (January 4, 1909): 13–17, here 13. He concluded, using the available statistics, that Jews were degenerating both physically and spiritually.

3. Gur Alroey, "Demographers in the Service of the Nation: Liebmann Hersch, Jacob Lestschinsky, and the Early Study of Jewish Migration," *Jewish History* 20 (2006): 265–282, here 267.

4. A. P. Subotin, *V cherte evreiskoi osedlosti: Otryvki iz ekonomicheskikh issledovanii v Zapadnoi i Iugo-Zapadnoi Rossii za leto 1887 g.,* vols. 1 and 2 (St. Petersburg: Izd-vo Ekonomicheskogo zhurnala, 1888–1890).

5. Liebmann Hersch, *Di yudiche emigratsye: Ir grois, aeygenarthnkeyt, sibot, bedeyting un regulirung* (Vilna: Di welt, 1914). For more on Hersch, see Alroey, "Demographers in the Service of the Nation."

6. Paul Glikson, "Jacob Lestschinsky: A Bibliographical Survey," *Jewish Journal of Sociology* 9, no. 1 (1967): 48–57, here 48.

7. Jacob Lestschinsky, "Statistika shel ayara ahat," *Ha-shiloah* 12 (1902/1903–1903/1904): 87–96, here 87.

8. The best analysis of Lestschinsky's science is offered in Gennady Estraikh, "Jacob Lestschinsky: A Yiddishist Dreamer and Social Scientist," *Science in Context* 20, no. 2 (2007): 215–237. Harriet Murav offers pertinent observations on Lestschinsky's loss of faith in the Soviet project as a solution to the "productivization" of Jewish labor under the influence of the pogroms in Ukraine in 1919 and in the 1920s. The pessimistic "negative statistics" of the pogromed Jewish communities ("cells") had undermined his earlier optimistic evolutionary Marxism. I am very grateful to Harriet Murav, who kindly shared with me the unpublished manuscript of her chapter "Archive of Violence: Part II, Documentation."

9. M. Sherman, "Otkrytoe pis'mo k evreiskim studentam," *Evreiskii meditsinskii golos* 1 (1909): 28.

10. A detailed analysis of the Russian tradition of students' self-censuses and other statistical projects is offered in A. E. Ivanov, *Studencheskaia korporatsiia Rossii kontsa XIX–nachala XX veka: Opyt kul'turnoi i politicheskoi samoorganizatsii* (Moscow: Novyi khronograf, 2004), 199–283.

11. For example, see V. V. Sviatlovskii, ed., *Studenchestvo v tsifrakh (Po dannym Iur'evskoi ankety 1907 g.)* (St. Petersburg: Izd. Obshchestva studentov, 1909).

12. M. A. Chlenov, *Polovaia perepis' moskovskogo studenchestva i ee obshchestvennoe znachenie* (Moscow: Studencheskaia med. izdatel'skaia komissiia, 1909), 28.

13. D. I. Sheinis, *Evreiskoe studenchestvo v tsifrakh (Po dannym perepisi 1909 g. v Kievskom universitete i politekhnicheskom institute)* (Kiev: Pechatnia Iosifa Shenfel'da, 1911); *K kharakteristike evreiskogo studenchestva (Po dannym ankety sredi evreiskogo studenchestva g. Kieva v noiabre 1910 g.)* (Kiev: Rabotnik, 1913); B. S-ii and M. K-r, *Odesskoe evreiskoe studenchestvo: Itogi ankety sredi evreev-studentov, proizvedennoi v 1911–1912 uchebnom godu* (Odessa: Tipogr. knigoiz-va M. S. Kozmana, 1913); D. I. Sheinis, *Evreiskoe studenchestvo v Moskve (Po dannym ankety 1913 g.)* (Moscow: Izd. Moskovskogo komiteta OPE, 1913).

14. Boris Cherniak, "Anketa mezhdu evreiami-studentami g. Kazani," *Rassvet* 28 (July 12, 1913): 13–16.

15. Sheinis, *Evreiskoe studenchestvo v Moskve,* 7. I used one of the few printed copies of the Moscow survey, available to scholars, preserved in GARF, op. 1, d. 65, ll. 1–36, esp. ll. 8 rev–9. All Jewish student surveys have been reproduced in the appendix in A. E. Ivanov, *Evreiskoe studenchestvo v Rossiiskoi imperii nachala XX veka: Kakim ono bylo?* (Moscow: Novyi khronograf, 2007), 165–422.

16. The Moscow survey was prepared by the Survey Committee of the Moscow chapter of the OPE under the chairmanship of I. S. Uryson. They used the 1909 Kiev questionnaire, the questionnaire for Jewish students composed in the statistical seminar at Moscow Commercial Institute, and questionnaires previously used in general student surveys. Sheinis, *Evreiskoe studenchestvo v Moskve,* 1.

17. Sheinis must have been a very able learner. Not only did he learn statistics by himself but after the revolution he became one of the founders of Soviet housing policy (*zhilishchnaia politika*). From 1922, Sheinis worked as head of the housing division of the Main Administration of Utilities of the NKVD (Soviet security service) of the Russian Soviet Federative Socialist Republic (RSFSR), and in 1930–1931 he became head of the Narkomat (Ministry) of Utilities of the RSFSR Council of People's Commissars. See A. N. Iakovleva, ed., *Lubianka: Organy VChK–OKGPU–NKVD–NKGB–MGB–KGB, 1917–1991; Spravochnik*, comp. A. I. Kokurin and N. B. Petrov (Moscow: MVD, 2003), 184–187. On Sheinis's role in the development of Soviet housing policy and his contribution to the idea of the "garden city," see M. G. Meerovich, *Gradostroitel'naia politika SSSR, 1917–1929: Ot goroda-sada k vedomstvennomu rabochemu poselku* (Moscow: NLO, 2017); Meerovich, "Rozhdenie i smert' goroda-sada: Deistvuiushchie litsa i motivy ubiistva," *Eurasica* 1 (2007): 118–166.

18. *K kharakteristike evreiskogo studenchestva*, 4 (introduction).

19. See materials of the 1909 survey in Ivanov, *Evreiskoe studenchestvo v Rossiiskoi imperii*, appendix, 165–422, here 168.

20. Sheinis, *Evreiskoe studenchestvo v Moskve*, i.

21. Sheinis, *Evreiskoe studenchestvo v Moskve*, 7–8.

22. Sheinis, *Evreiskoe studenchestvo v Moskve*, 43, 46.

23. Sheinis, *Evreiskoe studenchestvo v Moskve*, 43.

24. Ivanov, *Evreiskoe studenchestvo v Rossiiskoi imperii*, appendix, 168.

25. Steven Zipperstein, *Imagining Russian Jewry: Memory, History, Identity* (Seattle: University of Washington Press, 1999), 41–62.

26. Brian Horowitz, "The Return of the *Heder* among Russian Jewish Education Experts, 1840–1917," *Polin: Studies in Polish Jewry* 30 (2018): 181–193, here 181.

27. Horowitz, "The Return of the *Heder*," 181.

28. Horowitz, "The Return of the *Heder*," 188.

29. "Sovremennyi kheder: Po obsledovaniiu Obshchestva rasprostraneniia prosveshcheniia mezhdu evreiami v Rossii," *Vestnik evreiskogo prosveshcheniia* 17 (1912): 3–90.

30. J. Katsenelson, "Sovremennyi kheder kak ob"ekt issledovaniia," *Nedel'naia khronika Voskhoda* 12 (1895): 308.

31. "Sovremennyi kheder: Po obsledovaniiu obshchestva," 5.

32. K. Zhitomirskii, "Prakticheskie tseli evreiskoi shkoly," *Evreiskaia shkola* 12 (1904): 3–16, here 3.

33. M. Krefning, "K voprosu o fizicheskom vospitanii v shkole," *Evreiskaia shkola* 5 (1904): 3–11, here 3.

34. On the development of school medicine in the Russian Empire in general, see Andy Byford, "Professional Cross-Dressing: Doctors in Education in Late Imperial Russia (1881–1917)," *Russian Review* 65, no. 4 (2006): 586–616.

35. Alice Boardman Smuts, *Science in the Service of Children* (New Haven, CT: Yale University Press, 2006).

36. Andy Byford, *Science of the Child in Late Imperial and Early Soviet Russia* (Oxford: Oxford University Press, 2020), 2–3, 4.

37. See, for example, L. Rokhlin, "O fizicheskom razvitii uchashchikhsia v khederakh mestechka Krasnopol'ia Mogilevskoi gubernii," *Evreiskaia zhizn'* 8 (1905): 124–140, here 125.

38. Naum Vasil'evich Zak, *Fizicheskoe razvitie detei v sredneuchebnykh zavedeniikh g. Moskvy: Materialy dlia otsenki sanitarnogo sostoianiia uchashchikhsia* (Moscow: Tipogr. E. Gerbek, 1892). For another early work on Jewish children referenced in Zak, see S. N. Iashchinskii, "Antropologicheskie materialy k izucheniiu razvitiia, rosta, okruzhnosti grudi i vesa u poliakov i evreev v shkol'nom ikh vozraste," *Varshavskie universitetskie izvestiia* 3 (1889): 1–16; 4 (1889): 17–40.

39. Zak, *Fizicheskoe razvitie detei,* 99, 174.

40. Byford, *Science of the Child,* 12.

41. Michel Foucault, *The Birth of Biopolitics: Lectures at the Collège de France, 1978–1979,* ed. Michael Senellart, trans. Graham Burchell (Basingstoke, UK: Palgrave Macmillan, 2008).

42. Laura Engelstein, *The Keys to Happiness: Sex and the Search for Modernity in Fin-de-Siècle Russia* (Ithaca, NY: Cornell University Press, 1992); Daniel Beer, *Renovating Russia: The Human Sciences and the Fate of Liberal Modernity, 1880–1930* (Ithaca, NY: Cornell University Press, 2008).

43. Byford, "Professional Cross-Dressing," 602.

44. Andy Byford, "The Mental Test as a Boundary Object in Early-20th-Century Russian Child Science," *History of the Human Sciences* 27, no. 4 (2014): 42.

45. "K istorii OZE," RGIA, f. 1545, op. 1, d. 96, ll. 152–153.

46. "Komitet Obshchestva okhraneniia zdorov'ia evreiskogo naseleniia (Otchet za vremia s 1 avgusta 1914 goda po 1 avgusta 1916)," GARF, f. 9458, op. 1, d. 153, l. 11, ll. 11–11 rev.

47. D-r V. I. Binshtok, "Polozheniia k dokladu Sanitarno-statisticheskoi komissii 'Ob organizatsii Statisticheskoi chasti pri Obshchestve okhraneniia zdorov'ia evreiskogo naseleniia,'" GARF, f. 9458, op. 1, d. 153, l. 18.

48. "Proekt ustava Psikhonevrologicheskogo instituta," *Vestnik psikhologii, kriminal'noi anthropologii i gipnotizma* 1, no. 8 (1904): 711–715.

49. TsGA SPb, f. 115, op. 2, vol. 1, 1909–1917, d. 5120. The Russian government perceived the institute's curriculum as dangerous, leading to subversive views in youth. Several times it contemplated banning the educational activities of the institute and leaving it with research functions only.

50. Contacts between Jewish intellectuals and the institute existed at multiple levels. In the archive of the journal *Jewish Antiquity* (*Evreiskaia starina*) a letter is preserved from the Psychoneurological Institute's council signed by Bekhterev himself on September 22, 1911. In it, the council reports on the launching of the institute's own journal, *Herald of Psychology, Criminal Anthropology and Hypnotism,* and suggests exchanging publications. TsGIA SPb, f. 2134 (Redaktsiia zhurnala Evreiskaia starina), op. 1, d. 2 ("Perepiska redaktsii zhurnala 'Evreiskaia starina' s podpischikami za 1911"), l. 129.

51. "Protokol zasedaniia Komiteta OZE ot 15 fevralia 1919 g.," RGIA, f. 1545, op. 1, d. 168 ("Protokoly zasedanii Komiteta OZE za 1918–1919 gg.), ll. 88–89.

52. Andy Byford, "V. M. Bekhterev in Russian Child Science, 1900s–1920s: 'Objective Psychology'/'Reflexology' as a Scientific Movement," *Journal of the History of the Behavioral Sciences* 52, no. 2 (2016): 99–123, here 108. On alternative models of collecting children's medical statistics that also influenced Jewish statisticians, see Byford, "The Mental Test as a Boundary Object."

53. The original cards can be examined, for example, in DAKO, f. P-4018, op. 1, d. 16, l. 22.

54. "Doklad A. N. na Pervom Vseukrainskom soveshchanii deiatelei OZE," DAKO, f. P-4018, op. 1, d. 8, l. 2.

55. TsIAM, f. 1454, op. 1, d. 10, l. 19 rev.

56. "Polozheniia doklada d-ra M. S. Shvartsmana 'Nervno-psikhicheskoe zdorov'e evreev i zadachi ego okhraneniia,'" GARF, f. 9458, op. 1, d. 153, l. 14 rev.

57. "Tezisy k dokladu Ia. B. Eigera *Zadachi Obshchestva okhraneniia zdorov'ia evreiskogo naseleniia po zdravookhraneniiu detskogo i iunosheskogo vozrasta,*" GARF, f. 9458, op. 1, d. 155, ll. 17–18 rev., here l. 17.

58. These pre–World War I approaches were elaborated in "Doklad d-ra Patkanovoi-Kronkovskoi o defektivnykh detiakh," DAKO, f. P-4018, op. 1, d. 8 ("Pervoe Vseukrainskoe soveshchanie deiatelei OZE, 1–5 noiabra 1918"), l. 15–15 rev.

59. "Preniia," DAKO, f. P-4018, op. 1, d. 8, l. 15 rev.

60. "Preniia," DAKO, f. P-4018, op. 1, d. 8, l. 15 rev.

61. Henrika Kuklick, "Continuity and Change in British Anthropology, 1914–1919," in *Doing Anthropology in Wartime and War Zones: World War I and the Cultural Sciences in Europe,* ed. Reinhard Johler, Christian Marchetti, and Monique Scheer (Bielefeld, Germany: Transcript Verlag, 2010), 34.

62. Andrew Evans, "Science behind the Lines: The Effects of World War I on Anthropology in Germany," in Johler, Marchetti, and Scheer, *Doing Anthropology in Wartime and War Zones,* 99–122, here 114.

63. "Otchet gubernskogo podotdela doshkol'nogo vospitaniia za mai–iiun' 1919 goda. Prilozhenie 3: O roli vracha v rabote podotdela shkol'nogo vospitaniia," DAKO, f. P-4018, op. 1, d. 12, l. 19.

64. "Preniia," DAKO, f. P-4018, op. 1, d. 8, ll. 13–13 rev, 13 rev, 14.

65. According to an OZE questionnaire, in the cities of the Pale and the Kingdom of Poland, where in 1914 Jews composed 39.1 percent of the population, Jewish physicians constituted 30.4 percent of all physicians in these regions; the same ratio was true for medical assistants *(fel'dsher)* and midwives. "Po dannym ankety," *Vozrozhdenie* 6 (April 25, 1914): 14.

66. "Otchet o rabote i sostave detskogo ochaga v g. Zhitomere za 1917–1918 gg.," DAKO, f. P-4018, op. 1, d. 5, l. 7 rev. See also "Otchet o rabote Kamenets-Podol'skogo ochaga za 1917–1918 gg.," DAKO, f. P-4018, op. 1, d. 6, 14 ll.; "Otchet o rabote Rovenskogo detskogo ochaga za mai 1917 goda," DAKO, f. P-4018, op. 1, d. 4, 11 ll.; "Pasporta detskikh sadov v g. Kremenchuge i Lozovoi," DAKO, f. P-4018, op. 1, d. 4, ll. 3, 4.

67. S. Roizman, "Referat: D-r Haltrents. Dus Tuberkuloseproblem bei den Juden: Eine rassen- und sezialpoathologische Studie, 1925," in *Voprosy biologii i patologii evreev* [Questions of Jewish biology and pathology *(QJBP)*], vol. 2 (Leningrad: Izd. Evreiskogo istoriko-etnograficheskogo obshchestva, 1928), 243–248.

68. O. Maklakova, "Referat: D-r D. Eingorn: Predvaritel'nye dannye ob antropometricheskom issledovanii shkol'nikov g. Minska," in *QJBP,* vol. 2 (1928), 242–243.

69. "Fizicheskii oblik i zdorov'e russkikh evreev do mirovoi voiny," RGIA, f. 1545, op. 1, d. 96, l. 186.

70. "Fizicheskii oblik i zdorov'e russkikh evreev do mirovoi voiny," l. 186.

7. Civil War and the Biopolitics of Survival

1. Jeremy Adelman, *Sovereignty and Revolution in the Iberian Atlantic* (Princeton, NJ: Princeton University Press, 2006); Adelman, "An Age of Imperial Revolutions," *American Historical Review* 113, no. 2 (2008): 319–340.

2. Ilya Gerasimov, "The Great Imperial Revolution," *Ab Imperio* 18, no. 2 (2017): 21–44, here 22–24.

3. Pieter Judson, *The Habsburg Empire: A New History* (Cambridge, MA: Harvard University Press, 2016).

4. As Gerasimov reminds us, "Even the Grand Duchy of Finland, which had enjoyed legal status as a de facto nation-state within the Russian Empire and demonstrated highly developed and mass-scale nationalism, did not press for independence until several weeks after

the Bolsheviks' October coup. . . . Likewise, backed by massive national mobilization, the Ukrainian parliament—the Central Rada—insisted on its determination to see Ukraine as a part of democratic Russia. Despite the hostile attitude of the Provisional Government in Petrograd, which refused to recognize the Rada's legitimacy, it was only after the Bolshevik coup that the Central Rada proclaimed the sovereign Ukrainian People's Republic in federation with Russia (almost simultaneously with Finland's decision to attain independence). Other territories and national movements followed suit, as the trigger for disintegration was pulled not by "separatists," but by the central (technically, still "imperial") government. It was on November 2 (15), 1917, that the Bolshevik-led government issued the 'Declaration of the Rights of the Peoples of Russia,' which urged for 'decisive and irrevocable' emancipation of all the peoples of Russia in all social spheres, including the right to secede and form separate states." Gerasimov, "The Great Imperial Revolution," 24–25.

5. "Deiatel'nost' obshchestva v 1920–1921 gg.," RGIA, f. 1545, op. 1, d. 96, l. 48.

6. "Obshchii obzor sobytii za vremia voiny," RGIA, f. 1545, op. 1, d. 96, l. 40.

7. On Shteingauz's work with the children's program, see M. M. Shteingauz, *Letnie detskie kolonii* (Petrograd: Iz-vo OZE, 1918); Shteingauz, *Slovo, obraz i deistvie: God raboty s vos'miletkami* (Moscow: Tovarishchestvo "Mir," 1923). Shabad was a specialist on child speech. See, for example, E. Iu. Shabad, *Zhivoe detskoe slovo* (Moscow-Leningrad: Novaia Moskva, 1925).

8. "Pervoe Vseukrainskoe soveshchanie 'OZE' 1-go–5-go noiabria 1918 goda," DAKO, f. P-4018, op. 1, d. 8, l. 4 rev. (quotation); on administration matters, ll. 3 rev.–25.

9. M. Gran, "O zadachakh evreiskogo zdravookhraneniia v Svobodnoi Rossii," *Izvestiia OZE* 1 (1917): 4–18, here 4–5. The meeting resolved to allocate 40 out of 485 seats in the future assembly to relief networks, including the OZE, whereas the rest had to be determined by elections planned for the fall of 1917. Simon Rabinovitch, *Jewish Rights, National Rites: Nationalism and Autonomy in Late Imperial and Revolutionary Russia* (Stanford, CA: Stanford University Press, 2014), 229–231. The assembly, however, never had a chance to meet due to the Bolshevik coup (October 25, 1917).

10. M. Gran, "Stat'i i vospominaniia: S. G. Frumkin kak obshchestvennyi deiatel' i evreiskii obshchestvennyi vrach," *Izvestiia OZE,* December 1918–1919, 6–10, here 9.

11. "Rol' deiatelei obshchestva okhraneniia zdorov'ia v organizatsii obshchin," *Izvestiia OZE,* March 1917, 1–3.

12. V. Binshtok, "Statistika v novoi obshchine," *Izvestiia OZE,* March 1918, 18–19.

13. In terms of historiography, this original perception finds substantiation in Zvi Gitelman's fundamental study of Evsektsiia that shows how it functioned as an agent of the Sovietization and suppression of "authentic" Jewish national life and institutions. Gitelman, *Jewish Nationality and Soviet Politics: The Jewish Sections of the CPSU, 1917–1930* (Princeton, NJ: Princeton University Press, 1972); Mordechai Altshuler, *Ha-yevsektsiya bi-vrit ha-mo'atsot: Beyn komunizm ve leumiyut* (Tel Aviv: Sifriyat po'alim, 1981); Benjamin Pinkus, *The Jews of the Soviet Union: The History of a National Minority* (Cambridge: Cambridge University Press, 1988).

14. Jeffrey Veidlinger, *The Moscow State Yiddish Theater: Jewish Culture on the Soviet Stage* (Bloomington: Indiana University Press, 2000); David Shneer, *Yiddish and the Creation of Soviet Jewish Culture, 1918–1930* (Cambridge: Cambridge University Press, 2004); Anna Shternshis, *Soviet and Kosher: Jewish Popular Culture in the Soviet Union, 1923–1939* (Bloomington: Indiana University Press, 2006). For a discussion of the inclusive and complex nature of Soviet Jewishness, see Elissa Bemporad, *Becoming Soviet Jews: The Bolshevik Experiment in Minsk* (Bloomington: Indiana University Press, 2013); Jarrod Tanny, *City of Rogues and Schnorrers: Russia's Jews and the Myth of Old Odessa* (Bloomington: Indiana University Press, 2011); Andrew Sloin, *The Jewish Revolution in Belorussia: Economy, Race,*

and Bolshevik Power (Bloomington: Indiana University Press, 2017). In the same category, see Ken Moss, *Jewish Renaissance in the Russian Revolution* (Cambridge, MA: Harvard University Press, 2009).

15. "Deiatel'nost' obshchestva v 1920–1921 gg.," RGIA, f. 1545, op. 1, d. 96, l. 50. Where OZE committees could appeal to state authorities, primarily in big cities such as Minsk or Kiev, they retained property, but in the provinces this did not work.

16. "Deiatel'nost' obshchestva v 1920–1921 gg.," RGIA, f. 1545, op. 1, d. 96, ll. 53–54, 59.

17. Harriet Murav, "Archive of Violence: Part II, Documentation" (unpublished manuscript, consulted Fall 2021), 5. I am extremely grateful to the author for sharing this important text with me.

18. See, for example, GARF, f. R-1339, op. 1, d. 401 ("Statisticheskie tablitsy perepisi o sostave naseleniia v Kievskoi gubernii, 1920"), 188 ll; GARF, f. R-1339, op. 1, d. 402 ("Chernoviki perepisi evreiskogo naseleniia," 1920), 119 ll., esp. ll. 25–25 rev.

19. First, the Central United Committee of Jewish organizations (EKOPO, OZE, ORT, and OPE [Society for the Promotion of Culture among the Jews of Russia] [Obshchestvo rasprostraneniia prosveshcheniia mezhdu evreiami v Rossii]), whose role by that time had become largely symbolic, opened its bureau in Moscow in 1919. The Petrograd and Moscow OZE committees jointly decided to use it as their platform for negotiations with the central authorities.

20. "Deiatel'nost' obshchestva v 1920–1921 gg.," RGIA. f. 1545, op. 1, d. 96, ll. 63–81.

21. On JDC aid to Russian Jews in these years, see Michael Beizer, *Relief in Time of Need: Russian Jewry and the Joint, 1914–24* (Bloomington, IN: Slavica, 2015); Jaclyn Granick, "Humanitarian Responses to Jewish Suffering Abroad by American Jewish Organizations, 1914–1929" (PhD diss., Graduate Institute of International and Development Studies, Geneva, 2015).

22. Quoted in Michael Beizer, "The Emergence of the Society for the Protection of the Health of the Jewish Population (OZE): Historical Circumstances and Ideology of the Founders" (paper presented at the OZE [OSE-Union] Centennial Conference, Paris, June 2012), 1, http://pluto.huji.ac.il/~beizer/files/OZE%20100%20Paris.pdf.

23. "Deiatel'nost' obshchestva v 1920–1921 gg.," RGIA, f. 1545, op. 1, d. 96, l. 89, l. 92.

24. "Deiatel'nost' obshchestva v 1920–1921 gg.," RGIA, f. 1545, op. 1, d. 96, l. 108.

25. "Deiatel'nost' obshchestva v 1920–1921 gg.," RGIA, f. 1545, op. 1, d. 96, ll. 103–104.

26. Soviet and modern Russian manuals on statistics present Novosel'skii as a founder of Soviet demographic and sanitation statistics without ever mentioning his involvement with the OZE and its Jewish statistical project. Novosel'skii is usually credited for calculating, for the first time in the USSR, "characteristics of the stable population; [he] composed mortality tables for Leningrad and the first mortality tables for the population of the USSR and its regions for the years 1926–1927; he also composed brief mortality tables for specific social-professional groups. S. A. Novosel'skii is a founder of the academic school that carried out the multipart study of population health on the basis of the detailed analysis of demographic processes." V. P. Kornev, *Istoriia statistiki: Uchebnoe posobie* (Saratov: Saratov State Social-Economic University, 2013), 86–87. The connection of these interests of Novosel'skii with his prior experience with Jewish mortality statistics remains unknown.

27. In addition, in 1918 Gergel was appointed chairman of the bureau of the Jewish Ministry of Ukraine. After Hetman Skoropadskyi's coup in April 1918, Gergel effectively became the head of the Jewish Ministry of Ukraine. Nahum Gergel, "Di pogromen in Ukraine

1918–1921," *Shriftn far ekonomik un statistik* 1 (1928); the English translation of this article appeared more than twenty years later, as Nahum Gergel, "The Pogroms in Ukraine in 1918–1921," *YIVO Annual of Jewish Social Science* 6 (1951): 237–252. On the Civil War pogroms, see Oleg Budnitskii, *Russian Jews between the Reds and the Whites, 1917–1920*, trans. Timothy Portice (Philadelphia: University of Pennsylvania Press, 2012); L. B. Miliakova, ed., *Kniga pogromov: Pogromy na Ukraine, v Belorussii i evropeiskoi chasti Rossii v period Grazhdanskoi voiny, 1918–1922 gg.; Sbornik dokumentov* (Moscow: ROSSPEN, 2007); Elissa Bemporad, *Legacy of Blood: Jews, Pogroms, and Ritual Murder in the Lands of the Soviets* (New York: Oxford University Press, 2019). On mass rape during the Civil War pogroms, see Irina Astashkevich, *Gendered Violence: Jewish Women in the Pogroms of 1917 to 1921* (Boston: Academic Studies Press, 2018).

28. A. Vitlin, "Prirost evreiskogo naseleniia i nravstvennaia tsennost' ego," *Sibir-Palestina*, no. 4 (January 26, 1923): 2–3, here 2.

29. GARF, f. 1339, op. 1, d. 575, ll. 9–13.

30. GARF, f. 1339, op. 1, d. 575, ll. 7–7 rev.

31. GARF, f. 1339, op. 1, d. 575, 8–8 rev.

32. GARF, f. 1339, op. 1, d. 575, ll. 23–23 rev., l. 20.

33. "Dokladnaia zapiska o proizvedennoii i predpolagaemoi rabote 'OZE,'" [no year], SPbF ARAN, f. 282, op. 1, d. 176 (Materialy po evreiskomu voprosu), l. 735a.

34. As the head of the tuberculosis section of the Regional Medical Division (Gubzdravotdel), Bramson developed a system of clinical treatment for tuberculosis patients. His official Soviet biographies routinely omitted that in his medical work Bramson applied his rich experience in treating and studying Jewish tuberculosis and "narrow chests." He also prepared the project of the first Soviet Institute of Tuberculosis. See the official site of the Russian Federation's Ministry of Health, St. Petersburg Research Institute for Phthisiopulmonology, "Sozdanie Instituta tuberkuleza i ego deiatel'nost' v dovoennyi period," accessed March 21, 2022, https://www.spbniif.ru/about/sozdanie-instituta-tuberkuleza-i-ego-deyatelnost-v-dovoennyy-period.php.

35. "Doktoru A. M. Bramsonu, 1922," RGIA, f. 1545, op. 2, d. 28 (Perepiska s Berlinskim komitetom OZE ob okazanii pomoshchi evreiskomu naseleniiu), l. 35.

36. "Otchet o deiatel'nosti komissii po izucheniiu voprosov evreiskogo zdravookhraneniia 'OZE' za fevral' mesiats 1923 g.," RGIA, f. 1545, op. 2, d. 27 ("Perepiska s knigoizdatel'stvom 'Meditsinskii sovremennik,' kooperativnym izdatel'stvom i dr. ob izdanii meditsinskikh knig," 1922–1923), ll. 125–125 rev.

37. "Dokladnaia zapiska o proizvedennoii i predpolagaemoi rabote 'OZE,'" [n.d.], SPbF ARAN, f. 282, op. 1, d. 176 (Materialy po evreiskomu voprosu), l. 735a.

38. Beizer, *Relief in Time of Need,* 140.

39. Nadav Davidovitch and Rakefet Zalashik, whose work focuses on the interwar OZE history in East-Central Europe, believe that "the analysis of OZE as a transnational Jewish relief organization bears wider significances, as an example of international organizations originating from civil initiatives trying to promote minorities' health through field work and politics." Davidovitch and Zalashik, "'Air, Sun, Water': Ideology and Activities of OZE (Society for the Preservation of the Health of the Jewish Population) during the Interwar Period," *Dynamis* 28 (2008): 127–149, here 129. In 1924, the network included Latvian and Lithuanian national OZE organizations; Vilna and Bialystok organizations; Romanian, Bukovinian, Staro-Romanian, and Bessarabian national organizations; Danzig, Berlin, Paris, London, and Palestine (Institut Pasteur) organizations; and Maccabi—the world Jewish gymnastic union. The Russian OZE was not officially listed. "Kratkaia zapiska o deiatel'nosti 'OZE,'" LSHA, f. 6511 (Obshchestvo okhrany zdorov'ia evreev Litvy 'Oze'), op. 1, d. 2 (Protokoly zasedanii . . . Tezisy i rezolutsii . . . Otchety o deiatel'nosti), ll. 1–3, here 2. Eventually, the OZE

came to encompass forty-eight organizations from many countries. Albert Einstein was elected honorary president of the OZE–World Union.

40. In Poland, the TOZ and OZE collaborated as separate organizations (often as TOZ-OZE) until 1926. In 1927, the TOZ became the main organization and absorbed the OZE.

41. *Ustav OZE* (unpublished document, 1922). This document has a note stating, "The project was not implemented but is an example of the Statute on the basis of resolutions by the VTsIK [Vserossiysky Tsentral'ny Ispolnitelny Komitet—All-Russian Central Executive Committee] and SNK [or Sovnarkom, Soviet Narodnykh Komissarov—Council of People's Commissars] from August 8, 1922." TsGAM (Tsentral'nyi gosudarstvennyi arkhiv goroda Moskvy—Central State Archive of the city of Moscow), f. 1454, op. 1, d. 23, ll. 1–11, here l. 1.

42. Letter to Berlin, 1922, RGIA, f. 1545, op. 2, d. 28 (Perepiska s Berlinskim komitetom OZE ob okazanii pomoshchi evreiskomu naseleniiu), l. 33.

43. When the Civil War and the famine were over, Gran started his academic career—first in Moscow and after 1928 in Kazan, where he was appointed university chair in social hygiene. Gran finally retired in 1932 and returned to Moscow, where he continued his studies of medical statistics. V. Iu. Al'bitskii, M. E. Guryleva, N. Kh. Amirov, et al., eds., *Kazanskii gosudarstvennyi meditsinskii universitet (1804–2004): Zaveduiushchie kafedrami i professora; Biograficheskii slovar'* (Kazan: Magarif, 2004), 125–126; N. Kh. Amirov, L. M. Fatkhutdinova, and V. N. Krasnoshchekova, "V. M. Bekhterev i Kazanskaia gigienicheskaia shkola: Preemstvennost' pokolenii," in *V. M. Bekhterev i sovremennye Kazanskie meditsinskie shkoly*, ed. N. Kh. Amirov, M. F. Ismagilov, and D. M. Mendelevich (Kazan: Meditsina, 2007), 217–232; "Gran, Moisei Markovich," in *Bol'shaia biograficheskaia entsiklopediia* (Moscow: Online project by Academic.ru, 2009), http://dic.academic.ru/dic.nsf/enc_biography/39045/.

44. "Kruzhku Mozen v Moskve, 12 maia, 1922," RGIA, f. 1545, op. 2, d. 28, ll. 1 (quotation), 3.

45. Boris Bogen, *Born a Jew* (New York: Macmillan, 1930), 304–305.

46. Bogen, *Born a Jew*, 306.

47. Bogen, *Born a Jew*, 306.

48. Letter from Dr. Moisei Markovich Gran to Dr. Abram Moiseivich Bramson, April 6, 1922, RGIA, f. 1545, op. 2, d. 28, ll. 7–7 rev.

49. "Beseda s doktorom M. M. Granom," *Narodnaia mysl'*, no. 244 (September 28, 1924): 4.

50. *Der emes* (Truth) was the leading Yiddish newspaper of the Soviet Jewish socialist coalition (of the Jewish Social Democrats, Bolsheviks, and Left Socialist-Revolutionaries). From November 7, 1920, onward, it was the official organ of the Evsektsiia.

51. "Zadachi soiuzov 'OZE' i 'ORT': Beseda s predstavitelem Tsentral'nykh pravlenii 'Oze' i 'Ort' N. Iu. Gergelem," *Narodnaia mysl'*, no. 244 (September 28, 1924): 3.

52. "Intermeditzinish sanitare tetikheit fon 'OZE' in di Ukrain," *Buletin fon zentral Biro fon der geselshaft fur ferhiten di gezuntheit fon der idisher befolkerung "OZE,"* no. 2 (June 1923): 2. The Nansen Mission refers to an international humanitarian mission directed by the Norwegian explorer, scientist, and philanthropist Fridtjof Nansen. In 1921, Nansen was appointed the League of Nations' High Commissioner for Refugees, and in 1922 he was awarded the Nobel Peace Prize for his work for refugees and displaced victims of World War I and the Russian Civil War.

53. Letter to M. M. Gran, Berlin, December 11, 1923, RGIA, f. 1545, op. 2, d. 28, ll. 54–55.

54. Kamila Uzarczyk, "'Moses als Eugeniker'? The Reception of Eugenic Ideas in Jewish Medical Circles in Interwar Poland," in *"Blood and Homeland": Eugenics and Racial Nationalism in Central and Southeast Europe, 1900–1940*, ed. Marius Turda and Paul Weindling (Budapest: CEU Press, 2007), 283–298.

55. For example, see the report on OZE activities in Latvia in 1923–1926: "Doklad dr. Eliasberga," LSHA, f. 6511 (Obshchestvo okhrany zdorov'ia evreev Litvy 'Oze'), op. 1, d. 13, ll. 107–108 rev. A tradition of socioeconomic statistical research was resumed by the Yiddish Scientific Institute, YIVO, established in Vilna in 1925. Four departments were set up, including the Economic-Statistical Section headed by Lestschinsky. Gennady Estraikh, "Jacob Lestschinsky: A Yiddishist Dreamer and Social Scientist," *Science in Context* 20, no. 2 (2007): 226.

56. "Kratkaia zapiska o deiatel'nosti 'OZE,'" LSHA, f. 6511, op. 1, d. 2, l. 1–3, here 3.

57. Rakefet Zalashik, "Medical Welfare in Interwar Europe: The Collaboration between JDC and OZE-TOZ Organizations," in *The JDC at 100: A Century of Humanitarianism*, ed. Avinoam Patt, Atina Grossmann, Linda G. Levi, and Maud S. Mandel (Detroit: Wayne State University Press, 2019), https://books.google.com/books?id=6K5rDwAAQBAJ.

58. Zalashik, "Medical Welfare in Interwar Europe." In May 1925, Gran informed the JDC that they needed only $50,000 to reach their budget goal. Zalashik follows in detail the story of economic and policy conflicts between the OZE and the JDC.

59. Elissa Bemporad, "JDC in Minsk: The Parameters and Predicaments of Aiding Soviet Jews in the Interwar Years," in Patt et al., *The JDC at 100*.

8. The Gray Zone of Jewish Biopolitics and Its Place in Soviet Modernity

1. Michael Beizer, *Relief in Time of Need: Russian Jewry and the Joint, 1914–24* (Bloomington, IN: Slavica, 2015), 140.

2. B. Vishnevskii, "K antropologii evreev Rossii," *Evreiskaia starina* 11 (1924): 266–398.

3. Vishnevskii, "K antropologii evreev Rossii," 268.

4. On the ARA's principle to provide relief "without regard to race, politics, or religion," see H. H. Fisher, *The Famine in Soviet Russia 1919–1923: The Operations of the American Relief Administration* (New York: Macmillan, 1927), esp. 29–30. Fisher was chief of the ARA Historical Department and a lecturer in history at Stanford University.

5. Iu. A. Poliakov, "Demograficheskie posledstviia goloda 1921–1922 gg.," in *Naselenie Rossii v XX veke: Istoricheskie Ocherki*, ed. V. B. Zhiromskaia, vol. 1, *1900–1939 gg.* (Moscow: ROSPEN, 2000), 129, 131.

6. Bertrand M. Patenaude, *The Big Show in Bololand: The American Relief Expedition to Soviet Russia in the Famine of 1921* (Stanford, CA: Stanford University Press, 2002), 32.

7. "Mr. Hoover's Reply to Maxim Gorky's Appeal," *American Relief Administration Bulletin* 16 (September 1, 1921): 3–4.

8. See similar observations about the spirit of American philanthropy as practiced by the JDC and the Carnegie and Rockefeller Foundations, which "wanted to transfer the 'American way' to the European context through science, the management of projects and the rebuilding of communities," in Rakefet Zalashik and Nadav Davidovitch, "Taking and Giving: The Case of the JDC and OZE in Lithuania 1919–26," *Eastern European Jewish Affairs* 39, no. 1 (2009): 57.

9. Fisher, *The Famine in Soviet Russia*, 92–93, 124–125.

10. John P. Gregg, "The Tartar Socialist Soviet Republic (Formerly Kazan Government)," *American Relief Administration Bulletin* 18 (November 1, 1921): 26–29.

11. Voidinova was thirty-three in 1921. She graduated from the Moscow University Medical Department in 1917 and worked for a short time as a zemstvo physician in the Birsk district of Ufa province. Later she became a doctor at the Military Hospital in Kazan; simul-

taneously, she assumed positions as school and sanitary physician with the local Narkompros and Narkomzdrav. She was demobilized from the medical-sanitation department of the army in 1921, presumably after she had assumed her position with the ARA. See her job application in NART, f. p-41, op. 1, d. 70, ll. 63–63 rev.

The pelidisi test was designed by a Viennese medical doctor and professor, Clemens Pirquet, who served as chairman of the Austrian ARA public committee. Pirquet devised a formula for determining the degree of undernourishment in children up to the age of fifteen. The measurement was the cubic root of the tenfold weight of the body divided by that body's sitting height. For adults, the average would be 100, for children 94.5. Children with a pelidisi measurement of less than 94 were considered undernourished, and children with a measurement of 90 or less, seriously undernourished. "Dr. Clemens Pirquet, Founder of the NEM and Pelidisi Systems Explains Their Significance in Relation to Institutions and School Feeding, 3 November 1921," *American Relief Administration Bulletin* 18 (November 1, 1921): 33–38. All anthropometric measurements were taken according to the International Agreement on Anthropological Measurements: "Mezhdunarodnoe soglashenie dlia ob"edineniia antropologicheskikh izmerenii," *Russkii antropologicheskii zhurnal* 35–36 (1913): 103; Vishnevskii, "K antropologii evreev Rossii," 269.

12. Dozens of such handwritten, hard-to-read tables, composed according to the national principle and for internal usage, are collected in NART, f. p-41, op. 1, d. 6 (1921), 137 ll.

13. NART, f. p-41, op. 1, d. 6 (1921), ll. 47–48; ll. 43–44.

14. For example, "The list of children of Muslim nationality of the village of B. Achysypy of Shirdan *volost*' of district 2 of Sviiazhsk *kanton*, who had medical examinations according to the method of Professor Pirquet on October 27–28, 1921." NART, f. p-41, op. 1, d. 6 (1921), l. 38.

15. For example, "List of the children of workers of the uniforms manufactory of the Alafuzov brothers at Sennoi Market," 1921, NART, f. p-41, op. 1, d. 6, ll. 60–60 rev.

16. On the complexities of the institutionalization of anthropology in Russian imperial universities, where ethnography, geography, and anthropology had eventually been placed in the Departments of Physics and Mathematics, see Marina Mogilner, *Homo Imperii: A History of Physical Anthropology in Russia* (Lincoln: University of Nebraska Press, 2013), 34–53, esp. 43.

17. For Vishnevskii's personal file, which preserves materials highlighting the early stages of his career up to his arrest in 1937, see ARAN, f. 411, op. 6, d. 584, ll. 1 (3–3 rev).

18. NART, f. 1339, op. 2l, d. 1, ll. 41–42 rev. [Recommendation letter from professor Nikolai Fedorovich Katanov].

19. ARAN, f. 411, op. 6, d. 584, l. 4; NART, f. 1339, op. 2l, d. 1, ll. 1, 12, 54.

20. Mogilner, *Homo Imperii*, 360–363.

21. [B. Vishnevskii], "K voprosu o prepodavanii antropologii v Kazanskom universitete," *Zhurnal Kazanskogo mediko-antropologicheskogo obshchestva* 1 (1921): 270–271.

22. NART, f. 1963, op. 1, d. 12, l. 8 (June 25, 1922–May 26, 1923).

23. However, it could not reasonably demand that the Quakers, who acted as charity workers in Russia and collaborated with the ARA, bar their experienced staff of female relief workers.

24. Patenaude, *The Big Show in Bololand*, 50–51.

25. Beizer, *Relief in Time of Need*, 147.

26. "Agreement with the Joint Distribution Committee regarding Operations in the Ukraine and White Russia, August 14th, 1922," *American Relief Administration Bulletin* 28 (September 1922): 4–5.

27. NART, f. p-41, op. 1, d. 70, ll. 1–1 rev., 12, 21–22, 31, 34, 38, 49, 56, 60, 71, 80–84, 93–94, 118, 130–131, 140–145, 152, 159, 164.

28. "List of Collaborators of the Russian-American Committee of Help to the Hungering [original wording!—MM] Children," NART, f. p-41, op. 1, d. 6, ll. 3–6. The Russian equivalent of the document intended for the Soviet authorities is much more detailed, most probably composed a few months later than the English original, but it confirms the fact of staffing the Department of Medical Statistics exclusively with Jewish physicians. Unlike the English version, which gives only four "collaborators" (a literal translation of the Russian *sotrudnik*), the Russian document lists ten names. NART, f. p-41, op. 1, d. 6, ll. 1–5 rev.

29. L. K. Karimova, "Sotsial'naia struktura naseleniia Kazani po perepisiam 1923 i 1926 gg.," in *Sotsial'naia struktura i sotsial'nye otnosheniia v Respublike Tatarstan v pervoi polovine XX veka* (Kazan: Kheter, 2003), 230.

30. Karimova, "Sotsial'naia struktura naseleniia."

31. Peter Gatrell, *A Whole Empire Walking: Refugees in Russia during World War I* (Bloomington: Indiana University Press, 2005); Joshua A. Sanborn, *Imperial Apocalypse: The Great War and the Destruction of the Russian Empire* (Oxford: Oxford University Press, 2014).

32. On the mobilization of Jewish physicians during the late imperial and early Soviet period, see Marina Mogilner, "Toward a History of Russian Jewish 'Medical Materialism': Russian Jewish Physicians and the Politics of Jewish Biological Normalization," *Jewish Social Studies* 19, no. 1 (Fall 2012): 70–106.

33. On Luria's work with hunger statistics, see NART, f. p-4470, op. 1, d. 6, l. 270.

34. Liubov' Person was arrested in Kazan in February 1905 for supplying Jewish soldiers with revolutionary leaflets. S. Lifshits, *Kazan' v gody pervoi revoliutsii: 1905–1907 gg.* (Kazan: Nauchno-issledovatel'skii ekonomicheskii institut, 1930), 65.

35. A. Iu. Fedotova and N. A. Fedotova, *Pomoshch' golodaiushchemu naseleniiu TASSR sovetskimi i inostrannymi organizatsiami v 1921–1923 gg.* (Kazan: IITs UDP RT, 2013), 192.

36. RGIA, f. 1545, op. 1, d. 61 ("Ob otkrytii i rabote otdeleniia OZE v Kazani, 1916"), ll. 2, 3, 7–8.

37. Will Shafroth, "An Inspection Trip to the Volga: Extracts from the Notes Taken on Each Day's Work," *American Relief Administration Bulletin* 18 (November 1, 1921): 18.

38. NART, f. p-41, op. 1, d. 4, l. 13.

39. Anketa (application) no. 1160, NART, f. p-41, op. 1, d. 70, ll. 60–60 rev, 80–82 rev.

40. NART, f. p-41, op. 1, d. 6, ll. 1–137.

41. NART, f. p-41, op. 1, d. 5, l. 26 rev.

42. Anketa no. 82, NART, f. p-41, op. 1, d. 70, l. 82.

43. GARF, f. 1338, op. 1, d. 326, ll. 7, 9.

44. "Ochag v Kazani (Iz pis'ma G. I. Tsitovskoi)," *Izvestiia Obshchestva okhraneniia zdorov'ia evreiskogo naseleniia*, February 1918, 25–26. This report also claims that "local society accords the institution all possible attention and support" and that attendance rates at the *ochag* are quite high and stable. I was able to work with the complete set of *Izvestia OZE* for 1917, 1918, and 1919 at TsGAM. Here TsGAM, f. 1454, op. 1, d. 16, ll. 69–69 rev.

45. V. Iu. Al'bitskii, M. E. Guryleva, N. Kh. Amirov, et al., eds., *Kazanskii gosudarstvennyi meditsinskii universitet (1804–2004): Zaveduiushchie kafedrami i professora; Biograficheskii slovar'* (Kazan: Magarif, 2004), 125–126; N. Kh. Amirov, L. M. Fatkhutdinova, and V. N. Krasnoshchekova, "V. M. Bekhterev i Kazanskaia gigienicheskaia shkola: Preemstvennost' pokolenii," in *V. M. Bekhterev i sovremennye Kazanskie meditsinskie shkoly*, ed. N. Kh. Amirov, M. F. Ismagilov, and D. M. Mendelevich (Kazan: Meditsina, 2007), 217–232; "Gran, Moisei Markovich," in *Bol'shaia biograficheskaia entsiklopediia* (Moscow: Online

project by Academic.ru, 2009), http://dic.academic.ru/dic.nsf/enc_biography/39045/. Gran finally retired in 1932 and returned to Moscow, where he continued conducting studies of medical statistics.

46. RGIA, f. 1545, op. 1, d. 61, l. 28.

47. GARF, f. R1339, op. 1, d. 55, l. 200.

48. "Pelidisi Test: Statistical data proves that the Pelidisi method is more acceptable in Russia for children of school age than for younger children. In cases of acrofulosis and in rickets with increase in size of head and abdomen, the test is unreliable, the weight being much increased and the Pelidisi wanting in its correct and objective formula. The Pelidisi works to the disadvantage of children whose sitting height is 50 [centimeters], and to the advantage of children with a sitting height of 70. The test should be applied with great care in Russia and checked by general examination as quite frequently children with Pelidisi of 92–94 are refused because they are well nourished while others with 95 are accepted as medical examination reveals anemia and poor nutrition." Henry Beeuwkes, chief of the Medical Division, "Medical Examinations and Remarks on Pelidisi Test," *American Relief Administration Bulletin* 29 (October 1922): 19–43, here 33.

49. See TsGAM, f. 1454, op. 1, d. 11 ("Rezoliutsiia konferentsii OZE o rabote evreiskikh obshchin v oblasti zdravookhraneniia, spiski lekarstv, meditsinskikh otriadov i dr.; Materialy po vrachebno-sanitarnoi deiatel'nosti OZE, 1916–1921"), esp. ll. 10–14 ("Rukovodiashchie ukazaniia dlia stolovykh, ochagov, priiutov i vrachebno-sanitarnykh otriadov OZE po voprosu o kartinakh pitania").

50. GARF, f. 1339, op. 1, d. 575, ll. 1–2.

51. Patenaude, *The Big Show in Bololand,* 87.

52. L. Ia. Shternberg, "Problema evreiskoi natsional'noi psikhologii," *Evreiskaia starina* 11 (1924): 5–44.

53. M. M. Gran, "K voprosu o metodologii biologicheskogo izucheniia rasy i natsii," in *Voprosy biologii i patologii evreev,* vol. 2 (Leningrad: Izd. Evreiskogo istoriko-etnograficheskogo obshchestva, 1928), 7.

54. ARAN, f. 411, op. 6, d. 584.

55. For a typical example, see a quite detailed biography of Gran that even traces his and Nikolai Semashko's relationship back to their work in Samara zemstvo but gives no hint of Gran's Jewish activism before or after 1917: V. Iu. Al'bitskii, M. E. Guryleva, and N. Kh. Amirov, "Gran Moisei Markovich," in V. Iu. Al'bitskii et al., *Kazanskii gosudarstvennyi meditsinskii universitet (1804–2002),* 2–4. For a similar version of Bramson's Soviet professional biography, see the official site of the Russian Federation's Ministry of Health, St. Petersburg Research Institute for Phthisiopulmonology, "Sozdanie Instituta tuberkuleza," accessed March 21, 2022, https://www.spbniif.ru/about/sozdanie-instituta-tuberkuleza-i-ego-deyatelnost-v-dovoennyy-period.php.

56. Laura Engelstein, "Combined Underdevelopment: Discipline and the Law in Imperial and Soviet Russia," *American Historical Review* 98, no. 2 (1993): 338–353; Peter Holquist, "To Count, to Extract, and to Exterminate: Population Statistics and Population Politics in Late Imperial and Soviet Russia," in *A State of Nations: Empire and Nation-Making in the Age of Lenin and Stalin,* ed. Ronald Grigor Suny and Terry Martin (New York: Oxford University Press, 2001), 111–144; Daniel Beer, *Renovating Russia: The Human Sciences and the Fate of Liberal Modernity, 1880–1930* (Ithaca, NY: Cornell University Press, 2008).

57. Shimeliovich quoted in Andrew Sloin, *The Jewish Revolution in Belorussia: Economy, Race, and Bolshevik Power* (Bloomington: Indiana University Press, 2017), 83.

58. Sloin, *The Jewish Revolution in Belorussia,* 17.

59. Sloin, *The Jewish Revolution in Belorussia,* 134–135.

Conclusion

1. Published in *Biro-Bidzhan (Zemleustroistvo trudiashchikhsia-evreev v Rossii)* (Harbin: Kharbin Observer, 1930), 40–41, here 41. First published in *Izvestiia TsIK*, no. 265 (November 16, 1926).

2. Daniel E. Schafer, "Local Politics and the Birth of the Republic of Bashkortostan, 1919–1920," in *A State of Nations: Empire and Nation-Making in the Age of Lenin and Stalin,* ed. Ronald Grigor Suny and Terry Martin (New York: Oxford University Press, 2001), 165–190; Schafer, "Bashkir Loyalists and the Question of Autonomy: Gabdulkhai Kurban-galiev in the Russian Revolution and Civil War," in *Russia's Home Front in War and Revolution, 1914–1922,* bk. 1, *Russia's Revolution in Regional Perspective,* ed. Sarah Badcock, Liudmila Novikova, and Aaron Retish (Bloomington, IN: Slavica, 2016), 215–246.

3. Olivier Roy, *The New Central Asia: The Creation of Nations* (London: I. B. Tauris, 2000); Adeeb Khalid, *Making Uzbekistan: Nation, Empire, and Revolution in the Early USSR* (Ithaca, NY: Cornell University Press, 2015); Marco Buttino, *La rivoluzione capovolta: L'Asia centrale tra il crollo dell'impero zarista e la formazione dell'Urss* (Naples: Ancora del Mediterraneo, 2003).

4. In the course of this delimitation, some groups vanished from the Soviet map of Central Asia altogether, such as Sarts, who were reclassified as Uzbeks, or the distinct Central Asian Jews, who became generic Jews. On Sarts, see Sergei Abashin, *Natsionalizmy v Srednei Azii: V poiskakh identichnosti* (St. Petersburg: Aleteiia, 2007). On Jews, see Zeev Levin, *Collectivization and Social Engineering: Soviet Administration and the Jews of Uzbekistan, 1917–1939* (Leiden: Brill, 2015); Albert Kaganovitch, "Jews and Autonomy in Kokand, 1917–1918," *Jews in Eastern Europe* 38–39, no. 1–2 (1999): 74–87. Others were homogenized and denied recognition where they used to constitute important enclaves, especially in old urban centers now assigned to others' titular republics. See Matthias Battis, "The Aryan Myth and Tajikistan: From a Myth of Empire to One National Identity," *Ab Imperio* 17, no. 4 (2016): 155–183.

5. Terry Martin, *The Affirmative Action Empire: Nations and Nationalism in the Soviet Union, 1923–1939* (Ithaca, NY: Cornell University Press, 2001); Francine Hirsch, *Empire of Nations: Ethnographic Knowledge and the Making of the Soviet Union* (Ithaca, NY: Cornell University Press, 2005).

6. On the nomad Kazakhs, see Matthew Payne, "The Making of a Kazakh Proletariat? The Unexpected Russification of Nativization Policies on the Turksib," in Suny and Martin, *A State of Nations,* 223–252; Payne, *Stalin's Railroad: Turksib and the Building of Socialism* (Pittsburgh: University of Pittsburgh Press, 2001). On catching-up modernization and proletarization, see also Adrienne Lynn Edgar, *Tribal Nation: The Making of Soviet Turkmenistan* (Princeton, NJ: Princeton University Press, 2004); Douglas Northrop, *Veiled Empire: Gender and Power in Stalinist Central Asia* (Ithaca, NY: Cornell University Press, 2004); Marianne Kamp, *The New Woman in Uzbekistan: Islam, Modernity, and Unveiling under Communism* (Seattle: University of Washington Press, 2006). On the peoples of the Far North, see Igor Stas', "Vozrozhdennye narody: Natsional'noe grazhdanstvo indigennogo naseleniia Severa v pozdnem stalinizme," *Ab Imperio* 21, no. 3 (2020): 157–188; Yuri Slezkine, *Arctic Mirrors: Russia and the Small Peoples of the North* (Ithaca, NY: Cornell University Press, 1994), 95–386 (pts. 2–4).

7. V. I. Binshtok, A. M. Bramson, M. M. Gran, and G. I. Dembo, "O zadachakh nauchnykh sbornikov 'Voprosy biologii i patologii evreev,'" in *Voprosy biologii i patologii evreev* [Questions of Jewish biology and pathology (*QJBP*)], vol. 1 (Leningrad: Prakticheskaia meditsina, 1926), 3–6, here 5.

8. Warwick Anderson, *Colonial Pathologies: American Tropical Medicine, Race, and Hygiene in the Philippines* (Durham, NC: Duke University Press, 2006); Aro Velmet, *Pasteur's Empire: Bacteriology and Politics in France, Its Colonies, and the World* (New York: Oxford University Press, 2020).

9. The French anthropologist Marcel Mauss, who in 1934 introduced the expression "techniques of the body," defined it as "the ways in which from society to society men know how to use their bodies." Mauss, *Sociologie et anthropologie,* with an introduction by Claude Lévi-Strauss, 4th ed. (Paris: Presses universitaires de France, 1968), 364–386 ("Techniques of the Body," a lecture given at a meeting of the Société de psychologie, May 17, 1934).

10. Yohanan Petrovsky-Shtern, *Jews in the Russian Army, 1827–1917: Drafted into Modernity* (Cambridge: Cambridge University Press, 2009); Eugene M. Avrutin, *Jews and the Imperial State: Identification Politics in Tsarist Russia* (Ithaca, NY: Cornell University Press, 2010).

11. V. I. Binshtok, "K voprosu ob odarennosti evreev (Nekotorye statisticheskie materialy)," in *QJBP*, vol. 1 (1926), 7–29, here 26.

12. N. K. Kol'tsov, *Uluchshenie chelovecheskoi porody* (Petrograd: Vremia, 1923); T. I. Iudin, *Evgenika: Uchenie ob uluchshenii prirodnykh svoistv cheloveka* (Moscow: Sabashnikovy, 1925).

13. Binshtok et al., "O zadachakh nauchnykh sbornikov," 5.

14. Lev Trotskii, *Literatura i revoliutsiia* (Moscow: Krasnaia nov', 1923), 195–197.

15. Koltsov quoted in V. V. Babkov, *Zaria genetiki cheloveka: Russkoe evgenicheskoe dvizhenie i nachalo genetiki cheloveka* (Moscow: Progress-Traditsiia, 2008), 68.

16. M. M. Gran, "K voprosu o metodologii biologicheskogo izucheniia rasy i natsii," in *QJBP*, vol. 2 (Leningrad: Izd. Evreiskogo istoriko-etnograficheskogo obshchestva, 1928), 5–10, here 6. On the history of Soviet eugenics, see Mark B. Adams, "The Politics of Human Heredity in the USSR, 1920–1940," *Genome* 31 (1989): 879–884; Adams, "The Soviet Nature-Nurture Debate," in *Science and the Soviet Social Order*, ed. Loren R. Graham (Cambridge, MA: Harvard University Press, 1990), 94–138; Babkov, *Zaria genetiki cheloveka;* Loren R. Graham, "Science and Values: The Eugenics Movement in Germany and Russia in the 1920s," *American Historical Review* 82 (1977): 1133–1164; Nikolai Krementsov, "Eugenics in Russia and the Soviet Union," in *The Oxford Handbook of the History of Eugenics,* ed. Alison Bashford and Philippa Levine (Oxford: Oxford University Press, 2010), 413–429; Krementsov, "'From Beastly Philosophy' to Medical Genetics: Eugenics in Russia and the Soviet Union," *Annals of Science* 68 (2011): 61–92; Krementsov, *With and without Galton: Vasilii Florinskii and the Fate of Eugenics in Russia* (Cambridge: Open Book Publishers, 2018).

17. Binshtok, "K voprosu ob odarennosti evreev," 9.

18. Examples of population statistics: Kh. B. Braude, "Materialy po estestvennomu dvizheniiu evreiskogo naseleniia v Moskve za 48 let (1870–1917)," in *QJBP*, vol. 1 (1926), 64–68; L. G. Zinger, "Dvizhenie evreiskogo naseleniia SSSR," in *QJBP*, vol. 2 (1928), 125–139; Ia. Leshchinskii, "Dvizhenie evreiskogo naseleniia Rossii za 1897–1916," in *QJBP*, vol. 2 (1928), 140–204; G. D. Finkel'shtein, "Estestvennoe dvizhenie evreiskogo naseleniia g. Odessy," in *QJBP*, vol. 2 (1928), 205–216; Ts. O. Shabad, "Smertnost' i rozhdaemost' evreev v Vilne za 5-letie (1921–1925)," in *QJBP*, vol. 2 (1928), 217–227; S. A. Weissenberg, "Dvizhenie evreiskogo naseleniia Zinov'evska (Elizavetgrada) za 1901–1925," in *QJBP*, vol. 2 (1928), 189–204; G. O. Gol'dblat and I. L. Genkin, "K voprosu ob otnositel'no bol'shoi obrashchaemosti evreev za vrachebnoi pomoshchiiu," in *QJBP*, vol. 2 (1928), 98–101; E. Klionskii, "Rasovyi moment i techenie infektsionnykh boleznei (Avtoreferat)," in *QJBP*, vol. 2 (1928), 242. Examples of anthropometric studies: M. M. Patlazhan, "Fizicheskoe sostoianie podrostkov rabochikh g. Odessy v 1923 g.," in *QJBP*, vol. 2 (1928), 172–193; O. Maklakova, "Predvaritel'nye dannye ob antropologicheskom issledovanii shkol'nikov g. Minska (Iz

tsentral'noi detskoi ambulatorii g. Minska)," in *QJBP*, vol. 2 (1928), 247–248; M. S. Sofiev, "Materialy po obsledovaniiu tuzemnykh (bukharskikh) evreev g. Staraia Bukhara," in *QJBP*, vol. 3, no. 2 (Leningrad: Izd. Evreiskogo istoriko-etnograficheskogo obshchestva, 1930), 62–71.

19. I. Iakhinson, "Zadachi nauchno-pedologicheskoi raboty v evreiskoi srede," in *QJBP*, vol. 3, no. 1 (Leningrad: Izd. Evreiskogo istoriko-etnograficheskogo obshchestva, 1930), 107–111.

20. V. Rubashkin, "'Gruppy krovi' v antropologii," in *QJBP*, vol. 2 (1928), 107–111; L. A. Barinshtein, "Nekotorye dannye o krovianykh gruppakh u slavian i evreev," in *QJBP*, vol. 2 (1928), 19–28; M. S. Leichik, "Krovianye gruppy evreev-pereselentsev Odeshchiny (Iz nauchno-issledovatel'skoi kafedry eksperimental'noi i klinicheskoi meditsiny v Odesse privat-dotsenta E. Iu. Kramarenko), in *QJBP*, vol. 2 (1928), 29–51; S. S. Zabolotnyi, "Gruppy krovi karaimov i krymchakov," in *QJBP*, vol. 3, no. 1 (1930), 29–37; E. M. Semenskaia, "Krovianye gruppy sredi gruzinskikh evreev," in *QJBP*, vol. 3, no. 1 (1930), 38–46; N. P. Kevorkov and M. K. Martiukova, "Materialy k voprosu o rasovom biokhimicheskom indekse bukharskikh (tuzemnykh) evreev," in *QJBP*, vol. 3, no. 2 (1930), 60–61.

21. Editors, "From the Editors," in *QJBP*, vol. 2 (1928), 3–4, here 4.

22. M. M. Gran, "'Rodoslovnye v prilozhenii k izucheniiu biologii i patologii evreev," in *QJBP*, vol. 2 (1928), 102–117; See also G. D. Finkel'shtein, "Braki evreev v Odesse," in *QJBP*, vol. 3, no. 1 (1930), 99–108; I. Ia. Gamarnik, "K voprosu o semeinikh nevrodistrofiiakh u evreev," in *QJBP*, vol. 3, no. 2 (1930), 123–126.

23. In 1923–1925, the Commissariat of Health (Narkomzdrav) and the Central Statistical Bureau in Ukraine surveyed 2 percent of the Ukrainian rural population to understand the situation with human capital in the countryside after the Civil War. Jews were not targeted by this research, but the Odessa Regional Narkomzdrav sanctioned the JDC-OZE study as a part of the all-Ukrainian survey. For this, the Odessa physicians modified the Ukrainian survey household registers, prioritizing Jewish anthropometric and sanitation data, and medical statistics. The colonies they investigated were Bol'shaia Seideminukha and Malaia Seideminukha—the two oldest Jewish colonies in Kherson district—and Bogachevka and Manshurovo. M. M. Gran and G. S. Matul'skii, "Opyt sanitarnoi kharakteristiki evreiskoi derevni," in *QJBP*, vol. 1 (1926), 102–140.

24. S. I. Perkal', "Fizicheskoe razvitie evreev Ukrainy po noveishim antropometricheskim dannym," in *QJBP*, vol. 2 (1928), 252–255; A. M. Gabinskii, "Materialy k sravnitel'noi kharakteristike fizicheskogo razvitiia evreiskogo naseleniia Ukrainy," in *QJBP*, vol. 3, no. 2 (1930), 85–107.

25. Binshtok et al., "O zadachakh nauchnykh sbornikov," 5.

26. On "Soviet speak," see (the now classic) Stephen Kotkin, *Magnetic Mountain: Stalinism as Civilization* (Berkeley: University of California Press, 1995).

27. Sheila Fitzpatrick, ed., *Cultural Revolution in Russia, 1928–1931* (Bloomington: Indiana University Press, 1978).

28. A handwritten application to the EIEO requesting permission to establish a commission "for the study of Jewish psycho-physics" was composed and signed by Dr. E. Klionskii on September 9, 1923. SPbF ARAN, f. 282, op. 1, d. 130 ("Zametki i vypiski L. Ia. Shternberga po rasovomu voprosu"), l. 215. In the early 1920s, Efim Evseevich Klionskii moved to Petrograd from Vitebsk province and became the director and physician of the Jewish almshouse. At the same time, he worked as a research fellow at the Leningrad Tuberculosis Institute. He was among the contributors to the *QJBP*'s 1928 collection. Like many Jews of his generation, Klionskii later censored his past as a Jewish race scientist. After World War II, he became a professor at the Second Medical Institute in Leningrad and was a leading specialist

on tuberculosis. Limited information on Klionskii can be found in M. Beizer, *Evrei v Peterburge* (Jerusalem: Biblioteka Alia, 1990), 132–134.

29. A letter from Petrograd to the Berlin OZE Bureau, not signed, 1922, in RGIA, f. 1545, op. 2, d. 28 ("Perepiska s Berlinskim komitetom OZE ob okazanii pomoshchi evreiskomu naseleniiu"), l. 36.

30. Letter to Dr. A. M. Bramson (Petrograd) from L. I. Gurvich (Berlin), January 3, 1924, RGIA, f. 1545, op. 1, d. 189 ("Perepiska Komiteta OZE s berlinskoi i odesskoi evreiskimi organizatsiiami"), ll. 2–3.

31. Letter to Dr. A. M. Bramson (Petrograd) from L. I. Gurvich (Berlin), January 12, 1924, RGIA, f. 1545, op. 1, d. 189, ll. 14–15.

32. Letter to L. I. Gurvich (Berlin) from A. M. Bramson (Petrograd), January 6, 1924, RGIA, f. 1545, op. 1, d. 189 ("Perepiska Komiteta OZE s berlinskoi i odesskoi evreiskimi organizatsiiami"), l. 4.

33. For example, Iakov Eiger submitted an article on "the physical state of Jews." Efim Klionskii was preparing for submission a "very interesting work about the race factor in infectious diseases," for which he used the collection of the EIEO's Anthropology Museum. Letter to L. I. Gurvich (Berlin) from A. M. Bramson (Petrograd), January 10, 1924, RGIA, f. 1545, op. 1, d. 189, l. 12; on Klionskii: letter to L. I. Gurvich (Berlin) from A. M. Bramson (Petrograd), April 9, 1924, RGIA, f. 1545, op. 1, d. 189, ll. 51–52.

34. Letter to L. I. Gurvich (Berlin) from A. M. Bramson (Petrograd), May 22, 1924, RGIA, f. 1545, op. 1, d. 189, l. 67.

35. "Po povodu vypuska pervogo 'Nauchnogo sbornika' Soiuza 'OZE.' Polozheniia," RGIA, f. 1545, op. 1, d. 189, ll. 182–184, ll. 184–185.

36. Babkov, *Zaria genetiki cheloveka*, 173.

37. On scientific experiments of the 1920s and on their support by party leaders, see Alexander Etkind, *Eros of the Impossible: The History of Psychoanalysis in Russia* (Boulder, CO: Westview Press, 1997); A. I. Belkin and A. V. Litvinov, "K istorii psikhoanaliza v Sovetskoi Rossii," *Rossiiskii psikhoanaliticheskii vestnik* 2 (1992): 9–32; Andy Byford, *Science of the Child in Late Imperial and Early Soviet Russia* (Oxford: Oxford University Press, 2020), esp. the section "Revolutionizing the Human Science," 170–175; Kirill Rossiianov, "Beyond Species: Il'ya Ivanov and His Experiments on Cross-Breeding Humans with Anthropoid Apes," *Science in Context* 15, no. 2 (2002): 277–316; Nikolai Krementsov, *Revolutionary Experiments: The Quest for Immortality in Bolshevik Science and Fiction* (New York: Oxford University Press, 2014); Boris Groys and Michael Hagemeister, eds., *Die neue Menschheit: Biopolitische Utopien in Russland zu Beginn des 20. Jahrhunderts* (Frankfurt am Main: Suhrkamp, 2005); Nikolai Krementsov, *A Martian Stranded on Earth: Alexander Bogdanov, Blood Transfusions, and Proletarian Science* (Chicago: University of Chicago Press, 2011).

38. N. Semashko, "O biologicheskom podkhode k postanovke polovogo vospitaniia," *Zvezda* 5 (1924): 150–153, here 150.

39. Kendall Bailes, "Soviet Science in the Stalin Period: The Case of V. I. Vernadskii and His Scientific School, 1928–1945," *Slavic Review* 45, no. 1 (1986): 20–37; Yuri Slezkine, "The Fall of Soviet Ethnography, 1928–1938," *Current Anthropology* 32, no. 4 (1991): 476–484; Loren R. Graham, *Science in Russia and the Soviet Union: A Short History* (Cambridge: Cambridge University Press, 1993); Nikolai Krementsov, *Stalinist Science* (Princeton, NJ: Princeton University Press 1997), 31–53; Alexei Kojevnikov, "The Phenomenon of Soviet Science," *Osiris* 23, no. 1 (2008): 115–135.

40. N. Semashko, "Bibliografiia: Voprosy biologii i patologii evreev. Sb. 1," *Tribuna evreiskoi sovetskoi obshchestvennosti* 1, no. 3 (1927): 23. Also quoted in Gran, "K voprosu o metodologii," 5.

NOTES TO PAGES 255-257

41. Semashko, "Bibliografiia," 23.

42. Gran, "K voprosu o metodologii," 6, 7.

43. Gran, "K voprosu o metodologii," 7.

44. For the details of this transformation, see Krementsov, *Stalinist Science*, 11–92 (pt. 1, "The Making of Stalinist Science").

45. Krementsov, *Stalinist Science,* 35.

46. Quoted from M. Gran's letter to the *Kazan Medical Journal,* in which he tried to rebuff political accusations against him: Gran, "Pis'mo v redaktsiiu v otvet tov. D. Pravdinu," *Kazanskii meditsinskii zhurnal* 26, no. 11 (1930): 1165–1166.

47. See A. G. Hughes, "Jews and Gentiles, Their Intellectual and Temperamental Differences: A Psychological Study Which Reveals the Innate Superiority of London Jewish Children over Their Gentile School-Mates," *Eugenics Review* 20 (1928): 89–94; Mary Davies and A. G. Hughes, "An Investigation into the Comparative Intelligence and Attainments of Jewish and Non-Jewish School Children," *British Journal of Psychology* 18, no. 2 (October 1927): 134–146. Assuming the time for translation and a waiting period before the next collection was ready, the translation of Hughes's article must have started after the publication of the original in *Eugenic Review.*

48. A. G. Iuz [A. G. Hughes], "Sravnitel'noe issledovanie intellekta i psikhiki evreiskikh i neevreiskikh shkol'nikov Londona," *QJBP,* vol. 3, no. 1 (1930), 112–117, here 112.

49. Hughes understood his task as measuring differences, not building hierarchies. A footnote that was not included in the Russian text explained: "It should be pointed out that the term 'race' is used in this article in a very loose sense. Its use is not intended to imply more than that the Jews have been a relatively inbreeding group for a long space of time." Hughes, "Jews and Gentiles," 89.

50. The Jewish Health Organization conducted fundraising for the Berlin OZE Bureau as well as medical and sanitation work in London Jewish quarters among the Jewish refugees from the former Russian Empire. Todd Edelman, "The Jewish Health Organization of Great Britain in the East End of London," in *An East End Legacy: Essays in Memory of William J. Fishman,* ed. Colin Holmes and Anne Kershen (London: Routledge, 2017), 47–70; E. N. Levina, "V vikhre mirovykh sobytii ('O.Z.E.' v Vostochnoi i Tsentral'noi Evrope)" [In the whirlwind of world events (O.Z.E. in Eastern and Central Europe)], in *In Fight for the Health of the Jewish People (50 Years of OSE),* ed. Leon Wulman (New York: World Union OSE and American Committee of OSE, 1968), 20.

51. Karl Pearson and Margaret Moul, "The Problem of Alien Immigration into Great Britain, Illustrated by an Examination of Russian and Polish Jewish Children," *Annals of Eugenics* 1, no. 1 (October 1925): 5–54; Raphael Falk, *Zionism and the Biology of Jews* (Cham, Switzerland: Springer International, 2017), 114–115. Pearson established that the anthropometric measurements of children of Jewish immigrants deviated from those of English children; the amount of hemoglobin in their blood and, most significantly, their level of intelligence differed, too. Pearson and Moul attributed these differences not to the immigrants' difficult living conditions but to their race.

52. A. I. Iarkho, "Protiv idealisticheskikh techenii v rasovedenii SSSR," *Antropologicheskii zhurnal* 1 (1932): 9–23.

53. Slezkine, "The Fall of Soviet Ethnography"; T. D. Solovei, "'Korennoi perelom' v otechestvennoi etnografii," *Etnograficheskoe obozrenie,* no. 3 (2001): 101–121.

54. For the next decade, genetics took its place as a science preferred by the authorities, especially in application to agricultural studies, only to later become the next victim of ideological cleansing. Nikolai Krementsov, "Darwinism, Marxism, and Genetics in the Soviet Union," in *Biology and Ideology: From Descartes to Dawkins,* ed. Denis R. Alexander and Ronald L. Numbers (Chicago: University of Chicago Press, 2010), 215–246.

55. Mark B. Adams, ed., *The Wellborn Science: Eugenics in Germany, France, Brazil, and Russia* (New York: Oxford University Press, 1990); Paul Weindling, "German-Soviet Medical Co-operation and the Institute for Racial Research, 1927–c. 1935," *German History* 10, no. 2 (1992): 177–206; Eduard Kolchinskii, *Biologiia Germanii i Rossii—SSSR v usloviiakh sotsial'no-politicheskikh krizisov pervoi poloviny XX veka (Mezhdu liberalizmom, kommunizmom i natsional-sotsializmom)* (St. Petersburg: Nestor-Istoriia, 2006); Marina Mogilner, *Homo Imperii: A History of Physical Anthropology in Russia* (Lincoln: University of Nebraska Press, 2013), 347–373; Elazar A. Barkan, *The Retreat of Scientific Racism* (New York: Cambridge University Press, 1992).

56. Andy Byford, "Imperial Normativities and Sciences of the Child: The Politics of Development in the USSR, 1920s–1930s," *Ab Imperio* 17, no. 2 (2016): 71–123, here 111.

57. M. Kalinin, *Ob obrazovanii Evreiskoi avtonomnoi oblasti* (Moscow: Der emes, 1935), 6–7, 9, 11, 13. On the Birobidzhan project, see Robert Weinberg, *Stalin's Forgotten Zion: Birobidzhan and the Making of a Soviet Jewish Homeland* (Berkeley: University of California Press, 1998).

58. Kalinin, *Ob obrazovanii evreiskoi avtonomnoi oblasti,* 10.

59. For more on Soviet passports and nationality, see Albert Baiburin, "'The Wrong Nationality': Ascribed Identity in the 1930s Soviet Union," trans. Catriona Kelly, in *Russian Cultural Anthropology after the Collapse of Communism,* ed. Albert Baiburin, Catriona Kelly, and Nikolai Vakhtin (London: Routledge, 2012), 59–76. The following Soviet nationalities became victims of total deportations: Koreans, Germans, Ingria Finns, Karachays, Kalmyks, Chechens, Ingush, Balkars, Crimea Tatars, and Tatars-Meskhetians. Poles were deported before and during World War II, as were Lithuanians, Latvians, and Estonians.

60. The history of race in the USSR after World War II would require a multifactor analysis that would include the de facto failure of the Birobidzhan territorial project; the suppression of cultural and religious forms of transmission of Jewishness, which became ostensibly a matter of birth; the impact of the Holocaust; the establishment of the state of Israel and the new globalization of the Jewish problem; and, finally, the rise of Russian nationalism and state anti-Semitism in the postwar USSR. Another aspect of the same story is the institutionalization of the biosocial *etnos* concept in Soviet ethnography. On this, see David G. Anderson, Dmitry V. Arzyutov, and Sergei S. Alymov, eds., *Life Histories of* Etnos *Theory in Russia and Beyond* (Cambridge: Open Book Publishers, 2019).

61. A continuity narrative sensitive to historical and epistemological contextualizing has been attempted in Nadia Abu El-Haj, *The Genealogical Science: The Search for Jewish Origins and the Politics of Epistemology* (Chicago: University of Chicago Press, 2012), esp. 4, 222.

62. Derek J. Penslar, "Is Zionism a Colonial Movement?," in *Colonialism and the Jews,* ed. Ethan B. Katz, Lisa Moses Leff, and Maud S. Mandel (Bloomington: Indiana University Press, 2017), 275–300, here 276.

ACKNOWLEDGMENTS

In my entire experience as a researcher, this project stands apart as the most difficult, controversial, and long-lasting. It all started sometime in 2001, when I was browsing through the pages of a 1924 issue of the journal *Jewish Antiquity* (*Evreiskaia starina*) and saw an article analyzing the results of a 1921 anthropological examination of 289 Jewish children in the inner Russian city of Kazan—my home city. I was puzzled and began reading. The more I read the more I realized that the article referred to some completely unknown and even incomprehensible reality. I could not understand why, amid the ongoing Civil War and catastrophic famine, someone would deliver a group of poor, hungry, and malnourished Jewish children—World War I and Civil War refugees from Ukrainian and Belarusian lands—to a cold hall of Kazan University's main building, take their anthropometric measurements, and then prepare a research paper on the Jewish race. I was even more puzzled by multiple references in the paper to various studies by Russian scientists who, as it turned out, had for years explored races of the Russian Empire. These references suggested that the Russian Empire had had a rich tradition of academic and, in the Jewish case, semiacademic (judging by the venues in which the cited works appeared) race science. How come?, I asked myself. How come I had never heard about Russian race science, not to mention Russian Jewish race science? By that time, I was very interested in Russian imperial history and with a group of like-minded freshly minted PhDs was a member of the new imperial history collective. We had just founded the *Ab Imperio* quarterly to explore new paradigms of thinking and writing about the complex entangled past of the region formerly shorthanded as "Russia." In short, I knew the historiography of the Russian Empire quite well and found convincing the then dominant view that the biological category of race and the racialized science of the Other held a marginal place in Russian scholarly and intellectual traditions, broadly defined. Instead, a cultural and more flexible Völkisch category of ethnicity prevailed.

So, I decided to parse specific references in the Kazan article and to reconstruct the strange reality they concealed, which did not square with the old or new scholarship. If only I had known then that my investigation would take twenty years and bring me to archives and libraries in many countries, in addition to Russia (Moscow, Kazan, and St. Petersburg)—Ukraine, Finland, Lithuania, Latvia, the United States, and others. Further, the more I uncovered about Russian Jewish race science, the more I wondered about the presumably nonexistent race science in the Russian Empire. Was Jewish race science such an anomaly? But many Russian Jewish scientists obviously considered their non-Jewish colleagues in the empire as members of one common network. I was confused. The persistent silence about the phenomenon of race in the Russian Empire is particularly striking if one considers the general rise of interest in "race" in the 2000s. Having been conceptualized as an intersectional category, a multilayered dynamic space where subjecthood and citizenship were constructed and renegotiated, and a set of practices immanent to modernity and modern imperialism, "race" has graduated from a purely biological category to a nuanced social-cultural category. Against this background, insistence on the Russian Empire's peculiar indifference to "race" implied a *Sonderweg*-like view of Russian imperial experience and Russian modernity. My archives told a different story, but without having a general context for it I still could not properly explain and locate the reality behind the references in the Kazan article.

And thus, for the first and hopefully the last time in my career, I had to write another book as a "context" for my main project. In 2008, the book that told the story of Russian race science came out in Russian. In 2013, a substantially revised and expanded edition appeared in English as *Homo Imperii: A History of Physical Anthropology in Russia* (University of Nebraska Press). This work included a chapter on Jewish race science. Finally, I was ready to return to the project that became the manuscript for *A Race for the Future* in 2021.

As one can guess from this brief memoir, such a long project required the support of various institutions and, most importantly, people who shared my interest in "race" in the Russian imperial and Jewish contexts when this was hardly a popular topic, and who were ready to offer their advice and criticism as I progressed with my research. At its different stages, the project was financially supported by the following foundations and organizations: Gerda Henkel Stiftung (Germany, 2002–2004); International Center for Russian and East European Jewish Studies (Russian Federation, 2003–2004); Volkswagen Stiftung (Germany, 2006–2008); and the American Council of Learned Societies (2009). I was privileged to be a visiting fellow at Centre d'études des mondes russe, caucasien et centre-européen, CNRS-EHESS (France, 2010); National Research University "Higher School of Economics" (Moscow, Russia, 2016, 2017); University of Illinois at Chicago (UIC) Institute for the Humanities (United States, 2017–2018); and Ludwig-Maximilians-Universität, Graduate School for East and Southeast European Studies (Munich, summer 2018). All these foundations and institutions not only supported my archival expeditions but provided a welcome environment for presenting and discussing the results of my ongoing research with colleagues in different fields and from different countries, whereas the yearlong fellowship at the UIC Institute for the Humanities proved indispensable to my progress with the book's narrative. I am grateful to colleagues at all the foundations and institutions listed above and, at the UIC Institute for the Humanities, to its wonderful staff and my cohort of fellows for their criticism, good questions, and encouragement.

When one works on a book for as long as I did, many twists and turns alter one's life between the project's inception and its completion. In my case, my life took a most fortunate turn in 2013, when I joined the UIC history department, where I found a welcoming professional home. I am grateful to all my colleagues for their support from day one and for their consistent interest in my research. I am especially indebted to those who have read and commented on different parts of the manuscript or reached out to me after my talks with ideas and suggestions. This group of departmental colleagues is in no way limited to specialists on Russia, like my fellow Russianist, Jonathan Daly, or on Eastern Europe and Jews, like Keely Stauter-Halsted and Malgorzata Fidelis, to whom I am deeply indebted both personally and professionally. It also includes Americanists, or historians of South Asia or China, such as Lynn Hudson, Kevin Schultz, Rama Mantena, Loura Hostetler, and Jennifer Brier, all of whom pushed me to ask new questions and provide better answers. My friendship and professional collaboration with another UIC colleague, Julia Vaingurt, has been a constant source of inspiration. I am grateful for Julia's feedback on my writing as well as for her collaboration on the international conference "Thinking Race in the Russian and Soviet Empires" that we hosted together with a group of colleagues from the University of Chicago in March 2020, and on many other similar initiatives.

From the academic world outside of UIC, the list of people who have helped me throughout all these years of work on the project is quite long. Sergei Kan encouraged me to prepare the manuscript of *Homo Imperii* for an English-speaking audience and has been one of my most important readers ever since. When this leading historian of anthropology offered his professional approval of my research on Lev Shternberg, it gave me the confidence to pursue it further. In several fields and countries, different colleagues whose works I deeply respect and whose ideas and approaches have influenced my own thinking about race, empire, and Jewish science and biopolitics have generously shared their time, knowledge, and advice with me. My work has benefited immensely from the feedback and aid of Eugene Avrutin, Edyta Bojanowska, Jane Burbank, Frederick Cooper, Michael Gordin, Bruce Grant, Harriet Murav, Svetlana Natkovich, Riccardo Nicolosi, Dmitry Shumsky, Charles Steinwedel and Francesca Morgan, Darius Staliūnas, Vera Tolz, Jeffrey Veidlinger, and many others. My mentor, an excellent historian of the Russian Empire, the late Seymour Becker, and his wife, Alla, always believed in the success of my scholarly pursuits, and I could always count on them whenever their help was needed. From Seymour I learned many things about historical craft, the most important being a profoundly respectful and careful attitude to the ideas that people found relevant in the past, and the imperative of thinking globally and comparatively about them. I owe Seymour more than I was ever able to admit to him. Therese Malhame has for years been a friend and an ideal copy editor who not only polishes my imperfect English but teaches me the art of conveying my thoughts with precision and clarity. I am extremely grateful to Kathleen McDermott, my editor at Harvard University Press, for believing in this project and for her stewardship of it.

All these exceptional experts have contributed to making my book more complex, sophisticated, and interesting, and bear no responsibility for its mistakes and weaknesses. However, a small group of my colleagues do share such a responsibility with me, as they are my most immediate academic family—the people who share all my failures as well as achievements, and a long and challenging history of founding and coediting the *Ab Imperio* quarterly. Alexander Semyonov of the Higher School of Economics in St. Petersburg, Sergei Glebov of Smith and Amherst Colleges in Massachusetts, and Ilya Gerasimov, the executive editor of *Ab Imperio* and my most demanding, dedicated, and generous reader and partner in life and in all professional pursuits—this book would have not been possible without you. Ilya and our son Daniel, in particular, have lived with this project since around 2001. Ilya read and commented on all the iterations of this manuscript, in pieces and as a whole. One can hardly accomplish such a long-term project without the love and support that I received from my family.

This book is dedicated to my father, Boris Mogilner, who passed away in late January 2020. Like many protagonists of this book, he was a Russian Jewish physician and, quite amazingly, shared many of the predispositions of the generation of his fathers. This book is my tribute to him.

INDEX

Adesman, Srul'-Khaim, 172

Africa(ns), 24, 30, 42, 64, 67, 86, 87, 100, 110, 118, 121, 224. *See also* Negro/Black

All-Russian Jewish Assembly (or All-Russian Congress of Jewish Communities), 208, 209

America/American, 68, 95–97, 101, 115, 121, 123, 127, 128, 134–136, 137–138, 178, 181, 202, 213, 227, 228, 229, 232, 234, 237, 240, 259. *See also* Jew(s): American (and the United States); United States

American Anthropologist, 2, 80, 134

American Jewish Joint Distribution Committee (JDC or Joint), 9, 179, 213, 216, 220, 221, 222, 223, 224, 228, 232, 250

American Relief Administration (ARA): activities, 9, 220, 222, 223, 226, 227, 228, 229, 230, 231, 232, 239; *Bulletin,* 229, 234; Kazan Department of Medical Statistics, 233, 234, 235, 237

Andree, Richard, 63

An-sky, S., 25, 66, 68, 69–70, 71, 113, 117, 119, 120, 241. *See also* Rappoport, Shloyme Zanvil

Anuchin, Dmitrii, 50–51, 58, 74, 75, 76, 80, 81–82, 85, 86, 87, 88, 90, 91–93, 94, 98–99, 100, 103, 121, 128, 230–231

Armenia/Armenians, 54, 55, 86, 89, 91, 196

Asia, 42, 86, 87, 110, 121, 280n54, 280n59; Central, 4, 5, 15, 74, 94, 95, 101, 105, 245; intellectuals, 7, 16; Minor, 78. *See also* Jew(s): Asian

assimilation: per Ernest Renan, 21, 23; into Jewishness, 56; of Jews, 96–98, 124, 134, 137, 138; of Jews into Georgian nation, 56–57, 281n61; postimperial, 5; in the Russian Empire, 18, 19, 29, 39, 68, 133, 143, 149, 188, 247; on Soviet terms, 242, 259; in the United States, 1–2. *See also* Jew(s): assimilated

autonomism, 21, 25, 28, 109, 187, 238

Bardakh, Iakov Iul'evich, 168, 169, 299n42

Bashkirs, 51, 53, 54, 89, 90, 245

Beilis, Menahem Mendel, 79, 112, 114

Bekhterev, Vladimir, 174, 195, 217, 306n50

Belarus, 259, 323. *See also* Belorussia(ns)

Belorussia(ns), 44, 89, 90, 161, 163, 206, 223, 259. *See also* Belarus

Berlin, 69, 102, 142, 185, 219, 220, 221, 222, 223, 236, 252

Bernshtein, Alexander, 196

Bessarabia, 219, 223, 310n39

Białystok, 162, 163